Recent Developments in Markov Decision Processes

The Institute of Mathematics and its Applications Conference Series

---

Optimization, *edited by* R. Fletcher
Combinatorial Mathematics and its Applications, *edited by* D. J. A. Welsh
Large Sparse Sets of Linear Equations, *edited by* J. K. Reid
Numerical Methods for Unconstrained Optimization, *edited by* W. Murray
The Mathematics of Finite Elements and Applications, *edited by* J. R. Whiteman
Software for Numerical Mathematics, *edited by* D. J. Evans
The Mathematical Theory of the Dynamics of Biological Populations, *edited by* M. S. Bartlett and R. W. Hiorns
Recent Mathematical Developments in Control, *edited by* D. J. Bell
Numerical Methods for Constrained Optimization, *edited by* P. E. Gill and W. Murray
Computational Methods and Problems in Aeronautical Fluid Dynamics, *edited by* B. L. Hewitt, C. R. Illingworth, R. C. Lock, K. W. Mangler, J. H. McDonnell, Catherine Richards and F. Walkden
Optimization in Action, *edited by* L. C. W. Dixon
The Mathematics of Finite Elements and Applications II, *edited by* J. R. Whiteman
The State of the Art in Numerical Analysis, *edited by* D. A. H. Jacobs
Fisheries Mathematics, *edited by* J. H. Steele
Numerical Software—Needs and Availability, *edited by* D. A. H. Jacobs
Recent Theoretical Developments in Control, *edited by* M. J. Gregson
The Mathematics of Hydrology and Water Resources, *edited by* E. H. Lloyd, T. O'Donnell and J. C. Wilkinson
Mathematical Aspects of Marine Traffic, *edited by* S. H. Hollingdale
Mathematical Modelling of Turbulent Diffusion in the Environment, *edited by* C. J. Harris
Mathematical Methods in Computer Graphics and Design, *edited by* K. W. Brodlie
Computational Techniques for Ordinary Differential Equations, *edited by* I. Gladwell and D. K. Sayers
Stochastic Programming, *edited by* M. A. H. Dempster
Analysis and Optimisation of Stochastic Systems, *edited by* O. L. R. Jacobs, M. H. A. Davis, M. A. H. Dempster, C. J. Harris and P. C. Parks
Numerical Methods in Applied Fluid Dynamics, *edited by* B. Hunt
Recent Developments in Markov Decision Processes, *edited by* R. Hartley, L. C. Thomas and D. J. White

In Preparation

Power from Sea Waves, *edited by* B. Count

Recent Developments in Markov Decision Processes

Based on the proceedings of an International Conference on Markov Decision Processes organised jointly by the Department of Decision Theory of the University of Manchester and the Institute of Mathematics and its Applications and held at the University of Manchester on July 17th to 19th, 1978

Edited by

R. HARTLEY
L. C. THOMAS
D. J. WHITE

Department of Decision Theory
The University of Manchester
Manchester, England

1980

ACADEMIC PRESS

A Subsidiary of Harcourt Brace Jovanovich, Publishers

London New York Toronto Sydney San Francisco

ACADEMIC PRESS INC. (LONDON) LTD.
24/28 Oval Road,
London NW1

United States Edition published by
ACADEMIC PRESS INC.
111 Fifth Avenue
New York, New York 10003

British Library Cataloguing in Publication Data
Recent developments in Markov decision processes.—
(Institute of Mathematics and its Applications;
Conference series).
1. Markov processes —Congresses
I. Hartley, R. II. Thomas, L. C. III. White,
Douglas John IV. Series
519.2'33 QA274.7 80–40529

ISBN 0-12-328460-0

Printed in Great Britain by
Whitstable Litho Ltd., Whitstable, Kent

CONTRIBUTORS

R.C.H. Cheng, *Department of Mathematics, UWIST, King Edward VII Avenue, Cardiff CF1 3NU.*

K.-J. Farn, *Institute of Applied Mathematics, National Tsing Hua University, Kuang Fu Road, Hsinchu, Taiwan 300, Republic of China.*

A. Federgruen, *Graduate School of Business, Columbia University, New York City, NY.*

N. Furukawa, *Department of Mathematics, Faculty of Science, Kyushu University, Japan.*

R. Hartley, AFIMA, *Department of Decision Theory, University of Manchester, Manchester M13 9PL.*

A. Hordijk, *Department of Mathematics, University of Leiden, Wassenaarseweg 80, P.O. Box 9512, 2300 Ra Leiden, The Netherlands.*

G. Hübner, Institut für Mathematische Stochastik, Universität Hamburg, Bundesstrasse 55, 2 Hamburg 13, Germany.

L.C.M. Kallenberg, Department of Mathematics, University of Leiden, Wassenaarseweg 80, P.O. Box 9512, 2300 Ra Leiden, The Netherlands.

H. Myoken, *Faculty of Economics, Nagoya City University, Mizuhocho Nizuhoku, Nagoya, Japan.*

E.L. Porteus, *Graduate School of Business, Stanford University, Stanford, California 94305, USA.*

D. Reetz, *Fachbereich Wirtschaftswissenschaft, Freie Universität, Berlin.*

K. Sawaki, *Nanzan University, 18 Yamazato-Cho, Showa-ku Nagoya, Japan.*

P.J. Schweitzer, *Graduate School of Management, University of Rochester, Rochester, NY.*

S.S. Sheu, *Institute of Applied Mathematics, National Tsing Hua University, Kuang Fu Road, Hsinchu, Taiwan 300, Republic of China.*

L.C. Thomas, *Department of Decision Theory, University of Manchester, Manchester M13 9PL.*

H.C. Tijms, *Vrije Universiteit, Actuariaat and Econometrie, 1007 MC Amsterdam, de Boelelaan 1105, Postbus 7161, Amsterdam, The Netherlands.*

F.A. Van der Duyn Schouten, *Vrije Universiteit, Actuariaat and Econometrie, 1007 MC Amsterdam, de Boelelaan 1105, Postbus 7161, Amsterdam, The Netherlands.*

K.M. van Hee, *Department of Mathematics, Eindhoven University of Technology, P.O. Box 513, Eindhoven, The Netherlands.*

C.C. White, III, *Department of Engineering Science and Systems, University of Virginia, Charlottesville 22901, USA.*

D.J. White, FIMA, *Department of Decision Theory, University of Manchester, Manchester M13 9PL.*

P. Whittle, *Department of Pure Mathematics and Mathematical Statistics, University of Cambridge, 16 Mill Lane, Cambridge CB2 1SB.*

PREFACE

The ideas involved in Markov Decision Proccesses were inherent in the work of Bellman on dynamic programming and Shapley's paper [4] on stochastic games. It was given a formal basis with the publication of Howard's book [5], and a series of papers by Blackwell [1,2]. Between them they established the validity of value iteration and policy iteration as techniques for solving Markov Decision Processes. It was obvious from the start that the two main optimisation criteria were discounted rewards over an infinite horizon, and the average reward per step.

In the intervening eighteen years since 1960, the theory of Markov Decision Processes has expanded at an increasing rate, until we have now a broad understanding of the general theory. Meanwhile, the potential areas of application have also increased, if at a much slower rate. The one disappointing feature has been that the practical use of Markov Decision Processes to these applications has been minimal - the main reason being the computational "size" of most practical problems. With the advent of "silicon chip" technology, this difficulty may soon be overcome. With this background, the conference on Markov Decision Processes held at the University of Manchester, in July 1978, looked at some recent theoretical developments in Markov Decision Processes and their applications.

A Markov Decision Process is defined on a state space S, where each state $i \in S$, has a set of feasible actions K_i. Choosing an action $k \in K_i$, results in an immediate reward (or cost) r_i^k and a probability p_{ij}^k of moving to state j at the next stage. The two main types of objective function are discounted rewards, or average rewards per stage. The former criterion is closely related to the total reward, where there is some condition that ensures the reward is finite.

Porteus's invited paper describes the iterative methods for solving the discounted problem. He concentrates on the efficiency of the algorithms, explains how the bounds they produce can be used to delete non-optimal actions, and describes some transformations that can speed up the algorithms. Hubner's short note describes another set of bounds which are useful for approximating longer horizon problems by shorter horizon ones. Whittle's paper is also centred on the relation between finite horizon discounted reward (or total reward) problems and the infinite horizon variety. He displays a very simple condition that ensures that the optimal reward in the n-period problem converges to that in the infinite horizon one, with a similar convergence of the solutions of the optimality equations. Most previous conditions which guaranteed this convergence required that the system kept returning to certain distinguished states infinitely often, but Whittle's condition only requires that the total reward under a policy in the infinite horizon case is less than a multiple of the optimal reward in a n-stage problem.

Reetz looks at the strange discount problem where the discount factor is greater than one. He introduces punctuated annuity as his optimisation criterion, which is the reward until first return to the starting state, suitably "discounted". D.J. White's paper is a bridge between finite and infinite state discounted problems. He suggests a finite state algorithm based on value iteration which converges to the optimal value of the infinite state problem. The idea is to introduce one new state at each iteration of the algorithm. Obviously one would introduce the "important" states at the start, and this seems a useful technique for very large finite problems as well as infinite ones.

In the average cost problems, the theory in the finite state case has been well established. In their survey article, Schweitzer and Federgruen, describe the behaviour of the value iteration algorithms for this type of problem. They describe the asymptotic behaviour of the values for this algorithm and of the "maximising" policies at each iteration. They also look at the bounds given by the algorithm, and the usefulness of data transformations to speed up the algorithms. Sheu looks at the existence of stationary optimal policies in the case of compact action spaces. These results are well known, even for infinite state spaces [4] under the usual "average reward" criteria, but he proves it under the criteria derived by taking the limit as β tends to 1 in the discounted problem. (For connections between these criteria see Flynn [3].) Hordijk and Kallenberg also look at the undiscounted finite state space, finite action space Markov Decision Process. They, however, use the linear programming approach, but instead of taking the usual linear programme (see Manne [6]), they look at its dual, and show that this leads to a simplified algorithm.

The third section of papers deals with the average cost problem when the state space is infinite. The invited paper by Tijms briefly reviews the results on the existence of optimal policies before turning to the way policy iteration can be applied in this case. Obviously the ordinary algorithm cannot be applied to an infinite state problem, but Tijms looks at a modification which involves only a finite number of operations at each step. He applies this to three examples and the algorithm converges in each case, though, as yet, the conditions which guarantee convergence to an optimal policy are unknown. Thomas's paper looks at the conditions under which the existence of an optimal policy has been proved for the countable state problem. He concentrates on the conditions that are required on the transition matrices and describes the implications and equivalences among them. Furukawa extends the countable state problem by allowing for vector valued reward functions. The optimisation criterion is then taken from a partial ordering on the rewards, so that he is looking for policies that give rise to "efficient" rewards. He shows that a type of policy improvement algorithm will always converge to a set of policies including all the optimal "efficient" policies.

The papers by Sawaki, van Hee and Myoken all deal with the problem of adaptive or partially observed Markov Decision Processes. The transition probabilities, and possibly the rewards, depend on unknown parameters, or the state is incompletely known, and, as the system develops, one tries not only to maximise the reward, but also to obtain better estimates of these parameters. The approach throughout is to use Bayesian updating techniques to obtain posterior distributions of the parameters after each state of the process. Myoken looks at the discounted case with max-max, max-min and the Bayes decision criterion. Using ideas from dual control theory, since one is trying to both maximise the reward and estimate the unknown parameters, he describes an algorithm for solving the problem. van Hee looks at the average cost adaptive problem and describes the expected result that, provided the process is returning to every state infinitely often, then there is an optimal policy. He describes "Bayes equivalent rules" and shows the optimal policy must be of that type. Sawaki's paper, though dealing with the problem of Markov Decision Processes over an arbitrary state space, is based on techniques used in adaptive problems by Smallwood and Sondik [8]. The idea is to exploit the piecewise linearity of the optimal rewards, as functions of the modified state space. Sawaki describes both a value iteration and policy improvement algorithm based on this idea, and shows both are computationally feasible algorithms.

C.C. White discusses the problem of structured policies and shows how, in a very general framework, assumptions about criterion definition result in isotone stationary strategies being optimal. Related results connecting criteria and types of optimal strategy are also deduced.

Hartley deals with an undiscounted infinite horizon stochastic control problem, where the cost function is unbounded but convex. By using a controllability condition and the convexity assumptions, he is able to show the existence of an optimal stationary policy. The proof follows the approach of Ross by looking at the discounted problem as the discount factor tends to one. Cheng's paper deals with the optimal control of continuous systems subject to random Markov jump disturbances. By discretising the state space, and the time axis, he is able to approximate these processes by Markov Decision Processes with finite state spaces. On refining the approximate Markov Decision Process by increasing the state space, the optimal solution converges to that of the continuous control process. Hordijk and van der Duyn Schouten also look at discrete approximations of continuous processes - continuous in time. By using stopping times, they show the discrete time problems converge in the weak probabilistic sense to the continuous versions. They are then able to describe the structure of the optimal policy for several different types of continuous time models.

The papers in this volume give an indication of the progress made in Markov Decision Processes in the last twenty years. There was a feeling throughout the conference that, in the immediate future, progress must be made in developing algorithms for specific types of problems, which will prove computationally efficient and will be used in "real" problems.

The editors would like to thank the Institute of Mathematics and its Applications, especially Catherine Richards and Cacs Hinds, for the efficient and friendly way they helped to run the conference. We must also thank the secretarial staff at the Institute's head quarters in Southend as well as Susan Lee and Kate Baker in the Department of Decision Theory, Manchester for their hard work in helping to prepare this volume for publication. Finally, our thanks to the authors of the papers, without whom there would have been no conference proceedings.

R. Hartley
L.C. Thomas
D.J. White

1. Blackwell, D. "Discrete dynamic programming", *Ann. Math. Statist.*, **33**, 719-726, (1962).

2. Blackwell, D. "Discounted dynamic programming", *Ann. Math. Statist.*, **36**, 226-235, (1965).

3. Flynn, J. "Conditions for the equivalence of optimality criteria in dynamic programming", *Ann. Stat.*, **4**, 936-953, (1976).

4. Hordijk, A. "Dynamic programming and Markov potential theory", Mathematical Centre Tracts No. 51, Amsterdam, (1974).

5. Howard, R.A. "Dynamic programming and Markov processes", M.I.T. Press, Cambridge, Mass., (1960).

6. Manne, A.S. "Linear programming and sequential decisions", *Mgmt. Sci.*, **6**, 259-267, (1960).

7. Shapley, L.S. "Stochastic games", *Proc. Nat. Aca. Sci.*, **39**, 1095-1100, (1953).

8. Smallwood, R.D., Sondik, E.J. "Optimal control of partially observable processes over the finite horizon", *Opns. Res.*, **21**, 1071-1088, (1973).

ACKNOWLEDGEMENTS

The Institute thanks the authors of the papers, the editors, Dr. R. Hartley, Dr. L. Thomas, Professor D.J. White (University of Manchester) and also Miss Janet Fulkes for typing the papers.

CONTENTS

Contributors v
Preface vii
Acknowledgements xii

OVERVIEW OF ITERATIVE METHODS FOR DISCOUNTED FINITE MARKOV AND SEMI-MARKOV DECISION CHAINS by E.L. Porteus 1

SEQUENTIAL SIMILARITY TRANSFORMATION: A PROMISING SEQUENTIAL APPROXIMATION METHOD FOR STATIONARY SUB-MARKOV DECISION PROBLEMS by G. Hübner 21

NEGATIVE PROGRAMMING WITH UNBOUNDED COSTS: A SIMPLE CONDITION FOR REGULARITY by P. Whittle 23

PUNCTUATED AND TRUNCATED ANNUITIES FOR EXPANDING MARKOVIAN DECISION PROCESSES by D. Reetz 35

FINITE STATE APPROXIMATIONS FOR DENUMERABLE STATE INFINITE HORIZON DISCOUNTED MARKOV DECISION PROCESSES: THE METHOD OF SUCCESSIVE APPROXIMATIONS by D.J. White 57

A SURVEY OF ASYMPTOTIC VALUE-ITERATION FOR UNDISCOUNTED MARKOVIAN DECISION PROCESSES by A. Federgruen and P.J. Schweitzer 73

A SUFFICIENT CONDITION FOR THE EXISTENCE OF A STATIONARY 1-OPTIMAL PLAN IN COMPACT ACTION MARKOVIAN DECISION PROCESSES by S.S. Sheu and K.-J. Farn 111

ON SOLVING MARKOV DECISION PROBLEMS BY LINEAR PROGRAMMING by A. Hordijk and L.C.M. Kallenberg 127

AN ALGORITHM FOR AVERAGE COSTS DENUMERABLE STATE SEMI-MARKOV DECISION PROBLEMS WITH APPLICATIONS TO CONTROLLED PRODUCTION AND QUEUEING SYSTEMS by H.C. Tijms 143

CONNECTEDNESS CONDITIONS FOR DENUMERABLE STATE MARKOV DECISION PROCESSES by L.C. Thomas 181

VECTOR-VALUED MARKOVIAN DECISION PROCESSES WITH COUNTABLE STATE SPACE by N. Furukawa 205

ADAPTIVE DUAL CONTROL APPROACH TO MARKOVIAN DECISION PROCESSES AND ITS APPLICATION by H. Myoken 225

MARKOV DECISION PROCESSES WITH UNKNOWN TRANSITION LAW: THE AVERAGE RETURN CASE by K.M. van Hee 227

PIECEWISE LINEAR MARKOV DECISION PROCESSES WITH AN APPLICATION TO PARTIALLY OBSERVABLE MODELS by K. Sawaki 245

THE OPTIMALITY OF ISOTONE STRATEGIES FOR MARKOV DECISION PROBLEMS WITH UTILITY CRITERION by C.C. White 261

DYNAMIC PROGRAMMING AND AN UNDISCOUNTED, INFINITE HORIZON, CONVEX STOCHASTIC CONTROL PROBLEM by R. Hartley 277

OPTIMAL CONTROL OF SYSTEMS SUBJECT TO RANDOM JUMP DISTURBANCES AND THEIR REPRESENTATION BY MARKOV DECISION PROCESSES by R.C.H. Cheng 301

WEAK CONVERGENCE OF DECISION PROCESSES by A. Hordijk and F.A. Van der Duyn Schouten 323

OVERVIEW OF ITERATIVE METHODS FOR DISCOUNTED FINITE MARKOV AND SEMI-MARKOV DECISION CHAINS

Evan L. Porteus

(Graduate School of Business, Stanford University)

1. INTRODUCTION

This paper will provide an overview of the role of iterative methods in obtaining solutions to infinite horizon stationary, finite state and action, discrete time parameter Markov and semi-Markov decision chains with the maximal discounted present value criterion. After setting the notation and terminology, a brief review of the basic foundational results will be given. An overview of the basic algorithms for obtaining numerical solutions is followed by a review of some basic bounds and their relevance. More detailed discussion of alternative iterative methods begins with addressing various ways of measuring the performance of such methods. The paper then reviews in some detail various transformations that can be used to convert one process into an equivalent one that may be easier to solve. Often transformations are made only implicitly, with the result being an alternative iterative method. Viewing the result as coming from a transformation can facilitate the measurement of its performance. After touching on the use of reordering the states, the paper discusses in some detail the use of various extrapolations that are designed to accelerate convergence. The paper ends with a discussion of some numerical illustrations.

2. NOTATION AND TERMINOLOGY

We use the following notation and conventions. N is the number of states; i and j are arbitrary states; D_i is the finite set of actions available at state i; k is an arbitrary decision; and r_i^k is the expected present value of the returns received when the state is i and the decision is k, during the time elapsing between when decision k is made and the next decision is made, evaluated at the time that decision k is made. See Jewell (1963) in the case of a semi-Markov decision chain for how to construct each r_i^k and p_{ij}^k, which is the "discounted probability" of moving from state i to state j under decision k. In the standard discounted Markov decision chain case, p_{ij}^k is simply the one period discount factor times the probability of moving from i to j under k. However, the construction is more involved in the semi-Markov case. A policy is a vector δ of decisions, one for each state, so that $\delta(i)$ is a decision for state i. A strategy is a vector $\pi = (\delta_1, \delta_2, \ldots)$ of policies, one for each decision, so that δ_n is the policy for the nth decision. A stationary strategy $\pi = (\delta, \delta, \ldots)$ is denoted by δ^∞. For each possible policy δ, r_δ is the vector of returns $[r_\delta]_i = r_i^{\delta(i)}$, P_δ is the transition matrix, consisting of discounted transition probabilities $[P_\delta]_{ij} = p_{ij}^{\delta(i)}$, H_δ is the one stage return (present value) operator $H_\delta v := r_\delta + P_\delta v$ where v is an arbitrary vector of values (one for each state), and A is the one stage optimal return (present value) operator $Av := \sup_\delta H_\delta v$ which can also be written as $[Av]_i = \max_k (r_i^k + \sum_j p_{ij}^k v_j)$. For $n = 1,2,\ldots$, $H_\delta^n v = H_\delta(H_\delta^{n-1} v)$, with $H_\delta^0 v$ defined to be v. Multiple applications of the operator A are denoted similarly.

Among compatibly dimensioned real N-vectors u and v, $u \le v$ means $u_i \le v_i$ for each i and $u < v$ means $u \le v$ and $u \ne v$. The vector of ones is denoted by e, and n denotes an arbitrary positive integer.

When examining the speed of convergence of various iterative methods, we will need the following terminology. Suppose v is an arbitrary real N-vector. Then $\|v\|$ denotes an arbitrary norm of v, $\|v\|_2 := \left(\sum_{i=1}^N v_i^2\right)^{1/2}$ denotes the L_2 norm of v, and $\|v\|_\infty := \max_i |v_i|$ denotes the L_∞ norm of v. If P is an arbitrary real $N \times N$ matrix, then the matrix norm of P is defined in terms of the corresponding vector norm using

$$\|P\| := \sup_{\|x\| = 1} \|Px\| .$$

In particular, $\|P\|_\infty := \max_i \sum_{j=1}^{N} |P_{ij}|$. Let $\bar{\alpha}_\delta$ denote the maximum row sum of P_δ for each δ and $\bar{\alpha} := \sup \bar{\alpha}_\delta$ $(= \sup_\delta \|P\|_\infty)$. Define $\underline{\alpha}_\delta$ and $\underline{\alpha}$ analogously for minimum row sums. In general, $\underline{\alpha} \le \underline{\alpha}_\delta \le \bar{\alpha}_\delta \le \bar{\alpha}$ for all δ, whereas equalities hold throughout in the Markov chain case. Let $\lambda_1, \lambda_2, \ldots, \lambda_N$ denote the eigenvalues of P and assume without loss that $|\lambda_1| \ge |\lambda_2| \ge \cdots \ge |\lambda_N|$. Then $\rho(P) := |\lambda_1|$ is the (spectral) radius of P and $\rho^*(P) := |\lambda_2|$ is the subradius of P. We assume henceforth that the process is discounted: $\rho := \sup_\delta \rho(P_\delta) < 1$.

To facilitate the exposition, $a \ge b$ will be written verbally, for example, as a exceeds b rather than a exceeds or equals b. This convention will apply to all analogous terminology, such as better, higher, lower, tighter, increasing, decreasing, etc. The term "strictly" will be added when appropriate.

If $\{v_n\}$ is a sequence of real N-vectors that converges to, say, v^* ($\|v_n - v^*\| \to 0$ as $n \to \infty$), then the sequence converges geometrically at the rate α if $0 \le \alpha < 1$ and there exists a real number M such that $\|v_n - v^*\| \le \alpha^n M$ for all n. The sequence converges geometrically nearly at the rate α if $0 \le \alpha < 1$ and the sequence converges geometrically at all rates strictly between α and 1.

3. REVIEW OF BASIC RESULTS

The criterion of interest is the vector of expected present values over an infinite horizon when following an arbitrary strategy π: $v(\pi) := r_{\delta_1} + P_{\delta_1} r_{\delta_2} + P_{\delta_1} P_{\delta_2} r_{\delta_3} + \cdots$. The object is to find the optimal return function $f := \sup_\pi v(\pi)$ and an optimal strategy π^* such that $v(\pi^*) = f$. The results in this section are due to Shapley (1953), Bellman (1957), Blackwell (1962, 1965), Denardo (1967), and Veinott (1969).

Under our assumptions $v(\pi)$ exists and is uniquely defined for all π. Thus $v_\delta := v(\delta^\infty)$, the return from the stationary strategy δ^∞ exists and is uniquely defined by $v_\delta = r_\delta + P_\delta r_\delta + P_\delta^2 r_\delta + \ldots$. Hence $v_\delta = r_\delta + P_\delta v_\delta = H_\delta v_\delta$, v_δ is the unique fixed point of the operator H_δ, $v_\delta = (I-P_\delta)^{-1} r_\delta$, and $H_\delta^n v \to v_\delta$ for any vector v. The optimal return function f exists and is uniquely defined; it is the unique fixed point of the operator A, and $A^n v \to f$ for any vector v. There exists an optimal stationary strategy $\delta^{*\infty}$, and δ^* is called an optimal policy. Finally, if $H_\delta f = Af$ then δ is an optimal policy (δ^∞ is an optimal strategy).

4. OVERVIEW OF THE BASIC ALGORITHMS

There are several fundamental computing steps used in the basic algorithms to be discussed. The value determination step starts with a given policy δ and computes v_δ. The policy improvement step starts with a given vector v and finds a policy δ and its one stage return $H_\delta v$ such that $H_\delta v = Av$, namely, δ yields the optimal one stage return when the value of continuing is v. The modified policy improvement step starts with a given vector v and policy γ and, if possible, finds a policy δ and its one stage return $H_\delta v$ such that $H_\delta v > H_\gamma v$. Of course, a policy improvement usually makes up a modified policy improvement step (unless γ is optimal and $v = f$). A value improvement step starts with a given vector v and policy δ and computes $H_\delta v$. Finally, a normalization step starts with a given vector v and subtracts v_N from each component ($v - v_N e$ is computed).

We now review the basic algorithms. Value iteration (Bellman (1957)), also called pre-Jacobi iteration, consists of successive policy improvement steps: $v_i^n = \max_k \left\{ r_i^k + \sum_j p_{ij}^k v_j^{n-1} \right\}$. That is, given v^0, $v^n = Av^{n-1} = A^n v^0$. Variants of value iteration include Jacobi iteration in which $v_i^n = \max_k \, (r_i^k + \sum_{j \neq i} p_{ij}^k v_j^{n-1})/p_{ii}^k$ and Gauss-Seidel iteration

(Kushner (1971)) in which $v_i^n = \max_k \left\{ r_i^k + \sum_{j<i} p_{ij}^k v_j^n + \sum_{j>i} p_{ij}^k v_j^{n-1} \right\} / p_{ii}^k$. Relative value iteration (White (1963)) carries out a normalization step after each policy improvement step. It applies in the Markov chain case where $P_\delta e = \alpha e$ for each δ. Policy iteration (Howard (1960)) alternates between value determination and policy improvement steps. Variants of value and policy iteration consist of performing multiple improvement steps between each value determination step (Denardo (1967)) and approximating the value determination step by multiple value improvement steps (Porteus (1971), Morton (1971a), van der Wal (1976), van Nunen (1976), and Puterman and Shin (1978)). Finally, the simplex method applied to the linear programming formulation of the problem (D'Epenoux (1960) and Manne (1960)) amounts to alternating between value determination and modified policy improvement steps, the latter carried out so that exactly one decision (for one state) is improved at each step. See Heilmann (1978) for a recent survey.

5. BASIC BOUNDS

Any algorithm that uses the policy improvement step may then use the following, easily computable bounds on the optimal return function: Let a and b denote the minimum and maximum difference between Av and v, respectively ($a = \min_i (Av-v)_i$ and $b = \max_i (Av - v)_i$). For the sake of illustration, suppose $a \geq 0$ and $\bar{\alpha} < 1$. Then

$$Av + \underline{\alpha} ae/(1 - \underline{\alpha}) \leq f \leq Av + \bar{\alpha} be/(1 - \bar{\alpha}).$$

A slightly weaker form of these bounds was given first by MacQueen (1966) for the Markov case. The current form is due to Porteus (1971), who gives the result in a more general setting. See Hübner (1977), Federgruen, Schweitzer, and Tijms (1978), and Waldmann (1978a) for recent developments and extensions. The bounds are useful in several ways. Firstly, they can be used for termination tests, so an algorithm can be stopped when a sufficiently close estimate of f is at hand. Secondly, they can be used to identify nonoptimal decisions, using the criterion developed by MacQueen (1967), which amounts to concluding that a decision for a given state is nonoptimal if the best it can produce is worse than the worst that can happen under some optimal policy. Thirdly, the bounds are useful for devising extrapolations, as we discuss later.

6. MEASURES OF PERFORMANCE OF ITERATIVE METHODS

The first measure of performance uses norms. If there exists $\alpha < 1$ such that $\|P_\delta\| \leq \alpha$ for all δ, then a standard application of the fact that A is a contraction mapping says that given v, there exists M such that $\|A^n v - f\| \leq \alpha^n M$ for all n. Thus, value iteration converges geometrically at the rate α in this case. Furthermore, M is easy to compute: $M = \|Av - v\|/(1 - \alpha)$ works, so bounds on f are readily available, especially when the L_∞ norm is used, when α is easy to identify. However, these bounds are never tighter than the basic bounds discussed in the previous section when the L_∞ norm is used.

The second measure of performance uses radii. Recall our assumption that $\rho := \sup_\delta \rho(P_\delta) < 1$. Veinott (1969) showed that value iteration converges geometrically at nearly the rate ρ. Since $\rho(P_\delta) \leq \|P_\delta\|$ always holds, this convergence rate is better than that guaranteed using norms. However, the constant M is not easy to compute, so bounds on f are not readily available.

The third measure of performance uses subradii. Morton and Wecker (1977) (see Doob (1953) and Hajnal (1956, 1958) for basic building blocks, Morton (1971a) for an application in this context, and van der Wal (1976), Hübner (1977), Schweitzer and Federgruen (1979), and Federgruen, Schweitzer, and Tijms (1978) for closely related results) showed, among other things, that in the Markov chain (equal row sums) case in our context, if there exists a unique optimal policy, say δ^*, then relative value iteration converges geometrically at nearly the rate $\rho^*(P_{\delta^*})$, the subradius of P_{δ^*}. Since the subradius of a matrix can be significantly smaller than the radius, this result can provide a substantially better convergence rate than that guaranteed using radii. However, the constant M is again not easy to compute. Furthermore, the conditions under which this result holds are not general ones.

The first three measures address the number of iterations required for convergence of iterative methods. This last measure is a general one consisting of the amount of computational effort required for a single iteration of a given method. It usually consists of the number of numerical operations of a given type (such as multiplications/divisions) required by an iteration. The main factor in determining this number is usually the total number of nonzero probabilities. Thus, we address whether the sparsity of a transition matrix is preserved when transformations discussed later are applied to it.

7. TRANSFORMATIONS

7.1 Introduction

The idea here is to transform a given process into an equivalent one that may be easier to solve. Several transformations were discussed in Porteus (1975) for this purpose. We review them here, with an outline of their effect on the measures of performance mentioned in the previous section.

A given transformation converts each return vector r_δ and transition matrix P_δ into new ones, denoted by $\tilde{r}_\delta$ and $\tilde{P}_\delta$.

7.2 Return transformation

This transformation leaves the transition matrices unchanged and creates $\tilde{r}_\delta = r_\delta - (I - P_\delta)v$, where v is some specified vector. In the Markov (equal row sum) case, this transformation can be used to make the obviously valid scaling change of subtracting a fixed constant from each immediate return.

7.3 Similarity transformation

This transformation was introduced in this context by Veinott (1969). The return vectors become $\tilde{r}_\delta = Qr_\delta$ and the transition matrices become $\tilde{P}_\delta = QP_\delta Q^{-1}$, where Q is a diagonal matrix with strictly positive diagonal elements. Radii, subradii, and sparsity are unchanged. However, norms change. Indeed, it is possible to use this transformation to obtain a process wherein $\|\tilde{P}_\delta\| \leq \alpha < 1$ for all δ, even if such was not the case with the original process. This transformation can be used implicitly by redefining the norm appropriately. Wessels (1977a) independently developed the use of such norms to obtain more general existence results in the denumerable state space case.

7.4 Pre-inverse transformation

Totten (1971) was the first to use a form of this transformation in this context. It creates $\tilde{r}_\delta = (I - Q_\delta)^{-1} r_\delta$ and $\tilde{P}_\delta = (I - Q_\delta)^{-1}(P_\delta - Q_\delta)$ for all δ, where Q_δ will usually be either diagonal or triangular, to facilitate computation. Jacobi (Gauss-Seidel) iteration arises from use of certain diagonal (triangular) Q_δ matrices. This transformation can also be used to convert a semi-Markov (unequal row sum) problem into a Markov (equal row sum) problem. (See Schweitzer (1971) and Lippman (1975).) This general transformation is very closely related to the use of stopping times to generate equivalent processes and successive approximation methods, as initiated by Wessels (1977b).

Increasing Q_δ, in the range $0 \leq Q_\delta \leq P_\delta$ decreases the L_∞ norm (of $\tilde{P}_\delta$) if $\|P_\delta\| \leq 1$ and decreases the radius in general. The general effect on the subradius is as yet unknown. Moving from pre-Jacobi to Jacobi to Gauss-Seidel either retains or improves sparsity.

By applying two transformations of this type, it is possible to derive a variant of value iteration, called <u>VSOR</u> (variable successive overrelaxation) in Porteus (1978b). It is easier to explain the method when applied to the case of a single policy, so affixes associated with decisions can be suppressed. In this case the method prescribes

$$\tilde{v}_i^* = r_i + \sum_{j<i} p_{ij} v_j^n + \sum_{j>i} p_{ij} v_j^{n-1},$$

and

$$v^n = \omega_i \tilde{v}_i^n + (1 - \omega_i) v_i^{n-1},$$

where $\omega_i = 1/(1 - \sum_{j<i} p_{ij}\omega_j p_{ji})$ for each i, and we have assumed without loss that $p_{ii} = 0$ for all i. This method is analogous to SOR (successive overrelaxation) (see Varga (1962) and

Young (1971)) in which a single relaxation factor $\omega (1 < \omega < 2)$ is used. SOR can yield dramatically faster convergence than Gauss-Seidel (based on radii) in certain cases (such as when the transition matrix is symmetric) and when ω is chosen judiciously. However, such improvement is not guaranteed in general. Furthermore, its induced transition matrix has negative elements in it, so the basic bounds cannot be applied and the method is not guaranteed to work in an optimization context. Here is where VSOR fits in, as it was designed so that its induced transition matrix would be nonnegative. We briefly return to this method later when we discuss some numerical illustrations.

7.5 Post-inverse transformation

Totten (1971) first proposed the use of this transformation in this context. It retains r_δ and creates $\tilde{P}_\delta = (P_\delta - Q)(I - Q)^{-1}$, where we require that $(I - Q)^{-1} \geq 0$. Increasing Q, in the range $0 \leq Q \leq P_\delta$ for all δ, decreases the L_∞ norm of $\tilde{P}_\delta$ in certain cases, and decreases the radius in general. The general effect on the subradius is not known. No special use has yet been made of this transformation.

7.6 Pre-multiplication transformation

Here $\tilde{r}_\delta = (I + Q_\delta)r_\delta$ and $\tilde{P}_\delta = P_\delta - Q_\delta + Q_\delta P_\delta$. Using Q_δ in the range $0 \leq Q_\delta \leq P_\delta$ creates a smaller L_∞ norm if $\|P_\delta\|_\infty \leq 1$ and a smaller radius in general. The effect on the subradius is not known in general. The effect on sparsity depends on the choice of Q_δ and on whether the matrix $\tilde{P}_\delta$ is computed explicitly.

7.7 Post-multiplication transformation

Here, the returns are unchanged and $\tilde{P}_\delta = P_\delta - Q + P_\delta Q$, where we require that $Q \geq 0$. In certain cases, the L_∞ norm is reduced by using this transformation. Using Q in the range $0 \leq Q \leq P_\delta$ for all δ creates a smaller radius. The effect on the subradius is not known in general. However, there is an important special application for which much can be said. It applies in the Markov (equal row sum) case and it is called the column reduction. It consists of finding the minimum probability of moving into a state in one step, over all initial states and decisions, and subtracting that amount from each such probability (all those in the column corresponding to moving into that state). The L_∞ norm and radius (which coincide in this case) are decreased by the sum of the reductions made in each column. Brauer (1952) shows that such a transformation leaves the remaining eigenvalues unchanged, so, in particular, the subradius is unchanged. If we merely seek the infinite horizon return from a fixed policy, then the transformed matrix need not be nonnegative. In this case, we can select a constant to subtract from each column so that the sum of these constants equals the original row sum(s). The result is that the spectral radius of the transformed (transition) matrix equals the subradius of the original one. It is straightforward to see that relative value iteration can be viewed as applying exactly this procedure in this case of analyzing a fixed policy (the return from the last state is subtracted from all other returns and the discounted probabilities of moving out of the last state are subtracted from all others in the respective columns). This observation therefore provides a simple explanation for why relative value iteration converges geometrically at nearly the rate of the subradius of the optimal transition matrix whenever a single optimal policy is selected by every policy improvement step made after a certain number of iterations.

8. REORDERING

In the semi-Markov case or when the induced transition matrix has unequal row sums, reordering the states produces different results. Several reordering algorithms were proposed in Porteus and Totten (1978) and Porteus (1978b) in the context of analyzing a fixed policy. Two of them guarantee that if $\|P\|_\infty = 1$ and $\rho(P) < 1$, then $\|\tilde{P}\|_\infty < 1$ where $\tilde{P}$ is the induced Gauss-Seidel matrix. Such a reordering allows the basic bounds to be used. The general effect of these reorderings on the radii and subradii is not yet known.

9. EXTRAPOLATIONS

9.1 Introduction

We shall focus henceforth in this paper on the analysis of a fixed policy. Doing so allows the use of simple notation (we seek v^* such that $v^* = r + Pv^*$) and allows insights to be gained about the simplest form of the processes being studied.

This specialization also arises naturally when using some variant of policy iteration in which we seek to approximate the infinite horizon return from a fixed policy.

We also assume that a given iterative method, such as Jacobi, Gauss-Seidel, VSOR, etc., is being applied to approximate v^*. Each such method can be viewed as being of the form $v^n = r + Pv^{n-1}$, where r and P are the induced return vector and transition matrix, respectively. The idea of extrapolations here is to make an additive correction in the terminology of Settari and Aziz (1973). Formally, once we have computed $\tilde{v}_i^n = r_i + \sum_j p_{ij} v_j^{n-1}$ for all i, we set $v_i^n = \tilde{v}_i^n + c_i^n$, where c^n is the additive correction vector (we extrapolate from $\tilde{v}^n$ to $\tilde{v}^n + c^n$). There are numerous ways to select c^n. We shall review several of them.

9.2 Lower bound extrapolations

These extrapolations guarantee that $v^n \leq v^*$ for all n. They also provide monotone convergence to v^*, allowing them to be applied in more general settings. Although upper bound extrapolations are analogous in the fixed policy case, they are not in the general optimization context.

The first lower bound extrapolation is the <u>trivial lower bound extrapolation</u> which uses the lower bound on v^* provided by the basic bounds (in section 5). The second is the <u>optimal lower bound extrapolation</u> in which a basic direction vector d^n is specified and additional computation is carried out to find the largest multiple of the induced direction vector Pd^n that can be added to $\tilde{v}^n$ with the proviso that the result is still a lower bound on v^*. We require that $Pd^n \geq 0$ to guarantee convergence. Different direction vectors may result in different extrapolations, of course. The trivial lower bound extrapolation appears when e is used as the basic direction vector in the Markov chain case.

Totten (1971) was the first to discuss lower bound extrapolations in this context. Further discussion and results appear in Porteus (1975) and Porteus and Totten (1978).

9.3 Norm reducing extrapolations

Introduction

This approach was presented in Porteus and Totten (1978) and continued in Porteus (1978b). One or two basic direction vectors (more are theoretically possible) are specified for each iteration. Each of these is premultiplied by P to get the induced direction vectors. The problem is then to determine what multiples of these direction vectors should be added to $\tilde{v}^n$ (what distance should be moved in each direction). For lower bound extrapolations, one moves as far as possible while retaining a lower bound on v*. Here, we form the difference between the input and output of the iteration, and try to select the multiples (distances) to minimize the norm of that difference. In particular, if d_1 and d_2 are two given basic direction vectors and ξ_1 and ξ_2 are the multiples that we shall select, then the input of the iteration is $v^{n-1}+\xi_1 d_1+\xi_2 d_2$ and the output is $H(v^{n-1} + \xi_1 d_1 + \xi_2 d_2) = \tilde{v}^n + \xi_1 Pd_1 + \xi_2 Pd_2$. This method is guaranteed to converge under reasonable conditions.

L_2 Norm

The general problem when using this norm can be solved straightforwardly by differentiation. Only modest computation is required when no more than two directions are used. Three cases were worked out and illustrated in Porteus and Totten (1978). Although it appears that improved performance can be achieved using this approach, it is a theoretical nuisance that the convergence guarantee requires that $\|P\|_2 < 1$, which usually isn't known in this context.

L_∞ Norm

The problem of finding the optimal multiples when using this norm in general becomes a linear program, which is of doubtful computational value in this context. In the Markov case with the single direction vector e, this approach yields the trivial scalar extrapolation, which consists of the average of the upper and lower basic bounds. This extrapolation merely serves the role of scaling in this case, since it has no effect on subsequent iteration results (differences, basic bounds, etc.). However, there are two extrapolations introduced in

Porteus (1978b) that are based on using this norm and possess attractive properties.

The first is the row sum extrapolation. It starts with e as the single basic direction vector, so Pe, the vector of row sums, is the induced direction vector. It extrapolates by an amount that is analogous to the average of the upper and lower basic bounds, but using the average of the largest and smallest row sums. That is, $c^n = t^n Pe$, where $t^n = .5(a^n + b^n)/(1 - \beta)$, where $a^n = \min_i (\tilde{v}_i^n - v_i^{n-1})$, $b^n = \max_i (\tilde{v}_i^n - v_i^{n-1})$, and $\beta = (\underline{\alpha} + \overline{\alpha})/2$. A nice feature of this extrapolation is that it provides a guaranteed reduction in the norm of the difference. That is, it is shown in Porteus (1978b) that

$$\|H(v^{n-1} + t^n e) - (v^{n-1} + t^n e)\|_\infty \leq \|Hv^{n-1} - v^{n-1}\|_\infty - (1 - \overline{\alpha})|t^n|.$$

The second interesting extrapolation using the L_∞ norm is the delayed scalar extrapolation. Its derivation differs from the others. Rather than starting with a basic direction vector and finding an induced direction vector, we start with a specified "induced" direction vector, namely e, we bypass the issue of the existence and/or calculation of a basic direction vector that would yield e, and we focus on reducing bounds on the norm of the difference between the input and output at the next iteration. The expression yielding the extrapolation depends on $a^n + b^n$ (as defined above). The resulting guaranteed reduction is given by the following:

$$\|\tilde{v}^{n+1} - v^n\|_\infty \leq \overline{\alpha}\|\tilde{v}^n - v^{n-1}\|_\infty - (1 - \overline{\alpha})\|c^n\|_\infty$$

10. NUMERICAL ILLUSTRATIONS

Numerical illustrations of the use of iterative methods are given in numerous articles, including MacQueen (1967), Hitchcock and MacQueen (1970), Morton (1971b), Totten (1971), Porteus (1975), van Nunen (1976), Puterman and Shin (1977), Porteus and Totten (1978), and Porteus (1978b). For brevity, only the illustrations in the latter two are reviewed.

One 30 state Markov chain problem and one 30 state semi-Markov chain problem are used in Porteus and Totten (1978) to illustrate the use of reordering and various extrapolation types (excluding the row sum and delayed scalar extrapolations). They indicated that reordering can have a dramatic effect on required computational effort, and the trivial lower bound extrapolation used in conjunction with Gauss-Seidel iteration can cut the required effort in half or more, compared to relative value iteration.

Twenty Markov chain problems with 100 states each and one with 1000 states were used in Porteus (1978b) to illustrate the use of the row sum and delayed scalar extrapolations, VSOR, and three reordering methods. The results showed that relative value iteration can require dramatically less computational effort than that estimated by the use of norms and spectral radii, as would be predicted by the subradius rate of convergence results. The trivial lower bound extrapolation used in conjunction with Gauss-Seidel iteration and any one of two of the reordering methods cut the average effort required by relative value iteration by over a third. Substituting the delayed scalar extrapolation improved the performance and using the row sum extrapolation instead yielded the best average performance, with the effort required by relative value iteration cut by a little over one half. The average performance of VSOR was best when the row sum extrapolation and one of the reordering methods were used. This performance was better than relative value iteration but not as good as Gauss-Seidel as discussed above. However, VSOR did perform significantly better on one of the problems, which had characteristics that made VSOR significantly different from Gauss-Seidel in application to it.

It appears that iterative methods can be very efficient when used to find the infinite horizon values for a given policy. Relative value iteration performs well in the Markov case and that performance can be improved by use of reordering, Gauss-Seidel, and the row sum extrapolation. Furthermore, the latter approach applies in the general semi-Markov case, whereas relative value iteration does not.

It is not clear when iterative methods are preferable to direct methods, such as Gaussian elimination. Part of the difficulty lies in the fact that the computational effort depends on the data structures used to deal with fill-in. For example, if one took the naive approach of allowing for complete fill-in, then the back substitution step alone would take about half of the computational effort required to solve the 100 state problems in Porteus (1978b) by one of the iterative methods discussed therein. The Gaussian elimination (partial pivoting) steps, needed before the back substitution can be carried out, can require substantially more effort, again with the amount depending on the data structures used. (See Reid (1977).) Furthermore, direct methods will almost always require more storage space than iterative methods and possibly dramatically more. Future work is needed to clarify when iterative methods are preferable to well designed direct methods.

As the computational effort required to obtain solutions to large discounted Markov (and semi-Markov) decision processes is brought down by new theoretical developments, more attempts will be made to obtain numerical solutions to such models for their contextual, practical value. For some steps taken in this direction, see Lembersky and Johnson (1975) and MacCall (1978). These theoretical developments are not limited to the topics reviewed in this paper. One important approach is that of approximating a large problem by a smaller one, usually by partitioning the state space. Some basic convergence results were established by Fox (1971, 1973), bounds and further results were given by Whitt (1978, 1979), Larraneta (1978), and Waldmann (1978b), finite horizon results were obtained by Hinderer (1978), and related work appears in Zipkin (1977, 1978) and Mendelssohn (1978). Another approach is that of approximating the appropriate stochastic processes by deterministic ones. The initial work was given by Norman and White (1968). A weakness of the approach was pointed out by Morton (1969) and an adjustment designed to correct it was proposed by Porteus (1978a). Related work is given by Reetz (1977).

11. ACKNOWLEDGEMENTS

The author wishes to thank Professors Henk Tijms and Dieter Reetz for making helpful suggestions, particularly those leading to the relevance of the Brauer reference.

12. REFERENCES

Bellman, R. "Dynamic Programming", Princeton University Press, Princeton, (1957).

Blackwell, D. "Discrete Dynamic Programming", *Ann. Math. Statist.*, **33**, 719-726, (1962).

Blackwell, D. "Discounted Dynamic Programming", *Ann. Math. Statist.*, **36**, 226-235, (1965)

Brauer, A. "Limits for the Characteristic Roots of a Matrix IV: Applications to Stochastic Matrices", *Duke Math. J.*, **19**, 75-91, (1952).

Denardo, E. "Contraction Mappings in the Theory Underlying Dynamic Programming", *SIAM Rev.*, **9**, 165-177, (1967).

D'Epenoux, F. "Sur un Problème de Production et de Stockage dans l'Aléatoire", *Rev. Franc. Rech. Opérationelle*, **14**, 3-16; (1960), English translation in *Man Sci.*, **10**, 98-108, (1963).

Doob, J. "Stochastic Processes", J. Wiley, New York, (1953).

Federgruen, A., Schweitzer, P. and Tijms, H. "Contraction Mappings Underlying Undiscounted Markov Decision Problems", *J. Math. Anal. Appl.*, **65**, 711-730, (1978).

Fox, B. "Finite-state Approximations to Denumerable State Dynamic Programs", *J. Math. Anal. Appl.*, **34**, 665-670, (1971).

Fox, B. "Discretizing Dynamic Programs", *J. Optzn. Th. Appl.*, **11**, 228-234, (1973).

Hajnal, J. "The Ergodic Properties of Nonhomogeneous Markov Chains", *Proc. Cambridge Phil. Soc.*, **52**, 67-77, (1956).

Hajnal, J. "Weak Ergodicity in Nonhomogeneous Markov Chains", *Proc. Cambridge Phil. Soc.*, **54**, 233-246, (1958).

Heilmann, W.-R. "Solving Stochastic Dynamic Programming Problems by Linear Programming - An Annotated Bibliography", *Z. Opns. Res.*, **22**, 43-53, (1978).

Hinderer, K. "On Approximate Solutions of Finite-Stage Dynamic Programs", Dynamic Programming and Its Application, M. Puterman (ed.), Academic Press, London, New York, 207-220, (1978).

Hitchcock, D. and MacQueen, J. "On Computing the Expected Discounted Return in a Markov Chain", *Naval Res. Logist. Quart.*, **17**, 237-241, (1970).

Howard, R. "Dyanmic Programming and Markov Processes", John Wiley, New York, (1960).

Hübner, G. "Contraction Properties of Markov Decision Models with Applications to the Elimination of Non-optimal Actions", Dynamische Optimierung, Bonner Math. Schriften 98, 57-65, (1977).

Jewell, W. "Markov-renewal Programming. I: Formulation, Finite Return Models", *Opns. Res.,* **11**, 938-948, (1963).

Kushner, H. "Introduction to Stochastic Control", Holt, Rinehart and Winston, New York, (1971).

Larraneta, J. "Approaches to Approximate Markov Decision Processes", Dept. of Ind. Org., Univ. of Sevilla, Sevilla, Spain, (1978).

Lembersky, M. and Johnson, N. "Optimal Policies for Managed Stands - An Infinite Horizon Markov Decision Process Approach", *Forest Sci.,* **21**, 109-122, (1975).

Lippman, S. "Applying a New Device in the Optimization of Exponential Systems", *Opns. Res.,* **23**, 687-710, (1975).

MacCall, A. "Population Models for the Northern Anchovy (Engraulis Mordax)", Symposium on the Biological Basis of Pelagic Fish Stock Management, Rapp. P. - v. Réun Cons. int Explor. Mer., (1978).

MacQueen, J. "A Modified Dynamic Programming Method for Markovian Decision Problems", *J. Math. Anal. Appl.,* **14**, 38-43, (1966).

MacQueen, J. "A Test for Suboptimal Actions in Markovian Decision Problems", *Opns. Res.,* **15**, 559-561, (1967).

Manne, A. "Linear Programming and Sequential Decisions", *Man. Sci.,* **6**, 259-267, (1960).

Mendelssohn, R. "Improved Bounds for Aggregated Linear Programs", S.W.F.C. Admin. Report, 16H, National Marine Fisheries Service, Honolulu, Hawaii, (1978).

Morton, T. "A Critique of the Norman-White Dynamic Programming Approximation", *Opns. Res.,* **17**, 751-753, (1969).

Morton, T. "On the Asymptotic Convergence Rate of Cost Differences for Markovian Decision Processes", *Opns. Res.,* **19**, 244-248, (1971a).

Morton, T. "Undiscounted Markov Renewal Programming via Modified Successive Approximations", *Opns. Res.*, **19**, 1081-1089, (1971b).

Morton, T. and Wecker, W. "Discounting, Ergodicity, and Convergence for Markov Decision Processes", *Man. Sci.*, **23**, 890-900, (1977).

Norman, J. and White, D. "A Method for Approximate Solutions to Stochastic Dynamic Programming Problems Using Expectations", *Opns. Res.*, **16**, 296-306, (1968).

van Nunen, J. "A Set of Successive Approximation Methods for Discounted Markovian Decision Problems", *Z. Opns. Res.*, **20**, 203-208, (1976).

Porteus, E. "Some Bounds for Discounted Sequential Decision Processes", *Man. Sci.*, **18**, 7-11, (1971).

Porteus, E. "Bounds and Transformations for Finite Markov Decision Chains", *Opns. Res.*, **23**, 761-784, (1975).

Porteus, E. "An Adjustment to the Norman-White Approach to Approximating Dynamic Programs", to appear in *Opns. Res.*, (1978a).

Porteus, E. "Improved Iterative Computation of the Expected Discounted Return in a Semi-Markov Chain", Research Paper No. 443, GSB, Stanford University, Stanford, (1978b).

Porteus, E. and Totten, J. "Accelerated Computation of the Expected Discounted Return in a Markov Chain", *Opns. Res.*, **26**, 350-358, (1978).

Puterman, M. and Shin, M. "Modified Policy Iteration Algorithms for Discounted Markov Decision Problems", *Man. Sci.*, **24**, 1127-1137, (1978).

Reetz, D. "Approximate Solutions of a Discounted Markovian Decision Process", Dynamische Optimierung, Bonner Math. Schriften 98, 77-92, (1977).

Reid, J. "Sparse Matrices", The State of the Art in Numerical Analysis, D. Jacobs (ed.), Academic Press, London, New York, 85-148, (1977).

Schweitzer, P. "Iterative Solution of the Functional Equations of Undiscounted Markov Renewal Programming", *J. Math. Anal. Appl.*, **34**, 495-501, (1971).

Schweitzer, P. and Federgruen, A. "Geometric Convergence of Value-iteration in Multichain Markov Decision Problems", *Adv. Appl. Prob.*, **11**, 188-217, (1979).

Settari, A. and Aziz, K. "A Generalization of the Additive Correction Methods for the Iterative Solution of Markov Equations", *SIAM J. Numer. Anal.*, **10**, 506-521, (1973).

Shapley, L. "Stochastic Games", *Proc. Nat. Acad. Sci.*, **39**, 1095-1100, (1953).

Totten, J. "Computational Methods for Finite State Finite Valued Markovian Decision Problems", ORC 71-9, Operations Research Center, Univ. of Calif., Berkeley, (1971).

Varga, R. "Matrix Iterative Analysis", Prentice-Hall, Englewood Cliffs, (1962).

Veinott, A., Jr. "Discrete Dynamic Programming with Sensitive Discount Optimality Criteria", *Ann. Math. Statist.*, **40**, 1635-1660, (1969).

van der Wal, J. "A Successive Approximation Algorithm for an Undiscounted Markov Decision Process", *Computing*, **17**, 157-162, (1976).

Waldmann, K.-H. "A Natural Extension of the MacQueen Extrapolation", Preprint No. 436, Tech. Hoch. Darmstadt, Germany, (1978a).

Waldmann, K.-H. "On Approximation of Dynamic Programs", Preprint No. 439, Tech. Hoch. Darmstadt, Germany, (1978b).

Wessels, J. "Markov Programming by Successive Approximations with Respect to Weighted Supremum Norms", *J. Math. Anal. Appl.*, **58**, 326-335, (1977a).

Wessels, J. "Stopping Times and Markov Programming", Trans. 7th Prague Conf. 1974, Academia, Prague, 575-585, (1977b).

White, D. "Dynamic Programming, Markov Chains and the Method of Successive Approximations", *J. Math. Anal. Appl.*, **6**, 373-376, (1963).

Whitt, W. "Approximations of Dynamic Programs I", *Math. Opns. Res.*, **3**, 231-243, (1978).

Whitt, W. "Approximations of Dynamic Programs, II", *Math. Opns. Res.*, **4**, 179-185, (1979).

Young, D. "Iterative Solution of Large Linear Systems", Academic Press, New York, (1971).

Zipkin, P. "Bounds on the Effect of Aggregating Variables in Linear Programs", Research Paper No. 211, Graduate School of Business, Columbia Univ., New York, (1977).

Zipkin, P. "Bounds for Row-Aggregation in Linear Pgoramming", Research Paper No. 105A, Graduate School of Business, Columbia Univ., New York, (1978).

SEQUENTIAL SIMILARITY TRANSFORMATION: A PROMISING SEQUENTIAL APPROXIMATION METHOD FOR STATIONARY SUB-MARKOV DECISION PROBLEMS

G. Hübner

(University of Hamburg)

ABSTRACT

The value V^N of a stationary Markov decision problem with large horizon N may be obtained approximately by extrapolation from a small horizon n. The "classical" bounds of MacQueen (1966) and Porteus (1971) use constant extrapolation terms which are not always satisfactory. These bounds may be improved by using the similarity transformation of Veinott (1969) and - according to an idea of Schellhaas (1974) - by choosing the differences $d_n := V^n - V^{n-1}$, or d_{n-1}, as transforming functions. This is possible if d_n is positive for all states (or negative for all states), and all n.

If S and A are the state and action spaces, $Q^{sa}(.)$ is the transition law bounded by $\sup_{sa} Q^{sa}(S) = \beta < \infty$, $r(s,a)$ is the one-stage reward and $V^O(s)$ the final reward, then the resulting bounds read (see Hübner (1975), (1978/80).

$$V^n(s) + \sum_{k=1}^{N-n} (\gamma_n^-)^k d_n(s) \leqq V^N(s) \leqq V^n(s) + \gamma_n^+ \sum_{k=1}^{N-n} \gamma_n^k d_{n-1}(s)$$

with

$$O \leqq \gamma_n^- := \inf_s \frac{d_n(s)}{d_{n-1}(s)} \leqq \gamma_n^+ := \sup_s \frac{d_n(s)}{d_{n-1}(s)} \leqq \gamma_n := \sup_{s,a} \frac{Q^{sa} d_{n-1}}{d_{n-1}(s)}$$

These bounds worked very well in some small numerical problems, but it may be expected that they will be good for most unequal row sum problems since γ_n^- is a lower bound to the rate of geometrical convergence (if any) and γ_n^+ tends to an upper bound at least if there is a finite "turnpike horizon".

Similar bounds may be obtained if the value of $\lim_{n\to\infty} V^n$ is available - or more generally, a fixed point of the maximal reward operator.

Both types of bounds may be applied to eliminate nonoptimal actions at early stages (see Hübner (1979)).

REFERENCES

Hübner, G. "Extrapolation und Ausschlissung suboptimaler Aktionen in endlich-stufigen stationären Markoffschen Entscheidungsmodellen", Habilitationsschrift, Universität Hamburg, (1975).

Hübner, G. "Bounds and good policies in stationary finite-stage Markovian decision problems", Reprint (1978). To appear in *Adv. Appl. Prob.*, **12**, (1980).

Hübner, G. "Sequential similarity transformations for solving finite-stage sub-Markov decision problems", *Operations Research Verfahren*, **33**, 197-207, (1979).

MacQueen, J. "A modified dynamic programming method for Markovian decision problems", *J. Math. Anal. Appl.*, **14**, 38-43, (1966).

Porteus, E. L. "Some bounds for discounted sequential decision processes", *Management Science*, **18**, 7-11, (1971).

Schellhaas, H. "Zur Extrapolation in Markoffschen Entscheidungsmodellen mit Diskontierung", *Z. Operations Res.*, **18**, 91-104, (1974).

Veinott, A. F. "Discrete dynamic programming with sensitive discount optimality criteria", *Ann. Math. Statist.*, **40**, 1635-1660, (1969).

NEGATIVE PROGRAMMING WITH UNBOUNDED COSTS: A SIMPLE CONDITION FOR REGULARITY

P. Whittle

(University of Cambridge)

SUMMARY

A simple condition (the "bridging condition" (10)) is given for a Markov decision problem with non-negative costs to enjoy the regularity properties enunciated in Theorem 1. These are, roughly, that the s-horizon minimal loss function converge to the infinite horizon minimal loss as $s \to \infty$ for all terminal loss functions within some class, and that this limiting loss function be the unique solution in some class of the equilibrium form (11) of the dynamic programming equation. The bridging condition is sufficient for regularity, and is not far from being necessary, in a sense explained in section 2.

1. INTRODUCTION

We consider the usual Markov decision process with additive losses. The problem is framed in discrete time t ($t = \ldots -2,-1,0,1,2\ldots$), the values of state variable and decision variable at time t being denoted x_t and u_t. The decision u_t at time t is to be chosen as a function of current observable $W_t = \left\{x_s, u_{s-1};\ s \leq t\right\}$ alone. The distribution of x_{t+1} conditional upon W_t, u_t is conditioned by x_t, u_t alone. We also suppose this conditional distribution independent of t, so that the process is time-homogeneous.

The loss function over a horizon s is defined as

$$C_s = \sum_{r=1}^{s-1} \beta_1\beta_2\cdots\beta_{r-1}g_r + \beta_1\beta_2\cdots\beta_{s-1}K_s, \qquad (1)$$

where

$$g_t = g(x_t, u_t, x_{t+1})$$
$$\beta_t = \beta(x_t, u_t, x_{t+1}) \qquad (2)$$
$$K_t = K(x_t)$$

for given loss functions g, β, K. Here g is to be regarded as a current loss, K a terminal loss, and β a state-and action-dependent discount factor. We assume β non-negative, but not necessarily bounded above.

In defining an infinite horizon loss one generally sets K = O:

$$C_\infty = \sum_{r=1}^{\infty} \beta_1\beta_2 \cdots \beta_{r-1} g_r \qquad (3)$$

We shall suppose all the quantities (2) to be non-negative, so that we restrict ourselves to the area generally known as negative programming ("negative" because positive costs are negative rewards). In this case, C_∞ exists (by monotonicity of C_s if K = O). We shall not suppose the loss functions bounded above, however, so the limit C_∞ can well be infinite.

Strategies (i.e. admissible rules for specifying the u_t) are denoted by π. We define the expected losses

$$V_s^\pi(W_1) = E(\pi)(C_s|W_1)$$
$$V^\pi(W_1) = E(\pi)(C_\infty|W_1) \qquad (4)$$

where $E(\pi)(\cdot|W_1)$ denotes the expectation conditional on W_1 if strategy π is used. We define the infima

$$F_s^K = \inf_\pi V_s^\pi$$
$$F = \inf_\pi V^\pi . \qquad (5)$$

We have used the notation (5) because it is useful to make the dependence of F_s upon the terminal loss function K explicit. It follows then from the theory of Markov decision processes that F_s^K is a function of x_1 and s alone, denoted $F_s^K(x_1)$, which obeys the dynamic programming (DP) equation

$$F_s^K(x_1) = \inf_{u_1} E\left[g_1 + \beta_1 F_{s-1}^K(x_2) \mid x_1, u_1\right] \qquad (6)$$

$$(s > 1)$$

with terminal condition

$$F_0^K = K. \qquad (7)$$

(We refer to this as a terminal condition rather than an initial condition because (6) is a recursion backwards in time). We shall write (6) as

$$F_s^K = L\, F_{s-1}^K \qquad (s > 1) \qquad (8)$$

so defining the operator L. One can properly deduce from (7), (8) that

$$F_s^K = L^s K \qquad (9)$$

for finite s. For the case when K = 0 we shall denote F_s^0 simply as F_s, so that

$$F_s = L^s(0).$$

There are now a number of properties which one would expect and hope of the sequence $\{F_s^K\}$ and of F. The more important of these are listed in the enunciation of Theorem 1 in the next section: they are that F_s^K should converge to F as s increases for all K in some class, and that F should be the unique solution of $F = LF$ within some class. We shall refer to these conditions collectively as "regularity". One of the important applications of these conditions in particular problems is that one can often demonstrate that some particular property of K is also shared by all the iterates (9), and one would like to then conclude that it is also shared by F. Such conclusions often have implications for the form of an infinite-horizon optimal strategy.

"Regularity" is known to hold in the case when x and u can take only finitely many values, when g and K are uniformly bounded, and when discounting is strict (i.e. $0 \leq \beta \leq 1 - \delta < 1$). As soon as one departs from these assumptions, then there are examples to show that any of the desired properties can fail.

One would like weak and simply verifiable conditions which ensure regularity. A number of sufficient conditions have been proposed: Harrison (1972) makes some rather strong assumptions on the cost functions; Hinderer (1970) and Schäl (1975) make a deep and general attack on the problem, and derive sufficient conditions on the cost and transition functions (involving compactness and continuity demands). These conditions are however, far from necessary, and not very transparent. Derman and Veinott (1967) and Hordijk (1974) are more concerned with the case of average cost optimality, and phrase their conditions in terms of recurrence and recurrence time demands. Robinson (1976) is also concerned with average cost optimality, and obtains conditions (for uniqueness of solution of the appropriate DP equation) in terms of rate of growth of the solution. This is an approach which could plainly yield verifiable sufficient conditions in cases of some degree of generality; the argument is still one adapted to a rather special case, however.

The condition we shall set up is the following, that for some r and for some π there exists a finite constant a such that (10) (below) holds. This condition is sufficient for regularity, and is very nearly necessary, in a sense explained in the next section. It states simply that there is a policy π and a horizon r such that the infinite horizon loss V^{π} does not exceed a fixed multiple of the minimal loss over horizon r. The condition is easily verified as far as sufficiency goes, because one can often quite easily obtain an upper bound to V^{π} and a lower bound to F_r which are sufficient to establish (10). The condition (10) does not appeal to any very specific characteristic of the process, but this is its virtue, since there are so many types of special condition which will ensure regularity. Verification is bound to be problem-dependent past a certain point, and the art is to locate that point in simple, general terms.

2. THE BRIDGING CONDITION

We formulate the Bridging Condition: There exists a Markov policy π, a horizon r and a finite constant a such that

$$V^{\pi} \leq aF_r. \tag{10}$$

The description "bridging" refers to the fact that (10) relates infinite-horizon (but possibly non-minimal) losses to minimal (but possibly finite-horizon) losses. We also make the following Definition. Let K be a function from state-space $\mathcal{X}$ to the non-negative real line. Then $\mathcal{C}(K)$ is the set of such functions ϕ for which $\phi \leq aK$ for some finite a.

The bridging condition is thus a statement that $V^{\pi} \in \mathcal{C}(F_r)$.

Theorem 1. Suppose that g, K, $\beta \geq 0$ and that either (a) the bridging condition holds for some finite r or (b) the bridging condition holds for r infinite, and $F_{\infty} = \mathcal{L}F_{\infty}$. Then the decision process is regular, in that the following properties hold:

(i) F is a function of x_1 alone

(ii) The limit F_{∞}^{K} exists and equals F for all K in $\mathcal{C}(F_r)$

(iii) F satisfies

$$F = \mathcal{L}F \tag{11}$$

and is the unique solution of this equation in $\mathcal{C}(F_r)$.

The proof of the theorem is essentially a simple one, but is best distributed over a number of lemmas. We shall set these out in the next section.

The bridging condition is very nearly necessary as well as sufficient, because the property

$$F = F_{\infty}.$$

implied in (ii) can be written

$$\inf_{\pi} V^{\pi} = \lim_{r} F_r, \tag{12}$$

which could be viewed as a limiting form of the bridging condition. However, it does not necessarily imply the actual bridging condition for a particular admissible π and finite r. It is possible that there is an $\in$-version of the bridging condition which is both necessary and sufficient, but in such a version the condition would lose its attractive simplicity.

3. PROOF OF THEOREM 1

Recall that L is a monotone operator: $\phi \geq \psi$ implies that $L\phi \geq L\psi$. From this it follows that, if $LK \geq K$, then F_s^K is monotone increasing in s, and hence has a limit F_∞^K. This will be true in particular for $K = 0$. Correspondingly, if $LK \leq K$ then F_s^K decreases monotonely to a limit F_∞^K. Furthermore, both F_s^K and F_∞^K (if it exists) will be increasing in K.

Recall, furthermore, that

$$F \geq F_\infty$$

and

$$LF_\infty \geq F_\infty$$

and that cases can be found where both inequalities are strict for some values of x.

For a given Markov π let us write

$$V_s^{\pi} = G_s(\pi) + E_s(\pi)K$$

where the two terms are the contribution to $E(\pi)C_s$ from the sum over discounted g-values and from the discounted terminal loss in (1) respectively. The effect of the operator $E_s(\pi)$ on K is thus in general to apply a history-dependent discounting, and then take an expectation conditional on initial state.

Suppose that an optimal finite horizon policy exists, in that for all x_1 and all finite s the infimum in (6) is attained. Denote a policy thus generated by π^K, to emphasise its dependence upon the terminal loss function. Then

$$\begin{aligned} L^s K &= G_s(\pi^K) + E_s(\pi^K)K \\ &\leq G_s(\pi) + E_s(\pi)K \end{aligned} \tag{13}$$

for every other π.

Our arguments are simpler when such optimal finite-horizon policies exist; we shall then suppose this to be true. In the absence of this hypothesis the proofs still hold in an ϵ-version described in Whittle (1979).

If M is any other terminal loss function, then (13) implies that

$$G_s(\pi^M) + E_s(\pi^M)K \geq G_s(\pi^K) + E_s(\pi^K)K. \tag{14}$$

Setting $M = L^r K$ in (14) we deduce that

$$F^K_{r+s} - F^K_s \geq E_s(\pi^{L^r K})\left[L^r K - K\right]. \tag{15}$$

Replacing M by K and K by $L^r K$ in (14) we deduce that

$$F^K_s - F^K_{r+s} \geq E_s(\pi^K)\left[K - L^r K\right]. \tag{16}$$

These two inequalities give useful bounds for the righthand members in the cases where $L^s K$ is respectively increasing and decreasing in s. They hold as stated for r finite. If r is infinite then the presumption is that F^K_∞ exists. Both $L^r K$ and F^K_{r+s} can then be identified with this quantity if

$$F^K_\infty = L F^K_\infty \tag{17}$$

If equality does not hold in (17) then one can have the phenomenon that $L^s K$ and $L^s F^K_\infty$ can have different limits as $s \to \infty$, which would invalidate conclusions we would wish to draw.

Lemma 1 If $K \in \mathcal{C}(F_r)$ where either (a) r is finite, or (b) r is infinite and $F_\infty = LF_\infty$, then

$$\lim_{s\to\infty} E_s(\pi^{F_r})K = 0 \tag{18}$$

for all x such that $F_\infty(x) < \infty$.

Proof Set K = 0 in (15); thus

$$F_{r+s} - F_s \geq E_s(\pi^{F_r})F_r \geq 0.$$

This implies that

$$E_s(\pi^{F_r})F_r \to 0 \tag{19}$$

as s increases for all x such that F_{r+s} and F_s have a common finite limit. The conditions of the Lemma ensure this, and (19) will then imply (18) for all K in $\mathcal{C}(F_r)$. ■

Lemma 2 Under the conditions of Theorem 1 the limit F_∞^K exists and equals F_∞ for all x.

Proof

We have

$$\begin{aligned} F_s \leq F_s^K \leq G_s(\pi^{F_r}) + E_s(\pi^{F_r})K \\ = F_{r+s} + E_s(\pi^{F_r})\left[K - F_r\right]. \end{aligned} \tag{20}$$

By Lemma 1, the final E_s term will tend to zero with increasing s if x is such that F_∞ is finite. We thus deduce that, for such x, F_∞^K exists and

$$F_\infty^K = F_\infty . \tag{21}$$

If x is such that F_∞ is infinite, then (21) holds trivially, by the first inequality of (20). ■

Proof of Theorem 1 Let us, for simplicity, denote V^π just by V. The bridging condition is a statement that V belongs to $\mathcal{C}(F_r)$. It follows thus, from Lemma 2, that

$$\mathcal{L}^s V \to F_\infty \leq F.$$

On the other hand

$$F \leq \mathcal{L}^s V$$

since the right-hand member is the return from a particular policy (that in which π is preceded by s optimal steps). From these two inequalities we deduce that

$$F = F_\infty$$

and that F is also identifiable as the common limit of $\mathcal{L}^s K$ for any K in $\mathcal{C}(F_r)$. This establishes assertions (i) and (ii).

That F satisfies (11) is well known (see Hinderer, (1970)); that F belongs to $\mathcal{C}(F_r)$ follows from $0 \leq F \leq V \in \mathcal{C}(F_r)$.

Hence, if ϕ belongs to $\mathcal{C}(F_r)$ and

$$\phi = \mathcal{L}\phi \tag{22}$$

then

$$\phi = \mathcal{L}^s\phi \to F$$

which establishes assertion (iii). ■

4. AN EXAMPLE OF NON-UNIQUENESS OF THE NOTATION OF THE EQUILIBRIUM DYNAMIC PROGRAMMING EQUATION

In Whittle (1979) we extend this treatment to the case when optimal policies may not exist, deduce a number of corollaries to Theorem 1, and give an example of the application of the bridging condition. We also consider the situation where equation (22) has multiple solutions, ϕ_ν, say.

Such multiple solution is evidence of possible dependence of

F_∞^K on K, for (22) implies that

$$F_\infty^\phi = F_s^\phi = \phi.$$

We can then define the domain of attraction of ϕ_ν, denoted A_ν, as the set of terminal loss functions K for which

$$F_\infty^K = \phi_\nu$$

and in Whittle (1979) we deduce some properties of these domains of attraction. These properties are suggested, in part, by some particular examples in which (22) has several solutions. The examples given there are rather deterministic in character: different rates of deterministic passage of the state variable to infinity are weighed against differing costs. In this section we give a stochastic example more akin to that quoted by Robinson (1976) as due to Bather, in which differing degrees of concentration of the equilibrium distribution are weighed against differing costs.

One might think of the state variable x in the example as measuring incidence of some infection in a population. Two differing strategies are possible: to control the infection or not. The cost of control is neglected in itself, but it is assumed that a control programme has the effect of decreasing infection level, but of making a given level of infection more costly (since, in the presence of control measures, the population loses its natural resistance to that infection). So, it may be cheaper not to control at all, and, with a zero terminal loss, this will be the optimal strategy at all horizons. However, if at any current or future time one incurs a cost function appropriate to the control situation, then it is optimal to control immediately at all infection levels. Essentially, because of loss of immunity, the population has become "hooked" on infection control measures, even if such loss of immunity is a distant future prospect. So, (22) has two solutions, corresponding to control in all states, or zero control in all states.

Suppose that the strategies of zero control and control are denoted by u = 0,1 respectively, that x takes values in {0,1,2,...}, and that the transition matrix of x is

$$\begin{bmatrix} a_u & b_u & 0 & 0 & 0 & \cdots \\ r_u & q_u & p_u & 0 & 0 & \cdots \\ 0 & r_u & q_u & p_u & 0 & \cdots \\ 0 & 0 & r_u & q_u & p_u & \cdots \\ \cdot & \cdot & \cdot & \cdot & \cdot & \cdots \end{bmatrix}$$

if control u is used. Let us set

$$\pi_u(z) = p_u z + q_u \ r_u z^{-1},$$

and determine a,b by

$$a_u = 1 - b_u = \frac{\theta - \pi_u(\theta)}{\theta - 1} \tag{23}$$

where θ is yet to be determined.

We suppose that

$$\frac{p_u}{r_u} < 1$$

so that the process is always recurrent, and that

$$p_0 > p_1$$

$$r_0 < r_1$$

$$g(x,u) = \begin{cases} 0 & (u = 0) \\ \theta^x & (u = 1) \end{cases}$$

Thus, introduction of control increases costs, but inhibits growth of x. We suppose that

$$\theta > \max_u \left(\frac{r_u}{p_u}\right)$$

which makes specification (23) legitimate, and also that

$$\beta\pi_1(\theta) < 1 < \beta\pi_o(\theta)$$

where β is a fixed discount factor. We then find that (22) has the two solutions

$$\phi(x) = 0$$

$$\phi(x) = \frac{\theta^x}{1 - \beta\pi_1(\theta)}$$

corresponding to non-application or application of control in all states, respectively.

5. REFERENCES

1. Derman, C. and Veinott, A.F. "A solution to a countable system of equations arising in Markov decision processes", *Ann. Math. Statist.*, **38**, 582-584, (1976).

2. Harrison, J.M. "Discrete dynamic programming with unbounded rewards", *Ann. Math. Statist.*, **43**, 636-644, (1972).

3. Hinderer, A. "Foundations of non-stationary dynamic programming with discrete time parameter", Springer, (1970).

4. Hordijk, A. "Dynamic programming and Markov potential theory", Mathematical Centre Tracts, no. 51, Amsterdam, (1974).

5. Robinson, D.R. "Markov decision chains with unbounded costs and applications to the control of queues", *Adv. Appl. Prob.*, **8**, 159-176, (1976).

6. Schäl, M. "Conditions for optimality in dynamic programming and for the limit of n-stage optimal policies to be optimal", *Z. Wahrscheinlichkeitstheorie verw. Geb.*, **32**, 176-196, (1975).

7. Whittle, P. "A simple condition for regularity in negative programming", *J. Appl. Prob.*, **16**, 305-318, (1979).

PUNCTUATED AND TRUNCATED ANNUITIES FOR EXPANDING MARKOVIAN DECISION PROCESSES

D. Reetz
(Fachbereich Wirtschaftswissenschaft, Freie Universität, Berlin)

1. BASIC MODELS AND SOLUTION CONCEPTS

In the following we shall investigate two types of average pay-off criteria, namely punctuated and truncated annuities for Markovian decision processes (MDP) with discount (or growth) factor $\rho \geq 1$. Punctuated annuities will be defined using a stopping rule, whereby the process breaks off at the N-th entry into a specified state. Such a criterion for the case N = 1 was introduced by Reetz [12]. Truncated annuities on the other hand are defined by employing a stopping rule, according to which the process is halted after T stages have elapsed. If $\rho = 1$ such a procedure yields the classical average pay-off criterion as presented for example in Derman [3] on pp 25-28. Three primary results of our analysis may be summarized as follows. In Section 2 we investigate various punctuated annuities, leading to the fundamental theorem for recurrent MDP (Theorem 3). This theorem concerns the existence of a stationary policy which is optimal with respect to punctuated annuities. A second result shows that in the case $\rho = 1$ truncated annuities may be approached by punctuated annuities. Thus, the existence of a stationary policy optimal with respect to both annuities (which are then equal) is proved in Theorem 4 without employing an Abelian theorem as in Derman [3], pp 25-28, pp 143-144. Finally, we demonstrate that for an important class of MDP the solution concept of truncated annuities cannot sensibly be extended to the case $\rho > 1$ (Theorem 5). This seems to indicate that the concept of punctuated annuities is more general than that of truncated annuities.

We consider an infinite stage MDP with finite state space $S = \{1,\ldots,i,j,\ell,m,\ldots,M\}$ and finite feasible decision spaces K_i. A decision $k \in K_i$ in state i determines an immediate reward $r_i(k)$, to be multiplied by ρ^t, so as to yield the pay-off of stage $t=0,1,2,\ldots$. The parameter $\rho > 0$ represents a discount factor. The transition to a new state j occurs with probability

$$p_{ij}(k) \geq 0, \quad \sum_{j\varepsilon S} p_{ij}(k) = 1 \quad (i\varepsilon S, k\varepsilon K_i) \tag{1}$$

A decision function f is defined by a vector (k_i) of decisions $k_i \varepsilon K_i$, $i\varepsilon S$. A policy x is a sequence of decision functions $\{f_t\}_{t=0,1,2,\ldots}$. If $f_t = f$ for all $t \geq 0$, then the policy $x = f^\infty$ is called stationary. The set of all decision functions is denoted by F, and that of all policies by X.

If $f = (k_i)$ let $p_{ij}(f)$ denote $p_{ij}(k_i)$. Furthermore let $p_{ij}^{(t)}(x)$ represent t-stage transition probabilities under a policy x for all $t \geq 0$ with $p_{ij}^{(0)}(x) = \delta_{ij}$ (Kronecker delta). Define $r_j(f_t)$ by $r_j(k_{jt})$ if f_t is the decision function $f_t=(k_{jt})$ at stage t. For a T-stage process the expected present value with an initial state ℓ is then given by

$$u_\ell^T(x) = \sum_{t=0}^{T-1} \rho^t \sum_{j\varepsilon S} p_{\ell j}^{(t)}(x)\, r_j(f_t) \tag{2}$$

The T-truncated annuity $d_\ell^T(x)$ of a policy x, given initial state ℓ, is defined as the solution of

$$\sum_{t=0}^{T-1} \rho^t \sum_{j\varepsilon S} p_{\ell j}^{(t)}(x)\, r_j(f_t) = \sum_{t=0}^{T-1} \rho^t d_\ell^T(x) \tag{3}$$

According to (3) the present value (2) is equivalent to the present value of the same process in which the varying pay-offs $r_j(f_t)$ have been replaced by a fixed pay-off $d_\ell^T(x)$. Cf Hax [7] pp 10-11. We note that the T-truncated annuities are bounded: $|d_\ell^T(x)| \leq \max_{i,k} |r_i(k)|$.

A second (equivalent) interpretation of T-truncated annuities can be introduced if we reinterpret our MDP somewhat differently. A decision k in state i at stage t = 0,1,2,... determines a pay-off $r_i(k)$ per unit of time τ to be received over an interval of $\rho^t > 0$ units of time. At the end of such an interval a transition to a new state j occurs. Schematically this may be represented as follows:

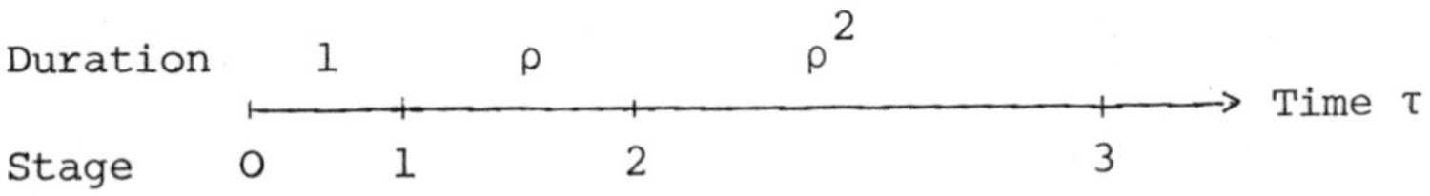

(4)

Solving (3) for $d_\ell^T(x)$ we observe that the T-truncated annuity is equal to the expected total pay-off divided by the duration of a process stopped after T stages have elapsed: $d_\ell^T(x) = u_\ell^T(x)/(1+\rho+\ldots+\rho^{T-1})$. In future definitions we shall always adhere to the second interpretation (4) of a MDP.

For $T \to \infty$ define <u>lower</u> and <u>upper truncated annuities</u> by

$$\begin{aligned} \underline{d}_\ell(x) &:= \liminf_{T \to \infty} \; d_\ell^T(x) \\ &\leq \limsup_{T \to \infty} \; d_\ell^T(x) =: \overline{d}_\ell(x) \end{aligned} \tag{5}$$

If lower and upper truncated annuities are equal, we speak of <u>truncated annuities</u> $d_\ell(x) = \underline{d}_\ell(x) = \overline{d}_\ell(x)$. The subscript ℓ is deleted if truncated annuities are independent of the initial state ℓ. A policy $x \in X$ is called <u>optimal with respect to truncated annuities</u> if for all initial states $\ell \in S$, and for all policies $y \in X$ every point of accumulation of $\{d_\ell^T(x)\}_{T=1,2,\ldots}$ is greater than or equal to every point of accumulation of $\{d_\ell^T(y)\}_{T=1,2,\ldots}$. This definition is equivalent to

$$\underline{d}_\ell(x) \geq \overline{d}_\ell(y) \quad (y \in X, \ell \in S). \tag{6}$$

If truncated annuities for an optimal policy x exist we write $d_\ell := \underline{d}_\ell(x) = \overline{d}_\ell(x)$ and delete the subscript if they are independent of ℓ.

In establishing the optimality criterion (6) for truncated annuities we have stopped the MDP at stage T, so as to define T-truncated annuities, and have subsequently let $T\to\infty$. In the following we shall derive an alternative optimality criterion using a different stopping rule in the same manner. Consider a process with initial state ℓ, which is to be stopped as soon as the state ℓ occurs for the N-th time ($N \geq 1$). If such a stopping rule is applied, then each realization may be split up into N renewal cycles as represented in the following diagram

$$\underbrace{\overbrace{\ell_0 \to\cdots\to \ell_1}^{N}}_{\substack{\text{Cycle}\\ 1}} \underbrace{\overbrace{\to\cdots\to \ell_2}^{N-1}}_{\substack{\text{Cycle}\\ 2}} \;\cdots\; \underbrace{\overbrace{\ell_{N-n} \to\cdots\to \ell_{N-n+1}}^{n}}_{\substack{\text{Cycle}\\ N-n+1}} \underbrace{\overbrace{\to\cdots\to \ell_{N-n+2}}^{n-1}}_{\substack{\text{Cycle}\\ N-n+2}} \;\cdots\; \underbrace{\overbrace{\ell_{N-1} \to\cdots\to \ell_N}^{1}}_{\substack{\text{Cycle}\\ N}} \tag{7}$$

Within a cycle the state ℓ does not occur. The subscript of ℓ represents the number of the cycle. The index n stands for the number of remaining cycles.

The case N=1 has been treated in detail by Reetz [12]. It is important for our investigation because it plays a representative role. With the stopping rule N=1 the state ℓ becomes absorbing. The resulting transformed MDP, which we call an ℓ-punctuated MDP possesses the transition probabilities

$${}_\ell p_{ij}(k) = \begin{cases} p_{ij}(k) & j \in S_\ell \\ 0 & j = \ell \end{cases} \quad (i,j\in S; k\in K_i) \tag{8}$$

with S_ℓ denoting the set $S - \{\ell\}$. For a given policy $x \in X$ the t-stage transition probabilities ${}_\ell p_{ij}^{(t)}(x)$ may be calculated by suitable matrix multiplication. As can be shown by induction on t, these transition probabilities are equal to the taboo probabilities

$$ {}_{\ell}p_{ij}^{(t)}(x) = \sum_{m_1 \varepsilon S_\ell} \cdots \sum_{m_{t-1} \varepsilon S_\ell} p_{im_1}(f_0) \cdots p_{m_{t-1}j}(f_{t-1}) \quad (9) $$

Hence the expected duration of a process under a policy x, starting in state i and stopping at the first entry into the specified state ℓ is given by the mean first passage time

$$ \mu_{i\ell}(x) = \sum_{t=0}^{\infty} \rho^t \sum_{j \varepsilon S} {}_{\ell}p_{ij}^{(t)}(x) \qquad (i \varepsilon S) \qquad (10) $$

Set ${}_{\ell}p_{ij}^{(0)}(x) = \delta_{ij}$. If the mean first passage times are finite for all stationary policies, $\mu_{i\ell}(f^{\infty}) < \infty \quad (i\varepsilon S, f\varepsilon F)$, we call ℓ a recurrent state. In this case the mean first passage times (10) for every policy $x \ \varepsilon \ X$ are finite, since the ℓ-punctuated MDP with transition probabilities (8) is transient. Cf. Blackwell [2] and Veinott [13]. If every state is recurrent we speak of a recurrent MDP. This obviously puts an upper bound $\bar{\rho}$ on ρ. In a recurrent MDP the punctuated present values

$$ u_{i\ell}(x) = \sum_{t=0}^{\infty} \rho^t \sum_{j \varepsilon S} {}_{\ell}p_{ij}^{(t)}(x) \ r_j(f_t) \qquad (i,\ell \varepsilon S) \qquad (11) $$

exist for all $x \ \varepsilon \ X$. Using (10) and (11) we may then define ℓ-punctuated annuities $a_\ell(x)$ as the expected total pay-off of one cycle in (7) starting and stopping in ℓ divided by the duration of this cycle: $a_\ell(x) = u_{\ell\ell}(x)/\mu_{\ell\ell}(x)$. Our first theorem is concerned with the existence and the determination of an optimal policy with respect to the punctuated annuities. A proof may be found in Reetz [12]. It coincides with the proof of Theorem 2 if N = 1.

Theorem 1

A recurrent MDP, satisfying the equal row sum condition (1), possesses a stationary policy $x = f^{\infty}$ which is optimal with respect to the punctuated annuities: $a_m(f^{\infty}) \geq a_m(y) \quad (m\varepsilon S, y\varepsilon X)$.

The optimal policy $x = f^{\infty}$ and the optimal punctuated annuities $a_m = a_m(f^{\infty})$ $(m \varepsilon S)$ can be obtained by first solving the following ℓ-punctuated system (12) - (13) for a_ℓ, $v_{i\ell}$ $(i \varepsilon S)$, f_ℓ and then using the transformations $f = f_\ell$, $a_m = a_\ell + (1-\rho) v_{m\ell}$ $(m \varepsilon S)$. The quantities a_ℓ and $v_{i\ell}$ $(i \varepsilon S)$ of (12) - (13) are unique.

$$v_{i\ell} + a_\ell = \overset{f_\ell}{\underset{k \varepsilon K_i}{\overset{\uparrow}{\max}}} \left\{ r_i(k) + \rho \sum_{j \varepsilon S} p_{ij}(k)\, v_{j\ell} \right\} \quad (i \varepsilon S) \tag{12}$$

$$v_{\ell\ell} = 0 \tag{13}$$

The maximization in (12) may yield more than one decision function. In order to rule this out, we shall adopt the convention of selecting (unique) optimal decision functions with the smallest indices k throughout the paper. A few remarks can be made about the above theorem: 1) the condition (1) can be replaced by some other conditions, which are however not easy to verify. We have omitted them for lack of space. 2) If $\rho = 1$, then the optimal punctuated annuities are independent of the specific punctuation and we write $a = a_\ell$ $(\ell \varepsilon S)$. 3) The assumptions of Theorem 1 do not allow us to claim that $v_{i\ell}$ are relative present values. On the other hand, there exist conditions (for example those in Morton/Wecker [10]) which guarantee that the $v_{i\ell}$ are relative present values. These same conditions may however yield negative a_ℓ, given positive $r_i(k)$, so that an interpretation of a_ℓ as an average criterion is questionable. We shall discuss this problem in a future paper.

Let us next consider a MDP which stops at the N-entry into the state ℓ as represented in (7). In order to prove a simple result we require the following complicated model. For mathematical reasons we wish to allow for policies differing in each renewal cycle and we therefore construct an <u>extended MDP</u> with state space

$$S^N = \{(i,n) \mid i \in S, \quad 1 \le n \le N\} \tag{14}$$

An <u>extended state</u> is a pair (i,n), with i representing the (original) state and n the number of <u>remaining</u> renewal cycles. The feasible decision spaces of the extended MDP are independent of n and are given by $K_i^n = K_i$ $(i\epsilon S, 1\le n\le N)$. The stochastic behaviour of the extended MDP is determined by

$$p_{(i,n)(j,m)}(k) = \begin{cases} p_{i\ell}(k) & \text{if } i\epsilon S, j = \ell, m = n-1\ge 1 \\ p_{ij}(k) & \text{if } i\epsilon S, j\epsilon S_\ell, \ 1\le m=n\le N \\ 0 & \text{otherwise} \end{cases} \tag{15}$$

Note that (15) coincides with (8) if N = 1. The immediate rewards in the extended MDP are independent of n and are given by the original rewards: $r_{(i,n)}(k) = r_i(k)$. The discount factor ρ remains unchanged.

An N-policy $\bar{x}^N$ in the extended MDP is defined by the matrix

$$\bar{x}^N = \begin{bmatrix} f_0^N & f_1^N & \cdots & f_t^N & f_{t+1}^N & \cdots \\ \vdots & \vdots & & \vdots & \vdots & \\ f_0^n & f_1^n & \cdots & f_t^n & f_{t+1}^n & \cdots \\ \vdots & \vdots & & \vdots & \vdots & \\ f_0^1 & f_1^1 & \cdots & f_t^1 & f_{t+1}^1 & \cdots \end{bmatrix} = \begin{bmatrix} x^N \\ \vdots \\ x^n \\ \vdots \\ x^1 \end{bmatrix} \tag{16}$$

An element f_t^n in (16) represents a decision function, which is to be applied in stage t with n cycles remaining. Each row of the matrix (16) is obviously an element of the set X. Thus, the element x^n in right array of (16) is a policy (as defined in the original MDP) which is to be used if n cycles remain. The lower right submatrix of (16) enclosed by the dashed lines is an n-policy which we designate by $\bar{x}_t^n$. The rows of this submatrix are written as $x_t^n, \ldots, x_t^1$ (starting from the top). Let $\bar{X}^N$ denote the set of all N-policies. If the columns of (16) are equal, the N-policy is called <u>stationary</u> and we use the symbol $\bar{x}_*^N = [f^n]$ to designate this property. If, on the other hand, the matrix (16) is composed of equal row vectors, we speak of a <u>cyclically independent</u> N-policy and write $\bar{x}_{.}^N = [f^t]$. A stationary cyclically independent N-policy $\bar{x}_{*.}^N = [f]$ is a matrix (16) such that $f_t^n = f$ for all $t \geq 0$, $1 \leq n \leq N$. Such an N-policy may be considered isomorphic to the stationary policy f^∞ in the original MDP.

If ℓ is a recurrent state, then the ℓ-punctuated MDP with transition probabilities (8) is transient. The transience of (8), in turn, implies the transience of the triangular system (15). Consequently, <u>mean n-th passage times</u> from i to ℓ exist for all N-policies $\bar{x}^N$:

$$\mu_{i\ell}^n(\bar{x}^N) := \sum_{t=0}^{\infty} \rho^t \sum_{(j,m)\in S^N} p_{(i,n)(j,m)}^{(t)}(\bar{x}^N) \quad ((i,n)\in S^N) \tag{17}$$

The expressions $p_{(i,n)(j,m)}^{(t)}(\bar{x}^N)$ in (17) denote the t-stage transition probabilities of (15), given the N-policy (16). We set $p_{(i,n)(j,m)}^{(0)}(\bar{x}^N) = \delta_{(i,n)(j,m)}$. Thus, (17) represents the expected duration of a process starting in state i with n cycles remaining and stopping at the n-th entry into the specified state ℓ. Obviously, only the lowest n rows of (16) are used in calculating (17). The n-th passage times (17) may alternatively be defined using the following recursion. Let $\mu_{i\ell}^1(\bar{x}_t^1) = \mu_{i\ell}^1(x_t^1)$ for all $\bar{x}_t^1$ $(t \geq N-1)$ according to (10). If $\mu_{i\ell}^{n-1}(\bar{x}_t^{n-1})$ $(i \in S)$ is known for all $\bar{x}_t^{n-1}$ $(t \geq N-n+1)$ determine $\mu_{i\ell}^n(\bar{x}_t^n)$ for all $\bar{x}_t^n$ $(t \geq N-n)$ by

$$\mu_{i\ell}^{n}(\bar{x}_t^{n}) = \sum_{s=0}^{\infty} \rho^s \sum_{j\varepsilon S} {}_{\ell}p_{ij}^{(s)}(x_t^n) .$$

$$\left[1 + \rho\, p_{j\ell}(f_{t+s}^{n})\, \mu_{\ell\ell}^{n-1}(\bar{x}_{t+s+1}^{n-1})\right] \tag{18}$$

$$= \mu_{i\ell}(x_t^n) + \sum_{s=1}^{\infty} \rho^s\, {}_{\ell}p_{i\ell}^{(s)}(x_t^n)\, \mu_{\ell\ell}^{n-1}(\bar{x}_{t+1}^{n-1})$$

If we continue in this manner until n = N we obtain $\mu_{i\ell}^{N}(\bar{x}^{N})$. The reader may wish to compare (18) with (11) and (10). The term in the brackets of (18) corresponds to the expression r_j in (11).

If ℓ is a recurrent state, then there also exist n-th punctuated present values

$$u_{i\ell}^{n}(\bar{x}^{N}) := \sum_{t=0}^{\infty} \rho^t \sum_{(j,m)\varepsilon S^N} p_{(i,n)(j,m)}^{(t)}(\bar{x}^{N})\, r_{(j,m)}(f_t^m) \tag{19}$$

Let $r_{(j,m)}(f_t^m) = r_j(f_t^m)$. The term $u_{i\ell}^{n}(\bar{x}^{N})$ represents the present value of a process controlled by the last (lower) n rows of (16), starting in state i with n remaining renewal cycles until break-off in state ℓ. If N=1 then (19) agrees with (11). The present values (19) may also be determined recursively in a manner analogous to (18), the details of which we omit. If the MDP is recurrent we may now define ℓ-punctuated N-annuities for all $\ell \varepsilon S$ as the ratio

$a_{\ell}^{N}(\bar{x}^{N}) := u_{\ell\ell}^{N}(\bar{x}^{N})/\mu_{\ell\ell}^{N}(\bar{x}^{N})$ using (18) and (19). $a_{\ell}^{N}(\bar{x}^{N})$ is equal to the expected total pay-off of N cycles in (7) starting and stopping in ℓ, divided by the expected duration of these N cycles if the N-policy $\bar{x}^{N}$ is applied. An N-policy $\bar{x}^{N}$ is called optimal with respect to punctuated N-annuities if $a_{\ell}^{N}(\bar{x}^{N}) \geq a_{\ell}^{N}(\bar{y}^{N})$ for all $\ell \varepsilon S$ and $\bar{y}^{N} \varepsilon \bar{X}^{N}$. The term $a_{\ell}^{N} = a_{\ell}^{N}(\bar{x}^{N})$

denotes the optimal ℓ-punctuated N-annuity. The existence of a stationary cyclically independent optimal N-policy $\bar{x}^N_{*\cdot}$ will be shown in Theorem 2.

Let us consider the case $N \to \infty$. We define an <u>∞-policy</u> $\bar{x}^{\infty}$ as an infinite matrix, whose elements are decision functions. Let $\bar{X}^{\infty} := \{\bar{x}^{\infty}\}$. In accord with previous definitions we characterize an ∞-policy as <u>stationary</u> if the column vectors are equal and as <u>cyclically independent</u> if the row vectors are equal. We denote such ∞-policies by $\bar{x}^{\infty}_{*}$ and $\bar{x}^{\infty}_{\cdot}$, respectively. The elements of a stationary cyclically independent ∞-policy $\bar{x}^{\infty}_{*\cdot} = [f]$ are all alike, so that this ∞-policy is isomorphic to the stationary policy f^{∞}.

For each N=1,2,... let $\bar{x}^N$ denote the matrix (N-policy) composed of the first (top) N rows of $\bar{x}^{\infty}$. Define $a^N_{\ell}(\bar{x}^{\infty}) = {}^N_{\ell}(\bar{x}^N)$. In analogy to the optimality definition of truncated annuities (6) we call an ∞-policy $\bar{x}^{\infty}$ <u>optimal</u> with respect to punctuated ∞-annuities if for all $\ell \in S$ and $\bar{y}^{\infty} \in \bar{X}^{\infty}$

$$\begin{aligned}\underline{a}^{\infty}_{\ell}(\bar{x}^{\infty}) &:= \liminf_{N \to \infty} a^N_{\ell}(\bar{x}^{\infty}) \\ &\geq \limsup_{N \to \infty} a^N_{\ell}(\bar{y}^{\infty}) =: \bar{a}^{\infty}_{\ell}(\bar{y}^{\infty})\end{aligned} \tag{20}$$

If such an ∞-policy exists, we call $a^{\infty}_{\ell} = a^{\infty}_{\ell}(\bar{x}^{\infty}) = \underline{a}^{\infty}_{\ell}(\bar{x}^{\infty}) = \bar{a}^{\infty}_{\ell}(\bar{x}^{\infty})$ the <u>optimal ℓ-punctuated ∞-annuity</u>. Theorem 3 demonstrates the existence of a stationary cyclically-independent policy $\bar{x}^{\infty}_{*\cdot}$ satisfying (20).

2. PUNCTUATED ANNUITIES

As a first step in this section we prove the existence of a stationary cyclically independent N-policy, which is optimal with respect to punctuated N-annuities. Such an N-policy may be obtained by solving the functional equation (12) - (13). Thus the optimization of punctuated N-annuities may be reduced to the optimization of punctuated annuities (N=1).

Theorem 2

A recurrent MDP, satisfying the equal row sum condition (1), possesses a stationary cyclically independent N-policy $\bar{x}^N_{*\cdot} = [f]$, which is optimal with respect to the punctuated N-annuities: $a^N_m(\bar{x}^N_{*\cdot}) \geq a^N_m(\bar{y}^N)$ for all $m \in S$ and $\bar{y}^N \in \bar{X}^N$.

The optimal N-policy $\bar{x}^N_{*\cdot} = [f]$ and the optimal punctuated N-annuities $a^N_m = a^N_m(\bar{x}^N_{*\cdot})$ $(m \in S)$ can be obtained by solving the ℓ-punctuated system (12) - (13) for a_ℓ, $v_{i\ell}$ $(i \in S)$, f_ℓ and using the transformations $f = f_\ell$, $a^N_m = a_\ell + (1-\rho)v_{m\ell}$ $(m \in S)$.

Proof: Let ℓ be a fixed state and a an arbitrary real parameter. Since ℓ is a recurrent state, parametric present values

$$v^n_{i\ell}(\bar{x}^N, a) := \sum_{t=0}^{\infty} \rho^t \sum_{(j,m)\in S^N} p^{(t)}_{(i,n)(j,m)}(\bar{x}^N) \left[r_{(j,m)}(f^m_j) - a \right] \tag{21}$$

$$= u^n_{i\ell}(\bar{x}^N) - a\, \mu^n_{i\ell}(\bar{x}^N) \qquad ((i,n) \in S^N)$$

exist. Cf. (17) and (19). Let $a^N_\ell(\bar{x}^N)$ be the unique zero of the function $v^N_{\ell\ell}(\bar{x}^N, a)$. We denote by

$$v^n_{i\ell}(a) := \sup_{\bar{x}^N \in \bar{X}^N} v^n_{i\ell}(\bar{x}^N, a) \qquad ((i,n) \in S^N) \tag{22}$$

the upper envelope of all $v^n_{i\ell}(\bar{x}^N, a)$. Each $v^n_{i\ell}(a)$ is obviously piecewise linear and continuous in a, and monotonically decreasing from $+\infty$ to $-\infty$, thus possessing a unique zero. Let a^N_ℓ denote the zero of $v^N_{\ell\ell}(a)$. Since $v^N_{\ell\ell}(a) \geq v^N_{\ell\ell}(\bar{x}^N, a)$,

a_ℓ^N dominates all of the zeros $a_\ell^N(\bar{x}^N)$, i.e. $a_\ell^N \geq a_\ell^N(\bar{x}^N)$.

Since ℓ is a recurrent state, the extended MDP with transition probabilities (15) is transient and we may appeal to fundamental properties of such MDP as stated in Veinott [13]. Thus, for each parameter a there exists a unique solution $v_{i\ell}^n(a)$ of the functional equation

$$v_{i\ell}^n(a) = \underset{k\varepsilon K_i}{\overset{f_\ell^n(a)}{\overset{\uparrow}{\max}}} \left\{ \left[r_i(k) - a\right] + \rho\ p_{i\ell}(k)\ v_{\ell\ell}^{n-1}(a) + \rho \sum_{j\varepsilon S_\ell} p_{ij}(k)\ v_{j\ell}^n(a) \right\} \quad (i\varepsilon S,\ 1 \leq n \leq N) \tag{23}$$

Set $v_{\ell\ell}^0(a) = 0$ in (23). The solution of (23) is equal to the upper envelope as defined in (22). Let $\bar{x}_*^N(a) = \left[f_\ell^n(a)\right]$ denote a stationary N-policy resulting from the maximization in (23). (The subscript ℓ denotes the fixed state ℓ and not the stage index t.) According to a fundamental theorem for transient MDP, this stationary N-policy is optimal with respect to parametric present values. For all (i,n) $\varepsilon\ S^N$ and $\bar{y}^N \varepsilon\ \bar{X}^N$ we thus have

$$v_{i\ell}^n(a) = v_{i\ell}^n(\bar{x}_*^N(a),\ a) \geq v_{i\ell}^n(\bar{y}^N,\ a) \qquad (i\ \varepsilon\ S) \tag{24}$$

By choosing a as the unique zero a_ℓ^N of $v_{\ell\ell}^N(a)$, the system (23) is transformed into the system

$$v_{i\ell}^n + a_\ell^N = \underset{k\varepsilon K_i}{\overset{f_\ell^n}{\overset{\uparrow}{\max}}} \left\{ r_i(k) + \rho\ p_{i\ell}(k)\ v_{\ell\ell}^{n-1} + \rho \sum_{j\varepsilon S_\ell} p_{ij}(k)\ v_{j\ell}^n \right\} \tag{25}$$

$$(i\ \varepsilon\ S,\ 1 \leq n \leq N)$$

$$v_{\ell\ell}^{N} = 0, \ v_{\ell\ell}^{0} = 0 \tag{26}$$

We have set $v_{i\ell}^{n}(a_{\ell}^{N}) = v_{i\ell}^{n}$ and $f_{\ell}^{n}(a^{N}) = f_{\ell}^{n}$. Since the solution of (23) is unique and there exists a unique zero a_{ℓ}^{N} of $v_{\ell\ell}^{N}(a)$, the system (25) - (26) possesses a unique solution. Let $\bar{x}_{*}^{N} = [f_{\ell}^{n}]$ denote the stationary N-policy determined by (25) - (26). If we substitute $a = a_{\ell}^{N}$ into (24) and select $i = \ell$ and $n = N$ we obtain as a consequence $a_{\ell}^{N} = a_{\ell}^{N}(\bar{x}_{*}^{N}) \geq a_{\ell}^{N}(\bar{y}^{N})$ for all $\bar{y}^{N} \in \bar{X}^{N}$. Thus we have shown the existence of a stationary N-policy which is optimal with respect to ℓ-punctuated N-annuity and which may be obtained by solving (25) - (26).

It is not difficult to see that $\bar{x}_{*}^{N}$ is also cyclically independent. Indeed a comparison of the unique solutions of systems (25) - (26) and (12) - (13) yields $v_{i\ell}^{n} = v_{i\ell}$, $a_{\ell}^{N} = a_{\ell}$ and $f_{\ell}^{n} = f_{\ell}$. Hence $\bar{x}_{*}^{N}. = [f_{\ell}]$.

Finally we show that $\bar{x}_{*}^{N}.$ does not depend on the specific punctuation ℓ. We compare the ℓ-punctuated system (12) - (13) with a corresponding m-punctuated system, both of which possess unique solutions. Using condition (1) this yields $a_{m} = a_{\ell} + (1 - \rho)\ v_{m\ell}$, $v_{im} = v_{i\ell} - v_{m\ell}$ $(i \in S)$ and $f_{m} = f_{\ell}$. Thus $\bar{x}_{*}^{N}.$ is independent of ℓ and can be obtained by solving (12) - (13). Simultaneously, all optimal m-punctuated N-annuities can be determined by $a_{m}^{N} = a_{m} = a_{\ell} + (1 - \rho)\ v_{m\ell}$ $(m \in S)$.

It is worthwhile to note the following immediate consequence of Theorem 2. Let $f = (k_{i})$ be an arbitrary decision function and set $K_{i} = \{k_{i}\}$. Then $a_{\ell}(f^{\infty}) = a_{\ell}^{N}(\bar{x}_{*}^{N}.)$ with $\bar{x}_{*}^{N}.$ given by $[f]$. For a nonstationary policy $x = \{f_{t}\}$, however, annuities and N-annuities will no longer be equal. Thus $a_{\ell}(x) \neq a_{\ell}^{N}(\bar{x}^{N}.)$ with $\bar{x}^{N}. = [f_{t}]$ in the general case $\rho \neq 1$.

Using Theorem 2 it is not difficult to prove the following fundamental theorem for recurrent MDP.

Theorem 3

A recurrent MDP, satisfying the equal row sum condition (1), possesses a stationary cyclically independent ∞-policy $\bar{x}^{-\infty}_{*\cdot} = \boxed{f}$, which is optimal with respect to the punctuated ∞-annuities: $\underline{a}^{\infty}_{m}(\bar{x}^{-\infty}_{*\cdot}) \geq \bar{a}^{-\infty}_{m}(\bar{y}^{-\infty})$ for all $m \in S$ and $\bar{y}^{-\infty} \in \bar{X}^{-\infty}$. The optimal ∞-policy $\bar{x}^{-\infty}_{*\cdot} = \boxed{f}$ and the optimal punctuated ∞-annuities $a^{\infty}_{m} = \underline{a}^{\infty}_{m}(\bar{x}^{-\infty}_{*\cdot}) = \bar{a}^{-\infty}_{m}(\bar{x}^{-\infty}_{\cdot})$ $(m \in S)$ can be obtained by solving the ℓ-punctuated system (12) - (13) for a_{ℓ}, $v_{i\ell}$ $(i \in S)$, f_{ℓ} and using the transformations $f = f_{\ell}$, $a^{\infty}_{m} = a_{\ell} + (1 - \rho)\, v_{m\ell}$ $(m \in S)$.

Proof: For each N=1,2,... let $x^{N}_{*\cdot} = \boxed{f}$ be the stationary cyclically independent optimal N-policy as specified in Theorem 2. Using the component f we define a stationary cyclically independent ∞-policy $\bar{x}^{-\infty}_{*\cdot} = \boxed{f}$. Let $\bar{y}^{-\infty}$ be an arbitrary ∞-policy. The first (top) N rows of $\bar{y}^{-\infty}$ define an N-policy $\bar{y}^{-N}$ for every N=1,2,... . According to Theorem 2 we have for all $m \in S$

$$a^{N}_{m}(\bar{x}^{-\infty}_{*\cdot}) = a^{N}_{m}(\bar{x}^{-N}_{*\cdot}) \geq a^{N}_{m}(\bar{y}^{-N}) = a^{N}_{m}(\bar{y}^{-\infty}) \qquad (27)$$

Letting $N \to \infty$ yields

$$\begin{aligned} \bar{a}^{-\infty}_{m}(\bar{x}^{-\infty}_{*\cdot}) &:= \limsup_{N \to \infty} a^{N}_{m}(\bar{x}^{-\infty}_{*\cdot}) \\ &\geq \limsup_{N \to \infty} a^{N}_{m}(\bar{y}^{-\infty}) =: \bar{a}^{-\infty}_{m}(\bar{y}^{-\infty}) \qquad (m \in S) \end{aligned} \qquad (28)$$

Since $a^{N}_{m}(\bar{x}^{-\infty}_{*\cdot}) = a^{N}_{m}(\bar{x}^{-N}_{*\cdot}) = a_{m}$ $(m \in S)$ by Theorem 2, we obtain $\bar{a}^{-\infty}_{m}(\bar{x}^{-\infty}_{*\cdot}) = \underline{a}^{\infty}_{m}(\bar{x}^{-\infty}_{*\cdot}) = a^{\infty}_{m}(\bar{x}^{-\infty}_{*\cdot}) = a_{m}$ on the left hand side of (28).

Thus there exists a stationary cyclically independent ∞-policy $\bar{x}^{\infty}_{*.} = [f]$ which is optimal with respect to ∞-annuities: $\underline{a}^{\infty}_{m}(\bar{x}^{\infty}_{*.}) \geq \bar{a}^{\infty}_{m}(\bar{y}^{\infty})$ for all $m \in S$ and $\bar{y}^{\infty} \in \bar{X}^{\infty}$. It suffices to solve (12) - (13) since $f = f_{\ell}$ and $a^{\infty}_{m} = a^{N}_{m} = a_{\ell} + (1 - \rho)\ v_{m\ell}$ for all $m \in S$, according to Theorem 2.

Theorems 2 and 3 demonstrate that optimal punctuated annuities, N-annuities and ∞-annuities are equal. For this reason we shall not consider punctuated N- or ∞- annuities in the next section. If f^{∞} is a stationary policy and $\bar{x}^{\infty}_{*.} = [f]$, then as a corollary to Theorem 3 we obtain $a_{\ell}(f^{\infty}) = a^{\infty}_{\ell}(\bar{x}^{\infty}_{*.})$. On the other hand let $x = \{f_t\}$ be a nonstationary policy and $\bar{x}^{\infty}_{.}$ the associated cyclically independent ∞-policy. Then in a recurrent MDP $a_{\ell}(x)$ will exist, but not in general $a_{\ell}(\bar{x}^{\infty}_{.})$.

3. TRUNCATED ANNUITIES

Using the results on punctuated annuities we investigate in this section the problem of existence of truncated annuities and the problem of determining optimal policies for them. We shall consider two classes of MDP: (i) single-chain MDP with $\rho = 1$, and (ii) single-chain aperiodic MDP with ρ lying in some interval $1 < \rho < \hat{\rho}$ to be specified. In our approach alternative proofs of well-known theorems for the first class are developed. The investigation of the second class uncovers a degeneracy, showing that the criterion of truncated annuities cannot sensibly be applied here.

We consider the first class first. Let $x = \{f_t\}$ be an arbitrary policy with components $f_t = (k_{it})$. A single-chain (irreducible) MDP with $\rho = 1$ and $K_i = \{k_{it}\}$ is obviously recurrent. The functional equation (12) - (13) is a linear system (29) - (30) in this case and possesses a unique bounded solution by Theorem 1. An alternative proof can be found in Derman/Veinott [5].

$$v_{i\ell}(f^{\infty}_{t},1) + a(f^{\infty}_{t},1) = r_i(f_t) + \sum_{j \in S} p_{ij}(f_t)\ v_{j\ell}(f^{\infty}_{t},1) \qquad (i \in S) \tag{29}$$

$$v_{\ell\ell}(f_t^{\infty},1) = 0 \tag{30}$$

Starting with (29) - (30) we have extended our notation by including the parameter ρ. The expression $a_\ell(x,\rho)$ for example represents an ℓ-punctuated annuity under the policy x if the discount factor is ρ. A similar convention holds for the other expressions.

Taking an appropriate T-stage average of both sides of (29) we obtain

$$\frac{\sum_{t=0}^{T-1} \rho^t \sum_{i\varepsilon S} p_{\ell i}^{(t)}(x)\; v_{i\ell}(f_t^{\infty},1)}{\sum_{t=0}^{T-1} \rho^t} + \frac{\sum_{t=0}^{T-1} \rho^t \sum_{i\varepsilon S} p_{\ell i}^{(t)}(x)\; a(f_t^{\infty},1)}{\sum_{t=0}^{T-1} \rho^t}$$

$$= \frac{\sum_{t=0}^{T-1} \rho^t \sum_{i\varepsilon S} p_{\ell i}^{(t)}(x)\; r_i(f_t)}{\sum_{t=0}^{T-1} \rho^t} + \frac{\sum_{t=0}^{T-1} \rho^t \sum_{j\varepsilon S} \sum_{i\varepsilon S} p_{\ell i}^{(t)}(x)\; p_{ij}(f_t) v_{j\ell}(f_t^{\infty},1)}{\sum_{t=0}^{T-1} \rho^t} \tag{31}$$

Letting

$$R_\ell^T(x,\rho) := \sum_{t=0}^{T-1} \rho^t \sum_{j\varepsilon S} \left[p_{\ell j}^{(t+1)}(x) - p_{\ell j}^{(t)}(x)\right] v_{j\ell}(f_t^{\infty},1) \tag{32}$$

we may rewrite (31) as

$$d_\ell^T(x,\rho) = \frac{\sum_{t=0}^{T-1} \rho^t\, a(f_t^{\infty},1)}{\sum_{t=0}^{T-1} \rho^t} - \frac{R_\ell^T(x,\rho)}{\sum_{t=0}^{T-1} \rho^t} \tag{33}$$

This equality is fundamental for the following presentation, since it relates truncated to punctuated annuities. In the next Lemma 1 we demonstrate the existence of truncated annuities for stationary policies in case $\rho = 1$ and show the equality with punctuated annuities.

Lemma 1:

In a single-chain MDP with $\rho = 1$ truncated annuities for stationary policies exist and are equal to punctuated annuities. Both annuities are independent of the initial state and specific punctuation:

$$d(f^{\infty},1) = a(f^{\infty},1) \qquad (f \in F) \tag{34}$$

Proof: If $\rho = 1$ and $x = f^{\infty}$ then (33) may be transformed into

$$d_{\ell}^{T}(f^{\infty}, 1) = a(f^{\infty}, 1) - \frac{1}{T} \sum_{j \in S} p_{\ell j}^{(T)}(f^{\infty})\, v_{j\ell}(f^{\infty}, 1) \tag{35}$$

Letting $T \to \infty$, we obtain (34).

Lemma 1 is used in the proof of Theorem 4. Note that the concept of optimality of truncated annuities is defined in (6), and that of punctuated annuities in Theorem 1.

Theorem 4:

In a single-chain MDP with $\rho = 1$ there exists a stationary policy which is optimal both with respect to truncated and punctuated annuities. An optimal policy f^{∞} and optimal annuities $d(f^{\infty}, 1) = a(f^{\infty}, 1)$ may be determined by solving the functional equations (12) - (13) with $\rho = 1$.

Proof: Letting $y = \{g_t\}_{t=0,1,2,\ldots}$ be an arbitrary policy and $\rho = 1$ in (33) we get

$$\begin{aligned} d_{\ell}^{T}(y,1) &= \frac{1}{T} \sum_{t=0}^{T-1} a(g_t^{\infty},1) - \frac{1}{T} R_{\ell}^{T}(y,1) \\ &\le \max_{g \in F} a(g^{\infty},1) - \frac{1}{T} \sum_{j \in S} p_{\ell j}^{(T)}(y)\, v_{j\ell}(g_{T-1}^{\infty},1) \end{aligned} \tag{36}$$

Using Theorem 1 and Lemma 1 with $T \to \infty$, inequality (36) implies

$$\begin{aligned} \overline{d}_\ell(y,1) &= \limsup_{T \to \infty} \; d_\ell^T(y,1) \\ &\leq \max_{g \varepsilon F} a(g^\infty,1) = a(f^\infty,1) \\ &= d(f^\infty,1) = d_\ell(f^\infty,1) = \underline{d}_\ell(f^\infty,1) \qquad (\ell \; \varepsilon \; S) \end{aligned} \tag{37}$$

Thus, the policy f^∞ determined by (12) - (13) is optimal with respect to punctuated and truncated annuities.

In the remainder of this section we consider the second class of single-chain aperiodic MDP with $1 < \rho < \hat{\rho}$. In order to specify $\hat{\rho}$ we introduce the following notation. For each decision function $f = (k_i)$ let $|\lambda_2(f)|$ be the <u>second largest modulus</u> for all eigenvalues $\lambda(f)$ of $P(f) = [p_{ij}(k_i)]$, i.e.

$$|\lambda_2(f)| = \max \{ |\lambda(f)| \; \Big| \; P(f)\, \nu(f) = \lambda(f)\, \nu(f), \quad \nu(f) \; \varepsilon \; \mathbb{R}^M, |\lambda(f)| \neq 1 \} \tag{38}$$

Set $|\lambda_2| = \max_{f \varepsilon F} |\lambda_2(f)|$ and $\hat{\rho} = 1 \,/\, |\lambda_2|$. For later use we require some auxiliary results in the form of Lemma 2, the proof of which may be found in Hadeler [6] and also in Bartmann [1].

<u>Lemma 2:</u>

Let $P = [p_{ij}]_{i,j \varepsilon S} \geq 0$ be an irreducible matrix with row sums equal to 1, $B = [b_{ij}]_{i,j \varepsilon S} \geq 0$ a matrix with identical rows: $b_{ij} = b_j$ ($i \; \varepsilon \; S$), and $Q = [q_{ij}]_{i,j \varepsilon S} := P - B$. Then every eigenvalue λ of P with the exception of $\lambda = 1$ is also an eigenvalue of Q. The remaining eigenvalue of Q is equal to $1 - \sum_{j \varepsilon S} b_j$.

If P is irreducible and aperiodic, 1 is the only eigenvalue of P with modulus 1 (Karlin [9], p. 101). If b_j are selected, so that $\sum_{j\varepsilon S} b_j = 1$, then Lemma 2 implies the following corollary.

Corollary:

Let $P \geq 0$ be an irreducible, aperiodic stochastic matrix and $b = (b_j)_{j\varepsilon S}$ a probability vector. Then the spectral radius of the matrix Q in Lemma 2 is equal to the second largest modulus of all eigenvalues of P, which is less than 1.

The above Corollary plays an essential role in the proof of the following Theorem 5 on the existence of truncated annuities with $1 < \rho < \hat{\rho}$.

Theorem 5:

In a single-chain aperiodic MDP with $1 < \rho < \hat{\rho}$, truncated annuities of stationary policies exist and are independent of the initial state ℓ and discount factor ρ, being equal to the truncated or punctuated annuities with $\rho = 1$:

$$d(f^{\infty},\rho) = d(f^{\infty},1) = a(f^{\infty},1) \tag{39}$$

Proof: We show that $R_{\ell}^{T}(x,\rho)$ of (32) is absolutely convergent if $x = f^{\infty}$ for some arbitrary $f \varepsilon F$ and ρ satisfies $1 < \rho < 1/|\lambda_2(f)|$. Suppressing the fixed parameter ℓ as well as the decision function f as far as possible we introduce the notation

$$\pi_i^{(t)} := p_{\ell i}^{(t)}(f^{\infty}) \qquad (i\varepsilon S,\ t = 0,1,2,...) \tag{40}$$

The sequence $\pi^{(t)}$ may be generated by the Chapman-Kolmogoroff recursion

$$\pi_j^{(t+1)} = \sum_{i\varepsilon S} \pi_i^{(t)} p_{ij} \qquad (j\varepsilon S,\ t = 0,1,2,...) \tag{41}$$

with p_{ij} denoting $p_{ij}(k_i)$, $f = (k_i)$. To demonstrate the geometric convergence of (41) the following auxiliary sequences (42) and (43) are defined, with $b = (b_j)$ as a given probability vector and $q_{ij} := p_{ij} - b_j$.

$$\eta_j^{(t+1)} = -b_j + \sum_{i\varepsilon S} \eta_i^{(t)} q_{ij} \qquad (j \varepsilon S) \tag{42}$$

$$\theta_j^{(t+1)} = \sum_{i\varepsilon S} \theta_i^{(t)} q_{ij} \qquad (j \varepsilon S) \tag{43}$$

If initial values are specified by $\eta_i^{(0)} = 0$, $(i\varepsilon S)$ and $\theta_i^{(0)} = \delta_{\ell i}$, $(i\varepsilon S)$, then $\pi_i^{(0)} = \theta_i^{(0)} - \eta_i^{(0)}$, $(i\varepsilon S)$. Since $\sum_j p_{ij} \equiv 1$, it can be shown by induction on t, that

$$\pi_i^{(t)} = \theta_i^{(t)} - \eta_i^{(t)} \qquad (i\varepsilon S,\ t = 0,1,2,\ldots) \tag{44}$$

If P(f) is single-chain (irreducible) and aperiodic with row sums equal to 1, then by our Corollary the spectral radius of Q(f) is equal to $|\lambda_2(f)|$. By a fundamental theorem (Ortega/Rheinboldt [11], p. 44) there exists a matrix norm $\| \ \|$ depending on f, such that $\|Q(f)\| \leq |\lambda_2(f)| + \varepsilon$ for arbitrarily small $\varepsilon > 0$. Applying an appropriate consistent vector norm to (44) we obtain

$$\begin{aligned} \|\pi^{(t+1)} - \pi^{(t)}\| &\leq \|\eta^{(t+1)} - \eta^{(t)}\| + \|\theta^{(t+1)} - \theta^{(t)}\| \\ &\leq \|Q(f)\|^t \{\|\eta^{(1)} - \eta^{(0)}\| + \|\theta^{(1)} - \theta^{(0)}\|\} \\ &\leq (|\lambda_2(f)| + \varepsilon)^t \{\|\eta^{(1)} - \eta^{(0)}\| + \|\theta^{(1)} - \theta^{(0)}\|\} \end{aligned} \tag{45}$$

Since $1 < \rho < 1/|\lambda_2(f)|$ an $\varepsilon > 0$ may be selected such that $\gamma := \rho\ (|\lambda_2(f)| + \varepsilon) < 1$ and therefore

$$\rho^t \|\pi^{(t+1)} - \pi^{(t)}\| \leq \gamma^t \{\|\eta^{(1)} - \eta^{(0)}\| + \|\theta^{(1)} - \theta^{(0)}\|\} < 1 \tag{46}$$

Consequently, $R_\ell^T(f^\infty,\rho)$ is absolutely convergent, implying

$$\lim_{T\to\infty} \frac{R_\ell^T(f^\infty,\rho)}{\sum_{t=0}^{T-1} \rho^t} = 0 \qquad (47)$$

The equalities (39) are in immediate result of (33), (34) and (47).

According to Theorem 5 truncated annuities do exist for stationary policies in single-chain aperiodic MDP with $1 < \rho < \hat{\rho}$, but they are indifferent with respect to the discount factor. Such a degeneracy speaks heavily against using the criterion of truncated annuities if $\rho > 1$. As we have seen in Section 2 the criteria of punctuated N- and ∞-annuities, on the other hand, can be used in this case if the MDP is recurrent $(1 \leq \rho < \bar{\rho})$.

4. REFERENCES

1. Bartmann, D. "Optimierung Markovscher Entscheidungsprozesse", Dissertation, Müchen, (1975).

2. Blackwell, D. "Discrete dynamic programming", *Ann. Math. Statist.*, **33**, 719-726, (1962).

3. Dermann, C. "Finite state Markovian decision processes", Academic Press, New York, (1970).

4. Derman, C. "Denumerable state Markovian decision processes-average cost criterion", *Ann. Math. Statist.*, **37**, 1545-1554, (1966).

5. Derman, C. and Veinott, A. F. Jr., "A solution to a countable system of equations arising in Markovian decision processes", *Ann. Math. Statist.*, **38**, 582-584, (1967).

6. Hadeler, K. P. "Abschätzungen für den zweiten Eigenwert eines positiven Operators", *Aeq. Math.*, **5**, 227-235, (1970).

7. Hax, H. "Investitionstheorie", 2. Aufl. Physica-Verlag, Würzburg-Wien, (1972).

8. Howard, R. A. "Dynamic Programming and Markov Processes", MIT Press Cambridge, (1960).

9. Karlin, S. "A first course in stochastic processes", Academic Press, New York-London, (1966).

10. Morton, T. E. and Wecker, W. E. "Discounting, ergodicity and convergence for Markov decision processes", *Management Science,* **23**, 890-899, (1977).

11. Ortega, J. M. and Rheinboldt, W. C. "Iterative solution of nonlinear equations in several variables", Academic Press, New York, (1970).

12. Reetz, D. "Average Payoff criteria for ρ-recurrent Markovian decision processes", *Math. Centre Tract.,* **93**, 95-103, Amsterdam, (1977).

13. Veinott, A. F. "Discrete dynamic programming with sensitive discount optimality criteria", *Ann. Math. Statist.,* **40**, 1635-1660, (1969).

FINITE STATE APPROXIMATIONS FOR DENUMERABLE STATE INFINITE HORIZON DISCOUNTED MARKOV DECISION PROCESSES: THE METHOD OF SUCCESSIVE APPROXIMATIONS

D.J. White

(Department of Decision Theory, University of Manchester)

1. INTRODUCTION

Fox [1] discusses a scheme for approximating the solution to the following problem. Let: $I = \{1,2,\ldots,\}$ be a denumerable set of states; $\Delta = \{\delta\}$ be a set of stationary policies, where $\Delta = \Pi_i K_i$, and K_i is the admissible decision set for state i; $r_\delta(i)$ $[r_k(i)]$ be the bounded immediate return associated with using policy δ [decision k] for state i; $Q_\delta = [q_\delta(i,j)]$ be a matrix with the contraction mapping property that $||H_\delta u - H_\delta v|| \leq c||u-v||$, $c < 1$, $u,v \in \ell_\infty$, where $H_\delta(u(i)) = r_\delta(i) + \sum_{j\in I} q_\delta(i,j)u(j)$, and where, for our purposes, $||\ \ ||$ is the usual supremum norm; H_δ be monotonic i.e. $u \geqslant v \rightarrow H_\delta u \geqslant H_\delta v$; $v^\delta(i) = \left[\sum_{t=0}^{\infty} Q_\delta^t r_\delta\right]_i$ (where the suffix i on the right hand side means "the ith component"); $v^*(i) = \sup_{\delta\in\Delta}[v^\delta(i)]$.

Fox's objective is to produce a finite state approximation scheme for $v^*(.)$, pointwise, where $v^*(.)$ is the unique solution in ℓ_∞ of the following equation, with the appropriate notational interpretation,

$$\underline{i\in I} \quad w(i) = \sup_{k\in K_i} \left[r_k(i) + \sum_{j\in I} q_k(i,j)w(j)\right]. \tag{1}$$

Fox's scheme is to define a sequence $\{v^{\delta,n}(.)\}$ as follows.

$$\underline{i \leq n} \quad v^{\delta,n}(i) = r_\delta(i) + \sum_{j \leq n} q_\delta(i,j) v^{\delta,n}(j)$$

$$\underline{i > n} \quad v^{\delta,n}(i) = 0.$$

Fox establishes, under certain weak conditions, that $\lim_n \sup_\delta [v^{\delta,n}(i)] = v^*(i)$, $i \in I$, and that, given $r_k(i) \geq 0$, $\forall\ i,k$, convergence is monotonic increasing. White [2] studies the bounds on the associated approximations, and, implicitly, gives a convergence result which is independent of the sign of $\{r_k(i)\}$, but with an extra condition on the tails of the probability distributions.

The purpose of this paper is to establish results for the following method of successive approximations for solving (1).

Let u(.) be an arbitrary member of ℓ_∞, whose admissible form will be studied later. Define a sequence $\{v_n(.)\}$ as follows.

$$\underline{n \geq 1} \quad \underline{i \leq n} \quad v_n(i) = \sup_{k \in K_i} \left[r_k(i) + \sum_{j \in I} q_k(i,j) v_{n-1}(j) \right] \quad (2)$$

$$\underline{i > n} \quad v_n(i) = u(i)$$

$$\underline{n = 0} \quad \underline{i \in I} \quad v_0(i) = u(i)$$

It is to be noted that u(.) has to be chosen in advance, and that, although the scheme is a finite state scheme in terms of the introduction of one state at a time in calculating $\{v_n(.)\}$, the right hand side of (2) may involve an infinite set of terms, in which case there are clearly some computational problems. One possibility is to put u(i) = 0, $i \in I$, in which case the computations involve only a finite set of additions at each stage. However, it was thought useful to keep the u(.) as general as possible at this stage and it will later be seen that the form of u(.) has some significance for the scheme.

Apart from using u(.) instead of 0 in this scheme, the essential difference between the method of Fox and the method in this scheme is that Fox requires us to solve the problem $\sup_{\delta}[v^{\delta,n}(i)]$ for each n and $i \leq n$, whereas in the proposed scheme $v_n(.)$ is calculated, using successive approximation methods, from the knowledge of $v_{n-1}(.)$.

We will first of all deal with situations in which the sequence $\{v_n(.)\}$ will exhibit monotonicity properties. We will drop the "$k \in K_i$" explicitly to simplify notational problems. Also the summation range for j will be I unless otherwise stated, as will suprema and infima.

2. MONOTONICITY ANALYSIS

From (2) we have

$$\underline{n \geq 2} \quad \underline{i \leq n-1} \quad \inf_k\Big[\sum_j q_k(i,j)(v_{n-1}(j)-v_{n-2}(j))\Big]$$

$$\leq v_n(i)-v_{n-1}(i) \leq \sup_k\Big[\sum_j q_k(i,j)(v_{n-1}(j) - v_{n-2}(j))\Big]. \tag{3}$$

We then have the following theorem.

<u>Theorem 1</u>

If condition (4) ((5)) holds the sequence $\{v_n(.)\}$ is monotonic decreasing (monotonic increasing).

$$u(i) \geq \sup_k\Big[r_k(i) + \sum_j q_k(i,j)u(j)\Big] \tag{4}$$

$$u(i) \leq \sup_k\Big[r_k(i) + \sum_j q_k(i,j)u(j)\Big]. \tag{5}$$

Proof

Using induction, monotonicity follows for $i \leq n-1$ in (3) since $v_1(i) \leq v_0(i)$ $(v_1(i) \geq v_0(i))$ if (4) ((5)) holds. For $i = n$

$$v_n(n) = v_{n-1}(n) = \sup_k \left[r_k(n) + \sum_j q_k(n,j) v_{n-1}(j) \right] - u(n). \qquad (6)$$

Now if $v_{n-1}(j) \leq u(j)$, $j \in I$, from (6) we obtain, using (4),

$$v_n(n) - v_{n-1}(n) \leq \sup_k \left[r_k(n) + \sum_j q_k(n,j) u(j) \right] - u(n) \leq 0. \qquad (7)$$

If $v_{n-1}(j) \geq u(j)$, $j \in I$ we obtain the reverse inequality in (7), using (5).

The case $i > n$ is trivial, and hence the theorem is established. Δ

(4) ((5)) is clearly established if $r_k(i) \leq 0$, $\forall$ i,k, $u(i) = 0$, $\forall$ i, $(r_k(i) \geq 0$, $\forall$ i,k, $u(i) = 0$, $\forall$ $i)$. If $u(.)$ corresponds to some policy $\delta \in \Delta$, (5) is automatically satisfied. If $u(.) = v^*(.)$, then both (4) and (5) are satisfied, and $v_n(.) = v^*(.)$, $\forall$ n.

We then have the following convergence theorem.

Theorem 2

If the sequence $\{v_n(.)\}$ is monotonic and if the following condition (8) is satisfied, the sequence $\{v_n(.)\}$ converges pointwise to $v^*(.)$.

$$\underline{i \in I} \quad \lim_{m \to \infty} \sup_k \left[\sum_{j>m} q_k(i,j) \right] = 0. \qquad (8)$$

<u>Remark</u> If the set K_i for $i\in I$ is finite, this condition is trivially satisfied.

<u>Proof</u>

Since $\{v_n(.)\}$ is bounded above and below, the monotonicity condition guarantees that the sequence $\{v_n(.)\}$ is pointwise convergent to some member $v(.)\in\ell_\infty$.

$$\underline{n\geq 1}\quad \underline{i\in I}\quad \Big|\sup_k\Big[r_k(i)+\sum_j q_k(i,j)v_{n-1}(j)\Big]-\sup_k\Big[r_k(i)+\sum_j q_k(i,j)v(j)\Big]\Big| \leq \sup_k\Big[\sum_{j\leq m} q_k(i,j)\,\big|v_{n-1}(j)-v(j)\big|\Big]+\sup_k\Big[\sum_{j>m} q_k(i,j)\,\big|v_{n-1}(j)-v(j)\big|\Big]. \quad (9)$$

By making m large enough, using (8) and the boundedness of $v_{n-1}(.),v(.)$, the second term on the right hand side of (9) can be made arbitrarily small. For such a large, but finite, m, using the pointwise convergence property, the first term on the right hand side can then be made arbitrarily small if n is large enough. Hence the left hand side of (9) can be made arbitrarily small if n is large enough. Since we may make $v_n(i)$ arbitrarily close to $v(i)$ if n is large enough, and since $v_n(i)$ satisfies (2) if n is large enough, we see that $v(i)$ satisfies (1). Hence, $v(.) = v^*(.)$. Δ

We could now consider the bounds on the approximation of $\{v_n(.)\}$ to $v^*(.)$. However, we will do this for the situations which may not be monotonic, and the monotonic case will be a special case of this.

3. NON-MONOTONICITY ANALYSIS AND BOUNDS

From (2) we obtain, for arbitrary $N \geq 1$,

$$\underline{n\geq 2}\quad \underline{i\leq n-1}\quad |v_n(i)-v_{n-1}(i)| \leq \sup_k\Big[\sum_{j=1}^{N+i-1} q_k(i,j)\,|v_{n-1}(j) - v_{n-2}(j)|\Big] + \sup_k\Big[\sum_{j=N+i}^{\infty} q_k(i,j)\,|v_{n-1}(j) - v_{n-2}(j)|\Big]. \qquad (10)$$

Let

$$\underline{i\in I}\quad \underline{r\geq 0}\quad \tau(i,r,N) = \sup_{n\geq rN+i}\Big[|v_n(i)-v_{n-1}(i)|\Big] \qquad (11)$$

$$\underline{n\geq 1}\quad A = \sup_{i,n}\Big[|v_n(i)-v_{n-1}(i)|\Big]. \qquad (12)$$

Then, since $1 \leq j \leq N+i-1$, $n \geq rN+i$ implies $n-1 \geq (r-1)N+j$, we have, from (10), (11), (12),

$$\underline{i\in I}\quad \underline{r\geq 1}\quad \tau(i,r,N) \leq \sup_k\Big[\sum_{j=1}^{N+i-1} q_k(i,j)\tau(j,r-1,N)\Big] + A\sup_k\Big[\sum_{j=N+i}^{\infty} q_k(i,j)\Big]. \qquad (13)$$

Let

$$\varepsilon(N) = \sup_{i,k}\Big[\sum_{j\geq N+i} q_k(i,j)\Big]. \qquad (14)$$

Then from (13), (14) and the contraction properties of $\{q_k(i,j)\}$, we have

$$\underline{i\in I}\quad \underline{r\geq 1}\quad \tau(i,r,N) \leq c \max_{1\leq j<N+i}\Big[\tau(j,r-1,N)\Big] + A\varepsilon(N). \qquad (15)$$

If we now repeat (15) we obtain

$$\underline{i\in I}\quad \underline{r\geq 1}\quad \tau(i,r,N) \leq ((1-c^r)/(1-c))A\varepsilon(N) + c^r \max_{1\leq j<rN+i}\Big[\tau(j,0,N)\Big]. \qquad (16)$$

Now

$$\tau(j,0,N) = \sup_{n\geq j}\left[\left|v_n(j)-v_{n-1}(j)\right|\right] \leq A. \qquad (17)$$

Hence, from (16), (17), and since it is trivial that the following result is true for $r = 0$, we have

$$\underline{i\in I}\quad \underline{r\geq 0}\quad \tau(i,r,N) \leq A\tau(r,N) \qquad (18)$$

where

$$\tau(r,N) = (c^r + ((1-c^r)/(1-c))\varepsilon(N)). \qquad (18)$$

(18) implies

$$\underline{n\geq rN+i}\quad \underline{r\geq 0}\quad \underline{i\in I}\quad v_n(i) - A\tau(r,N) \leq v_{n-1}(i) \leq v_n(i)$$

$$+ A\tau(r,N).$$

If we replace $v_{n-1}(i)$ by $v^*(i)$ in the above analysis, and use the following definitions, we obtain result (20).

$$\underline{i\in I}\quad \underline{r\geq 0}\quad \lambda(i,r,N) = \sup_{n\geq rN+i}\left[\left|v_n(i)-v^*(i)\right|\right]$$

$$B = \sup_{i,n}\left[\left|v_n(i)-v^*(i)\right|\right]$$

$$\lambda(i,r,N) \leq B\tau(r,N)$$

$$\underline{n\geq rN+i}\quad \underline{r\geq 0}\quad \underline{i\in I}\quad v_n(i) - B\tau(r,N) \leq v^*(i)$$

$$\leq v_n(i) + B\tau(r,N). \qquad (20)$$

As a consequence of (19), we have the following theorem.

Theorem 3

If condition (21) holds, the sequence $\{v_n(.)\}$ converges pointwise to $v^*(.)$.

$$\lim_{N\to\infty}\left[\varepsilon(N)\right] = 0. \tag{21}$$

Proof

From (19), (21), $\tau(r,N)$ tends to 0 as r,N tend to ∞. Δ

Now although (20) gives bounds for $v^*(.)$, we need to determine policies which will give rise to a value function within specified bounds. Let δ_n be a policy such that, given $\eta > 0$, δ_n brings the right hand side of (2) within an amount η of its supremum, for $i \leqslant n$, and is arbitrary for $i > n$. We then have the following.

$$\underline{n\geqslant 1}\quad \underline{i\leqslant n}\quad v_n(i) - v^{\delta_n}(i) = (v_n(i) - r_{\delta_n(i)}(i)$$

$$- \sum_j q_{\delta_n(i)}(i,j)v_{n-1}(j)) + (\sum_j q_{\delta_n(i)}(i,j)(v_{n-1}(j)$$

$$- v_n(j))) + (\sum_j q_{\delta_n(i)}(i,j)(v_n(j) - v^{\delta_n}(j))).$$

Let

$$\underline{r\geqslant 0}\quad \mu(i,r,N) = \sup_{n\geqslant rN+i}\left[\left|v_n(i) - v^{\delta_n}(i)\right|\right]$$

$$C = \sup_{i,n}\left[\left|v_n(i) - v^{\delta_n}(i)\right|\right].$$

Then, applying a similar analysis to the preceding one, we obtain the following.

$$\underline{r\geqslant 1}\quad \mu(i,r,N) \leqslant \eta + c\max_{1\leqslant j<N+i}\left[\tau(j,r-1,N)\right]$$

$$+ c\max_{1\leqslant j<N+i}\left[\mu(j,r-1,N)\right] + C\varepsilon(N) + A\varepsilon(N). \tag{22}$$

We have

$$\mu(j,0,N) = \sup_{n \geq j}\left[\left|v_n(j) - v^{\delta_n}(j)\right|\right] \leq C.$$

Hence, repeating (22), since the following result (23) is also clearly true for $r = 0$, we have, with appropriate definitions, result (23).

$$\underline{r \geq 0} \quad \mu(i,r,N) \leq A\mu(r,N) + C\mu'(r,N) + \eta\mu''(r,N)$$

where

$$\mu(r,N) = (((1-c^r)-(1-c)rc^r)/(1-c)^2)\varepsilon(N) + rc^r$$

$$\mu'(r,N) = ((1-c^r)/(1-c))\varepsilon(N) + c^r$$

$$\mu''(r,N) = (1-c^r)/(1-c)$$

$\underline{n \geq rN+i}$ $\underline{r \geq 0}$ $\underline{i \in I}$

$$v_n(i) - A\mu(r,N) - C\mu'(r,N) - \eta\mu''(r,N) \leq v^{\delta_n}(i) \leq v_n(i)$$

$$+ A\mu(r,N) + C\mu'(r,N) + \eta\mu''(r,N) \tag{23}$$

Combining (20), (23), we have, since $v^*(i) \geq v^{\delta_n}(i)$, $i \in I$, the following theorem.

<u>Theorem 4</u>

$\underline{n \geq rN+i}$ $\underline{r \geq 0}$ $\underline{i \in I}$

$$v^{\delta_n}(i) \leq v^*(i) \leq v^{\delta_n}(i) + A\mu(r,N) + B\tau(r,N)$$

$$+ C\mu'(r,N) + \eta\mu''(r,N). \tag{24}$$

●

If (21) is true, then the limit of (24) as r,N tend to ∞ is as follows, where $\lim_n\left[v^{\delta_n}(i)\right]$ is <u>any</u> limit point, which must exist because $\{v^{\delta_n}(.)\}$ is bounded above and below.

$$\underline{i \in I} \quad \lim_n\left[v^{\delta_n}(i)\right] \leq v^*(i) \leq \lim_n\left[v^{\delta_n}(i)\right] + \eta/(1-c).$$

If a policy δ_n exists for which $\eta = 0$, then $\lim_n [v^{\delta_n}(.)]$ must exist from (24).

4. UNIFORMLY NEAR OPTIMAL POLICIES

Now, given $\eta > 0$, the policy δ_n (depending on η) will give a $v^{\delta_n}(i)$ value within a specified distance of the optimal $v^*(i)$ <u>for the specified i</u>. It is of interest to see if a uniformly near optimal policy, δ, exists, such that $|v^{\delta}(i)-v^*(i)|$ is within a specified error range for all i. Let us define, for a specific r, $N(\geq 1)$, a policy δ as follows.

$$\underline{i\in I} \quad \delta(i) = \delta_{rN+i}(i).$$

We then have the following, where we write δ_i for δ_{rN+i}, r,N understood.

$$\underline{i\in I} \quad v^{\delta_i}(i) = r_{\delta_i}(i) + \sum_j q_{\delta_i}(i,j)v^{\delta_i}(j) \tag{25}$$

$$v^{\delta}(i) = r_{\delta_i}(i) + \sum_j q_{\delta_i}(i,j)v^{\delta}(j). \tag{26}$$

Hence from (25), (26) we have the following.

$$\underline{i\in I} \quad v^{\delta_i}(i) - v^{\delta}(i) = \sum_j q_{\delta_i}(i,j)(v^{\delta_i}(j)-v^{\delta}(j)). \tag{27}$$

Now, from (24), (27) we have the following, since $v^{\delta_i}(j) \leq v^*(j)$, $i,j\in I$,

$$\underline{i\in I} \quad v^*(i)-v^{\delta}(i) \leq A\mu(r,N)+B\tau(r,N)+C\mu'(r,N)+\eta\mu''(r,N)$$

$$+ \sum_j q_{\delta_i}(i,j)(v^*(j)-v^{\delta}(j)). \tag{28}$$

From (28) we automatically obtain the following theorem.

Theorem 5

$$\underline{i \in I} \quad 0 \leqslant v^*(i) - v^{\delta}(i) \leqslant (A\mu(r,N) + B\tau(r,N) + C\mu'(r,N) + \eta\mu''(r,N))/(1-c).$$

Hence we can make $|v^*(i) - v^{\delta}(i)|$ arbitrarily small, simultaneously for all $i \in I$, if r,N are large enough, and η is small enough.

In practice, if we commence with $i = i_0$, we would determine δ_{i_0}, and then calculate δ_j at any subsequent stage only when j was greater than the highest state entered into prior to that stage.

As a point of interest, if j(max) is the highest state to date at any stage, we might consider using $\delta_{j(\max)}$ for all subsequent states up to j(max) until we reach a state $j > j(\max)$. However, this would give a non-stationary policy and further analysis would be needed.

In the previous analysis the degrees of approximation have been related to three parameters A,B,C. Let us now consider how upper bounds for A,B,C may be determined.

5. UPPER BOUNDS FOR A,B,C

Let

$$M = \sup_{i,k} [|r_k(i)|]$$

$$m = \sup_{i} [|u(i)|]$$

$$\underline{n \geqslant 1} \quad x_n = \sup_{i} [|v_n(i)|]$$

$$z_n = \max_{i \leqslant n} [|v_n(i)|].$$

Then from (2) we have the following, with $z_0 = 0$.

$$\underline{n \geq 1} \quad z_n \leq \max\left[M + cz_{n-1}, M + cm\right]. \tag{29}$$

Let

$$\alpha_t = \left(\sum_{s=0}^{t-1} c^s\right)M + c^t m.$$

Then, repeating (29), we have the following.

$$\underline{n \geq 1} \quad x_n \leq \max\left[z_n, m\right] \leq \max\left[\max_{1 \leq t \leq n}\left[\alpha_t\right], m\right].$$

Let

$$x = \sup_{n \geq 1}\left[x_n\right].$$

We then have the following.

$$x \leq \max\left[\sup_{t \geq 1}\left[\alpha_t\right], m\right].$$

Since $\alpha_t - \alpha_{t-1} = c^{t-1}(M-(1-c)m)$ for $t \geq 2$, we see that

$$x \leq \max\left[M/(1-c), m\right]. \tag{30}$$

Hence, from (30) we have the following.

$$A \leq 2 \max\left[M/(1-c), m\right].$$

Since clearly $|v^*(i)| \leq M/(1-c)$, $i \in I$, we have the following.

$$B \leq \max\left[2M/(1-c), m + M/(1-c)\right].$$

Similarly

$$C \leq \max\left[2M/(1-c), m + M(1-c)\right].$$

If $\{v_n(.)\}$ is monotonic we obviously have the following.

$$A \leqslant \max[\bar{M}/(1-c), m]$$

$$B \leqslant \max[\bar{M}/(1-c), \bar{m}]. \qquad (31)$$

If $r_k(i) \geqslant 0$, $\forall$ i,k, or $r_k(i) \leqslant 0$, $\forall$ i,k we can get still better bounds.

Even without the monotonicity conditions, we can still get bound (31), which follows from the following theorem.

Theorem 6

If $u(i) \leqslant v^*(i)$, $i \in I$ ($u(i) \geqslant v^*(i)$, $i \in I$), we have $v_n(i) \leqslant v^*(i)$, $i \in I$ ($v_n(i) \geqslant u^*(i)$, $i \in I$), $n \geqslant 0$.

Proof

Let $u(i) \leqslant v^*(i)$, $i \in I$. Then we have the following.

$\underline{i>1}$ $\quad v_1(i) = u(i) \leqslant v^*(i)$

$\underline{i=1}$ $\quad v_1(1) = \sup_k [r_k(1) + \sum_j q_k(1,j)u(j)]$

$$\leqslant \sup_k [r_k(1) + \sum_j q_k(1,j)v^*(j)] = v^*(1).$$

Now assume $v_{n-1}(i) \leqslant v^*(i)$, $i \in I$. Then we have the following.

$\underline{i \leqslant n}$ $\quad v_n(i) = \sup_k [r_k(i) + \sum_j q_k(i,j)v_{n-1}(j)]$

$$\leqslant \sup_k [r_k(i) + \sum_j q_k(i,j)v^*(j)] = v^*(i)$$

$\underline{i \geqslant n}$ $\quad v_n(i) = u(i) \leqslant v^*(i).$

The case $u(i) \geqslant v^*(i)$, $i \in I$, follows similarly. The case $n = 0$ is trivial. Δ

The bounds so far calculated do not take into account the possible closeness of u(.) and v*(.) which may significantly influence the convergence rate. Let us now consider this.

Let

$$\sup_i \left[\left| u(i) - v^*(i) \right| \right] = \sigma$$

$$\underline{n \geq 1} \quad \alpha_n = \sup_{i \leq n} \left[\left| v_n(i) - v^*(i) \right| \right]$$

$$\underline{n \geq 0} \quad \beta_n = \sup_i \left[\left| v_n(i) - v^*(i) \right| \right].$$

Then, from (1) (with w(.) = v*(.)), and (2), we have the following.

$$\underline{n \geq 1} \quad \alpha_n \leq c\beta_{n-1}.$$

We then have the following.

$$\begin{aligned} \underline{n \geq 1} \quad \beta_n &\leq \max\left[\alpha_n, \sigma\right] \\ &\leq \max\left[c\beta_{n-1}, \sigma\right] \\ &\leq \max\left[c^n \beta_0, \sigma\right] \\ &\leq \max\left[c^n \sigma, \sigma\right] \leq \sigma. \end{aligned}$$

Hence we have the following.

$$A \leq 2\sigma$$

$$B \leq \sigma.$$

A better bound for A will obtain if we use either Theorem 6 or monotonicity conditions. We will then have

$$A \leq \sigma.$$

Now if $\delta_n(i)$ is arbitrary for $i > n$, this poses particular difficulties in determining C in terms of σ. Now let δ^*_n be a specialisation of δ_n where now, for <u>every</u> $i\in I$, $\delta_n^*(i)$ gets to within a distance η of the right hand side of (2). Then we have the following.

$$\underline{n\geq 1} \quad \underline{i\in I} \quad 0 \geq v^{\delta_n^*}(i)-v^*(i) = r_{\delta_n^*(i)}(i) + \sum_j q_{\delta_n^*(i)}(i,j)v^{\delta_n^*}(j) - v^*(i)$$

$$= (r_{\delta_n^*(i)}(i) + \sum_j q_{\delta_n^*(i)}(i,j)v_{n-1}(j) - v^*(i)) + (\sum_j q_{\delta_n^*(i)}(i,j)(v^{\delta_n^*}(j)-v^*(j))$$

$$+ (\sum_j q_{\delta_n^*(i)}(i,j)(v^*(j)-v_{n-1}(j))). \qquad (23)$$

From (32) we have the following.

$$\underline{n\geq 1} \quad \underline{i\in I} \quad 0 \geq v^{\delta_n^*}(i)-v^*(i) \geq -\eta-\sigma-c\sigma+\sum_j q_{\delta_n^*(i)}(i,j)(v^{\delta_n^*}(j)-v^*(j)). \qquad (33)$$

From (33) we easily derive the following.

$$\underline{n\geq 1} \quad \underline{i\in I} \quad |v^{\delta_n^*}(i)-v^*(i)| \leq (\eta+(1+c)\sigma)/(1-c).$$

Hence we have the following.

$$C \leq (\eta+(1+c)\sigma)/(1-c)+\sigma = (\eta+2\sigma)/(1-c).$$

In order to use δ_n^* we need to determine $\delta_n^*(i)$ for each i as it arises.

6. REFERENCES

1. Fox, B.L. (1971) "Finite state approximations to denumerable state dynamic programs", *J.M.A.A.*, **34**, 665-670.

2. White, D.J. (1977) "Finite state approximations for denumerable state infinite horizon discounted Markov decision processes", Notes in Decision Theory, No. 43, Department of Decision Theory, Manchester University.

A SURVEY OF ASYMPTOTIC VALUE-ITERATION FOR UNDISCOUNTED MARKOVIAN DECISION PROCESSES

A. Federgruen

(Graduate School of Business, Columbia University, New York City, NY)

and

P.J. Schweitzer

(Graduate School of Management, University of Rochester, Rochester, NY)

ABSTRACT

This paper reviews the asymptotic behaviour of undiscounted value-iteration in Markov Decision Problems with finite state and action spaces. The asymptotic results concern both the value functions and the sets of optimizing policies as the length of the planning period tends to infinity.

1. INTRODUCTION

The value-iteration equations for finite, stationary Markov decision processes (MDPs) are

$$v(n)_i = \max_{k \in K(i)} \left[q_i^k + \beta \sum_{j=1}^{N} P_{ij}^k v(n-1)_j\right]; \; 1 \leq i \leq N, \; n = 1,2,3,... \tag{1.1}$$

[Howard (1960)], where N is the finite number of states, K(i) is the finite, non-empty, set of actions available in state i, β is the one-period discount factor, and $v(0)_i$ denotes the <u>scrap value</u> of state i upon termination. We refer to $0 \leq \beta < 1$ and $\beta = 1$ as the <u>discounted</u> and <u>undiscounted</u> cases, respectively. If action $k \in K(i)$ is chosen upon entrance to state i, an immediate expected reward q_i^k is earned (where every q_i^k is finite) and P_{ij}^k denotes the probability that the next state will be j ($P_{ij}^k \geq 0$, $\sum_{j=1}^{N} P_{ij}^k = 1$). Equation (1.1) results from Bellman's principle of optimality [Bellman (1957)] with $v(n)_i$ denoting the maximum expected cumulative return in an n-period process starting from state i.

The following notation will be employed. We let $S_R = \{[f_{ik}] \mid f_{ik} \geq 0, \sum_{k \in K(i)} f_{ik} = 1\}$ denote the set of all randomized stationary policies. Here f_{ik} is the probability action k is chosen whenever entering state i. A pure (<u>non-randomized</u>) policy has each $f_{ik} = 0$ or 1, since it associates a single action $k = f(i)$ with each i. $S_p = X_i K(i)$ will represent the finite subset of pure (stationary) policies.

With each policy $f \in S_R$, we associate an N-component reward vector $q(f)$ and an N×N-transition probability matrix $P(f)$:

$$q(f)_i = \sum_{k \in K(i)} f_{ik} q_i^k \; ; \; 1 \leq i \leq N$$

$$P(f)_{ij} = \sum_{k \in K(i)} f_{ik} p_{ij}^k ; \; 1 \leq i,j \leq N. \tag{1.2}$$

We rewrite (1.1) as

$$v(n) = Tv(n-1) = T^n v(0) ; \; n = 1,2,... \tag{1.3}$$

where the operator $T:E^N \to E^N$ is defined by

$$Tx_i = \max_{k \in K(i)} \left[q_i^k + \beta \sum_j P_{ij}^k x_j\right], \; 1 \leq i \leq N \tag{1.4}$$

and where T^n represents the n-fold application of the T-operator.

Finally, for any $\varepsilon \geq 0$, let

$$S(n) = \{f \in S_p \mid v(n) = q(f) + \beta P(f) v(n-1)\}$$

$$S(n,\varepsilon) = \{f \in S_p \mid q(f) + \beta P(f) v(n-1) \geq v(n) - \varepsilon \underline{1}\}$$

denote, respectively, the set of pure policies which attain (or come within ε of) $v(n)$, when the remaining planning period consists of n periods. Here $\underline{1}$ represents the N-vector with all components equal to unity.

The present survey describes the undiscounted case, with emphasis on the basic properties of $\{v(n)\}_{n=1}^{\infty}$ (section 2), its dependence upon the scrap value vector v(0) (section 3), and the asymptotic behaviour of the sets of optimal and ε-optimal policies $\{S(n)\}_{n=1}^{\infty}$ and $\{S(n,\varepsilon)\}_{n=1}^{\infty}$ (section 4). Section 5 discusses Lyapunov functions and other techniques for bounding a solution pair to the optimality equations arising in this model, while section 6 describes uses of data-transformations to simplify computations in the infinite horizon case. In each section, differences between the undiscounted and discounted cases are pointed out. Some of the material in sections 2-4 overlaps our previous survey on value-iteration (cf. Federgruen and Schweitzer (1978a)), to which the reader is referred for additional details.

The following notation will be employed: For any $x \in E^N$, let

$$x_{max} = \max_i x_i \; ; \; x_{min} = \min_i x_i \quad \text{and}$$

$$|x|_{\infty} = \max_i |x|_i .$$

In addition, we will use the span quasi-norm $sp[x]$, where $sp[x] = x_{max} - x_{min}$ (cf. Bather (1973)).

2. THE ASYMPTOTIC BEHAVIOUR OF v(n)

The following additional notation is required for studying the asumptotic behaviour of v(n). For any pure or randomized policy f, let the N×N-matrix $\Pi(f)$ denote the Cesaro limit of the sequence $\{P^n(f)\}_{n=1}^{\infty}$, and let $R(f) = \{i | \Pi(f)_{ii} > 0\}$ represent the set of recurrent states for P(f). Finally, we associate with each $f \in S_R$ its gain rate vector

$$g(f) = \lim_{M\to\infty} \frac{1}{M} \sum_{n=0}^{M} P^n(f)q(f) = \Pi(f)q(f).$$

Denote the maximal gain rate vector by g^*, where

$$g_i^* = \sup_{f \in S_R} g(f)_i, \quad 1 \leq i \leq N. \tag{2.1}$$

It is known (cf. Derman (1970)) that a pure policy achieves the N maxima in (2.1) simultaneously. Accordingly, let

$$S_{RMG} = \{f \in S_R | g(f) = g^*\} \text{ and } S_{PMG} = \{f \in S_p | g(f) = g^*\}$$

denote the set of all randomized maximal gain policies and the set of all pure maximal gain policies, and define

$$R^* = \{i | i \in R(f) \text{ for some } f \in S_{RMG}\}.$$

It is known that $R^* = \{i | i \in R(f) \text{ for some } f \in S_{PMG}\}$, i.e., any state that is recurrent under a randomized maximal gain policy will also be recurrent under at least one pure maximal gain policy. It is also known (cf. Schweitzer and Federgruen (1978b)) that the set

$$S_{RMG}^* = \{f \in S_{RMG} | R(f) = R^*\} \neq \phi.$$

Note that randomization is essential here: there generally does not exist a pure maximal gain policy whose set of recurrent states is R^*. To illustrate this, consider the following 4-state example.

Example 1

$N = 4$; $K(1) = K(3) = K(4) = \{1\}$; $K(2) = \{1,2\}$.

$q_i^k \equiv 0$, $g^* = (0,0,0,0)$, $S_{RMG} = S_R$, $R^* = \{1,2,3,4\}$

i	k	P_{i1}^k	P_{i2}^k	P_{i3}^k	P_{i4}^k
1	1	0	1	0	0
2	1	1	0	0	0
	2	0	0	1	0
3	1	0	0	0	1
4	1	0	1	0	0

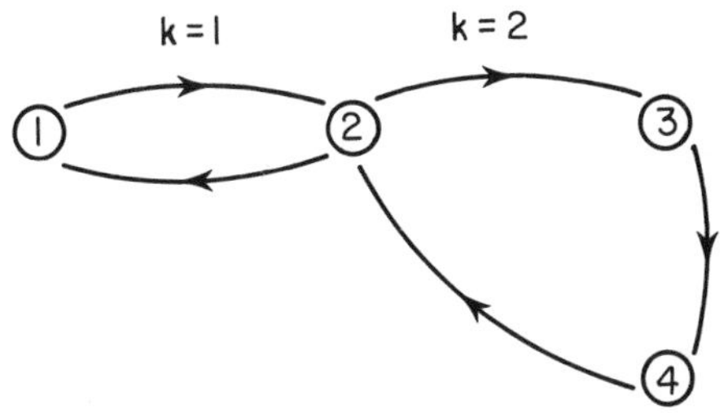

in which only state 2 has multiple alternatives.

Both pure policies are maximal gain, with recurrent states {1,2} if k = 1 is chosen and {2,3,4} if k = 2 is chosen. No pure policy has $R^* = \{1,2,3,4\}$ as its set of recurrent states, but any randomized policy which uses both alternatives in state 2 with positive probability will have R^* as its set of recurrent states. Randomization here plays the important role of coalescing the recurrent chains of the pure policies.

The historical account of the literature on the asymptotic behaviour of v(n) goes back to Bellman (1957) and Howard (1960). Bellman showed that if every $P_{ij}^k > 0$, then $v(n)_i \sim n\langle g^*\rangle$ where every $g_i^* = \langle g^*\rangle$. Howard gave examples where $\lim_{n\to\infty}\left[v(n) - ng^*\right]$ existed, and conjectured that this was always true. The conjecture is false if some of the (maximal gain) policies have tpm's† with periodic states. As an illustration, consider the following 2-state example with only one policy.

† "tpm" is henceforth used as an abreviation for "transition probability matrix"

Example 2

$$q = \binom{0}{0} \qquad P = \begin{pmatrix} 0 & 1 \\ 1 & 0 \end{pmatrix} \qquad g^* = (0,0)$$

$$v(n) = P^n v(0) = \begin{cases} (v(0)_1, v(0)_2) & n \text{ even} \\ (v(0)_2, v(0)_1) & n \text{ odd} \end{cases}$$

so that $\lim[v(n) - ng^*]$ exists if and only if $v(0)_1 = v(0)_2$. We remark here that there always exists choices of $v(0)$ such that $\lim[v(n) - ng^*]$ exists.

On the other hand, Brown (1965) showed that $\{v(n) - ng^*\}_{n=0}^{\infty}$ is always bounded in n so that

$$g_i^* = \lim_{n\to\infty} v(n)_i/n, \; 1 \le i \le N$$

(for any $v(0)$). This justifies the interpretation of g_i^* as the maximum average expected reward per transition when starting in state i. It remained to show when $\lim_{n\to\infty}[v(n) - ng^*]$ exists and the behaviour when the limit fails to exist. As mentioned above, non-existence of the limit is associated with the occurrence of maximal gain policies having periodic tpm's.

An elegant result by White (1963) states conditions which exclude periodicities, and ensure existence of the limit. Specifically, if there exists a state s, integer m and number $\alpha > 0$ such that

$$[P(f_1)P(f_2)\dots P(f_m)]_{is} \ge \alpha > 0, \; 1 \le i \le N \qquad \text{all } f_1, f_2, \dots, f_m \in S_p, \tag{2.2}$$

then $g_i^* = \langle g^* \rangle$ for all i, $v^* = \lim_{n\to\infty}[v(n) - n\langle g^*\rangle \underline{1}]$ exists for any choice of $v(0)$, and the approach to the limit is geometric.

To avoid the linear divergence of $v(n)$ with n, White proposed working with the relative values, obtained by setting (say) the Nth component of v^* equal to (say) zero. Under condition (2.2), the vector

$$v^{*rel} \equiv v^* - (v^*)_N \underline{1} \qquad (\text{with } v_N^{*rel} = 0)$$

is unique, and the relative value iteration scheme

$$y(n+1) = Ty(n) - \left[Ty(n)\right]_N \underline{1}, \qquad n = 0,1,2,\ldots \tag{2.3}$$

(where $y(n) = v(n) - v(n)_N \underline{1}$) will converge geometrically to v^{*rel}.

Brown (1965) and Lanery (1967) have shown, albeit with faulty proofs, that there exists an integer $J \geq 1$ such that $\lim_{n\to\infty}\left[v(nJ) - nJg^*\right]$ exists for every $v(0) \in E^N$. The present authors (cf. Schweitzer and Federgruen (1978a)) showed, in fact, that in the general case the process has to be observed every J^* steps, where

$$J^* = \min\{J \geq 1 \mid \text{there exists } f \in S^*_{RMG} \text{ such that } P(f)^J \text{ is an aperiodic tmp}\}$$

so as to overcome the irregular behaviour caused by the periodicities in the tpm's. Note that J^* is independent of the scrap values.

More specifically, we showed that

$\lim_{n\to\infty}\left[v(nJ + r) - (nJ + r)g^*\right]$ exists for all $v(0) \in E^N$ for some $r = 0,\ldots, J-1$, if and only if $J \geq 1$ is a multiple of J^*. (2.4)

For each $v(0) \in E^N$, there exists an integer J^0, which depends upon $v(0)$ and divides J^*, such that $\lim_{n\to\infty}\left[v(nJ + r) - (nJ + r)g^*\right]$ exists for any $r = 0,1,\ldots,J-1$ if and only if J is a multiple of J^0. Thus, in Example 2, $J^* = 2$ and $J^0 = 1$ or 2, depending on whether $v(0)_1 - v(0)_2 = 0$ or $\neq 0$. (2.5)

It follows from (2.4) that $\lim_{n\to\infty}\left[v(n) - ng^*\right]$ exists for all $v(0)$ if and only if $J^* = 1$, which holds if and only if there exists a (randomized) policy $f \in S^*_{RMG}$ with $P(f)$ aperiodic. Randomization is <u>essential</u> here because it serves to reduce the periods of the recurrent chains. As an illustration, the two pure policies in Example 1 have periods 2 and 3, while the randomized policy in S^*_{RMG} has period 1, so $J^* = 1$. Coalescing subchains and reducing periods appear to be among the few instances in MDPs where <u>randomized</u> policies play a central role. However, a finite procedure exists to calculate J^* from the periods of the <u>finite</u> set of pure maximal gain policies. (cf. Schweitzer and Federgruen (1978a).)

A sufficient condition for $J^* = 1$, hence for the existence of $\lim_{n\to\infty}\left[v(n) - ng^*\right]$ for every $v(0)$, is that every pure maximal gain policy (or every pure policy in S_p) has an aperiodic tpm. The motivation for seeking easily-verified sufficient conditions for $J^* = 1$ is given in section 4.

The present authors have also shown (cf. Schweitzer and Federgruen (1979)) that if $v(0)$ is such that $v^* = \lim_{n\to\infty}\left[v(n)-ng^*\right]$ exists, then the approach to the limit is ultimately <u>geometric</u>. That is, there exist scalars $c > 0$, and $0 \leq \lambda^* < 1$ such that

$$|v(n) - ng^* - v^*|_\infty \leq c(\lambda^*)^n; \quad n = 0,1,2,\ldots, \qquad (2.6)$$

where λ^* represents the ultimate average contraction factor per step, and is independent of $v(0) \in E^N$, and where $c > 0$ does depend upon $v(0)$. The geometric convergence result is achieved by showing the existence of an integer $M \geq 1$ (independent of $v(0)$) and an integer $n_0 \geq 1$ (dependent upon $v(0)$) as well as M-step contraction factor $\lambda(v(0)) < 1$, such that

$$sp\left[v(n+M) - (n+M)g^* - v^*\right] \leq \lambda sp\left[v(n) - ng^* - v^*\right] \quad \text{for all } n \geq n_0. \tag{2.7}$$

In fact, whenever $g^* = \langle g^* \rangle \underline{1}$, as happens to be the case in most real life applications, n_0 may be taken to be equal to one.

In the general multi-chain case, however, no good estimates are available for n_0, and likewise no good upper bounds exist for λ^*, except if the problem satisfies a simultaneous scrambling condition (cf. Federgruen, Schweitzer and Tijms (1978)), or if one is iterating a single fixed policy. In that case, it may be readily shown (cf. Morton (1977)) that λ can be taken as the modulus of the subdominant eigenvalue of the tpm.

The absence of estimates for λ is unfortunate because it precludes the use of (2.7) to estimate the deviation of $v(n) - ng^*$ from its limit. Consequently, (2.6)-(2.7) are not yet as useful as MacQueen's bounds in the discounted case (cf. MacQueen (1966) and also eq. (5.1)). Additional investigation is needed here to complete our understanding of the rate of contraction.

Finally, a generalization of these results is available for the <u>non-stationary</u> case, where, instead of having perfect knowledge of the parameters of the model, only approximations to the latter can be generated, or where it is computationally preferable to generate sequences of approximations for the parameters (cf. Federgruen and Schweitzer (1978b) for an enumeration of models in which this situation occurs). So in the non-stationary model we assume that we are able to generate sequences

$$\{q_i^k(n)\}_{n=1}^{\infty} \rightarrow q_i^k;\ 1 \leq i \leq N \text{ and } k \in K(i)$$

$$\{P_{ij}^k(n)\}_{n=1}^{\infty} \rightarrow P_{ij}^k;\ 1 \leq i,j \leq N \text{ and } k \in K(i)$$

$$\{K(i,n)\}_{n=1}^{\infty} \rightarrow K(i);\ 1 \leq i \leq N.$$

Moreover, in most cases, geometric convergence for the parameter approximations may be achieved, and the present authors showed (cf. Federgruen and Schweitzer (1978b)) that in this case the quantities of interest in our model can be approximated via the scheme

$$x(n+1)_i = \max_{k \in K(i,n)} \left[q_i^k(n) + \sum_j P_{ij}^k(n)x(n)_j\right],\quad 1 \leq i \leq N. \qquad (2.8)$$

In particular, the asymptotic behaviour of the sequence $\{x(n)\}_{n=1}^{\infty}$ is similar to that described for the stationary case:

(a) $\{x(n) - ng^*\}_{n=1}^{\infty}$ is bounded

(b) if $\lim_{n\to\infty}\left[x(n) - ng^*\right]$ exists, then the limit is approached geometrically

(c) If $J^* = 1$, then $\{x(n) - ng^*\}_{n=1}^{\infty}$ converges for every choice of $x(0) \in E^N$, and, if $J^* \geq 2$, then $\lim_{n\to\infty}\left[x(nJ^*+r)-(nJ^*+r)g^*\right]$ exists for all $x(0) \in E^N$ and $r = 0,\ldots,J^* - 1$.
However, unlike the stationary case (cf. section 3),

(d) $\lim_{n\to\infty}\left[x(n) - ng^*\right]$ may exist for every $x(0)$, even when $J^* \geq 2$

and

(e) $\lim_{n\to\infty}\left[x(n) - ng^*\right]$ may fail to exist for every $x(0)$, when $J^* \geq 2$.

Finally, examples in Federgruen (1978) show that, even with all of the tpm's of all of the policies being aperiodic, all kinds of irregular behaviour of the sequence $\{x(n)\}_{n=1}^{\infty}$ may be expected when the parameters of the model are approximated at a slower than geometric rate.

3. DEPENDENCE OF THE ASYMPTOTIC BEHAVIOUR UPON THE SCRAP VALUES

Let W denote the set of starting points for which the value iteration scheme converges, i.e.,

$$W = \{v(0) \in E^N \mid \lim_{n\to\infty}[v(n) - ng^*] \text{ exists}\}$$

$$= \{x \in E^N \mid \lim_{n\to\infty}[T^n x - ng^*] \text{ exists}\}.$$

If $J^* = 1$, then $W = E^N$, while if $J^* \geq 2$, W is a <u>non-empty strict</u> subset of E^N (cf. Schweitzer and Federgruen (1979)). For $v(0) \in W$, we let

$$L(v(0)) = \lim_{n\to\infty}[v(n) - ng^*].$$

This section summarizes the relatively few results that are known with respect to both W and $L(\cdot)$; moreover, we state some of our conjectures with respect to their properties.

The differences between the discounted and undiscounted cases deserve some emphasis here. In the discounted case, $\lim_{n\to\infty} v(n)$ <u>always</u> exists, and is <u>independent</u> of v(0). In the undiscounted case, $\lim_{n\to\infty}[v(n) - ng^*]$ <u>doesn't</u> always exist (except when $J^* = 1$), and when it does exist, it will explicitly depend upon v(0). (For instance, adding a constant to every component of v(0) adds the same constant to every component of v(n) and of $L(v(0))$.) These differences motivate our inquiry into the structure of the set W where the limit does exist, and the dependence of $L(\cdot)$ upon the scrap values.

The following notation will be needed. Suppose $v(0) \in W$ and $\lim_{n\to\infty}[v(n) - ng^*] = v^*$. Then $\{g^*, v^*\}$ satisfy the two coupled functional equations (cf. Howard (1960)):

$$g_i = \max_{k \in K(i)} \sum_j P_{ij}^k g_j, \qquad 1 \leq i \leq N \tag{3.1}$$

$$v_i = \max_{k \in L(i)} \{q_i^k - g_i + \sum_j P_{ij}^k v_j\}, \qquad 1 \leq i \leq N \tag{3.2}$$

where $L(i) = \{k \in K(i) | g_i = \sum_j P_{ij}^k g_j\}$. These equations determine g uniquely as g = g* in agreement with (2.1). But they determine v only tp to certain additive constants. We let

$$V = \{v \in E^N | (g^*, v) \text{ is a solution to (3.2)}\}. \tag{3.3}$$

In general, V may have a complicated structure, a characterization of which is given in Schweitzer and Federgruen (1978b). V is known to be closed, unbounded, connected but generally non-convex. The necessary and sufficient condition for the convexity of V was derived, and each of the following three conditions are easily verified sufficient conditions:

(a) $R^* = \{1,\ldots,N\}$

(b) for each $i \notin R^*$, L(i) is a singleton

(c) $n^* = 1$, where

$$n^* = \min\{n(f) | f \in S^*_{RMG}\} \tag{3.4}$$

with $n(f) \geq 1$ representing the number of subchains (closed, irreducible sets of states) of P(f), $f \in S_R$. $n^* = 1$ is in fact the necessary and sufficient condition for $v \in V$ to be unique up to a multiple of $\underline{1}$, i.e., under $n^* = 1$, V takes the simple structure $V = \{v^0 + c\underline{1} | -\infty < c < +\infty\}$ and in case $n^* \geq 2$, the set V can be shown to be an n^*-dimensional subset of E^N. A finite algorithm was given to determine the number n^*, as well as a triangular decomposition for the polyhedral set from which we may choose the n^* parameters (degrees of freedom), which determine $v \in V$ (cf. Schweitzer and Federgruen (1977a)). Finally, the condition $n^* = 1$ is trivially met if every pure or every pure maximal gain policy is unichained.

The multichain policy iteration algorithm (Howard (1960)) may be used to find an element of V. If $v(0) \in W$, then $\lim_{n\to\infty}[nv(n-1) - (n-1)v(n)] = \lim_{n\to\infty}[v(n) - ng^*] \in V$ so that value-iteration may also be employed to approximate an element of V, since several devices have been proposed to avoid the numerical difficulty of g^* being unknown and of $\{v(n)\}_{n=1}^{\infty}$ diverging linearly with n.

For later use we define for each $v \in V$, the set $S^*(v)$ of maximizing policies in (3.2):

$$S^*(v) = X_{i=1}^{N} L(i,v) \quad \text{where}$$
$$L(i,v) = \{k \in L(i) \,|\, v_i = q_i^k - g_i^* + \sum_j P_{ij}^k v_j\}, \qquad 1 \le i \le N. \tag{3.5}$$

We now consider the function $L(x)$, $x \in W$. The following properties are known to hold:

$$L(x + a\underline{1}) = L(x) + a\underline{1}, \text{ for any scalar a, i.e., } L(\cdot) \text{ is unbounded} \tag{3.6}$$

$L(\cdot)$ is continuous on W with Lipschitz norm of unity:

$$|L(x) - L(y)|_\infty \le |x-y|_\infty;$$
$$sp[L(x) - L(y)] \le sp[x-y]; \qquad x,y \in W \tag{3.7}$$

$L(\cdot)$ is a convex function on W:

$$L((1-a)x + ay) \le (1-a)L(x) + aL(y); \qquad x,y \in W \tag{3.8}$$

for $0 \le a \le 1$ such that $(1-a)x + ay \in W$

$$L(x) \in V \quad \text{if } x \in W \tag{3.9}$$

$$\text{if } x \in W, \text{ then } Tx \in W \text{ and } L(Tx) = L(x) + g^*. \tag{3.10}$$

In general, $L(x)$ is very difficult to display in closed form because it involves the (transient-type) decisions when termination of the process is near.

One simple example, patterned after Brown (1965), illustrates some of the structure in $L(\cdot)$:

Example 3: $N = 2$; $q_i^k = 0$ for all $1 \leq i \leq N$; $k \in K(i)$. Hence $g^* = \underline{0}$.

i	k	P_{i1}^k	P_{i2}^k
1	1	.4	.6
	2	.5	.5
2	1	.6	.4

$J^* = 1$; $W = E^2$, i.e.,

$L(x)$ exists for every $x \in E^2$

Note that

(a) $x_1 \leq x_2 \Rightarrow Tx = P(f^1)x$ with $(Tx)_1 \geq (Tx)_2$

(b) $x_1 \geq x_2 \Rightarrow Tx = P(f^2)x$ with $(Tx)_1 \leq (Tx)_2$

where $P(f^r)$ uses alternative r in state 1 ($r = 1,2$). Consequently, value-iteration will alternate between $P(f^1)$ and $P(f^2)$. This implies

$$v(2n) = \begin{cases} \left[\begin{pmatrix} .5 & .5 \\ .6 & .4 \end{pmatrix} \begin{pmatrix} .4 & .6 \\ .6 & .4 \end{pmatrix}\right]^n v(0) & \text{if } v(0)_1 \leq v(0)_2 \\ \\ \left[\begin{pmatrix} .4 & .6 \\ .6 & .4 \end{pmatrix} \begin{pmatrix} .5 & .5 \\ .6 & .4 \end{pmatrix}\right]^n v(0) & \text{if } v(0)_1 \geq v(0)_2 \end{cases}$$

Letting $n \to \infty$,

$$L(x) = \begin{cases} \left(\frac{24x_1}{49} + \frac{25x_2}{49}\right)\underline{1} & x_1 \leq x_2 \\ \\ \left(\frac{27x_1}{49} + \frac{22x_2}{49}\right)\underline{1} & x_1 \geq x_2 \end{cases} .$$

The following example illustrates more complex behaviour

Example 4: $N = 3$; $K(1) = K(3) = \{1\}$, $K(2) = \{1,2\}$

i	k	P^k_{i1}	P^k_{i2}	P^k_{i3}
1	1	0	1	0
2	1	1	0	0
	2	0	0	1
3	1	0	0	1

$$q^k_i \equiv 0; \quad g^* = (0,0,0)$$

$$V = \{v \in E^3 | v_1 = v_2 \geq v_3\} \text{ is convex.}$$

$$T^{2n}x = T^2x = \begin{bmatrix} \max[x_1, x_3] \\ \max[x_2, x_3] \\ x_3 \end{bmatrix} \qquad n = 1,2,3,...$$

$$T^{2n+1}x = T^3x = \begin{bmatrix} \max[x_2, x_3] \\ \max[x_1, x_3] \\ x_3 \end{bmatrix} \qquad n = 1,2,3,...$$

Note that $\lim_{n\to\infty} T^n x$ exists if and only if $T^2 x = T^3 x$, hence

$$W = \{x \in E^3 \mid \max[x_1, x_3] = \max[x_2, x_3]\}.$$

Note that W may be written as the union of two polytopes W_1 and W_2, where

$$W_1 = \{x \in E^3 \mid x_1 = x_2 \geq x_3\} = V$$

$$W_2 = \{x \in E^3 \mid x_3 \geq \max[x_1, x_2]\}.$$

In addition,

$$L(x) = \begin{cases} x_1 \underline{1}, & \text{for } x \in W_1 \\ x_3 \underline{1}, & \text{for } x \in W_2 \end{cases}.$$

We know (cf. Schweitzer and Federgruen (1979)) that W is always closed and unbounded; e.g., if $x \in W$, then $x + a\underline{1} \in W$ for all scalars a. W is connected in all cases examined by the authors, and it is conjectured that this holds in all generality. The above example shows, however, that W does not need to be convex, even if V is convex. The above examples suggest, in addition, that W may always be decomposed into a finite number of polytopes such that $L(\cdot)$ is linear on each of the polytopes and has directional derivatives in any feasible direction wherever two such polytopes join. These polytopes may have the structural form of cones. (See also Theorem 3.2.)

In general, it is very difficult to compute W, which depends sensitively upon the value-iteration decisions when termination is near. It is hard even to give an analytic characterization of W. The following theorem provides a step in that direction, but is not useful at present because the function $L^*(x)$ (defined below) is as poorly understood as $L(x)$ itself.

Define $L^*(x) \equiv \lim_{n\to\infty} [T^{nJ^*}x - nJ^* g^*]$. According to (2.4), $L^*(x)$ exists for every $x \in E^N$. Note that $L(x)$ agrees with $L^*(x)$ for $x \in W$; however, $L(x)$ is undefined for $x \in E^N\backslash W$.

Note that the J^*-step operator T^{J^*} may be interpreted as the value iteration operator in a J^*-step MDP, with $\{1,\ldots,N\}$ as its state space, and with action spaces, one step expected rewards and transition probabilities given by:

$$\tilde{K}(i) = \{(f^{J^*},\ldots,f^1) \mid f^r \in S_p,\ 1 \le r \le J^*\},\quad 1 \le i \le N$$

$$\tilde{q}_i^{\xi} = q(f^{J^*})_i + P(f^{J^*})q(f^{J^*-1})_i + \ldots + [P(f^{J^*})\ldots P(f^2)]q(f^1)_i$$

$$1 \le i \le N \text{ and } \xi = (f^{J^*},\ldots,f^1)$$

$$\tilde{P}_{ij}^{\xi} = P(f^{J^*})\ldots P(f^1)_{ij};\quad 1 \le i,j \le N;\ \xi = (f^{J^*},\ldots,f^1)\ .$$

As a consequence, the properties (3.6) and (3.7) carry over to $L^*(x)$, and property (3.8) shows $L^*(x)$ is a convex function everywhere on E^N.

Theorem 3.1: (characterization of W)

$$x \in W \text{ if and only if } L^*(x) \in V.$$

Proof: If $x \in W$, combine $L^*(x) = L(x)$ with $L(x) \in V$ to conclude $L^*(x) \in V$. Conversely, assume $L^*(x) \in V$ and define $L^{*,r}(x) = \lim_{n\to\infty} T^{nJ^*+r}x - (nJ^* + r)g^*$ $(r = 1,2,\ldots,J^*)$. Observe that for all $1 \le i \le N$:

$$T^{nJ^*+1}x_i - (nJ^* + 1)g_i^* = \tag{3.11}$$

$$\max_{k\in K(i)} \{nJ^*[\sum_j P_{ij}^k g_j^* - g_i^*] + q_i^k - g_i^* + \sum_j P_{ij}^k[T^{nJ^*}x - nJ^* g^*]\}$$

and note that for n sufficiently large, only $k \in L(i)$ achieve the maximum on the right of (3.11). Use this when letting n tend to infinity, to conclude for all $1 \le i \le N$:

$$L^{*,1}(x)_i = \max_{k \in L(i)} \{q_i^k - g_i^* + \sum_j P_{ij}^k L^*(x)_j\} = L^*(x)_i,$$

where the second inequality follows from $L^*(x) \in V$. Likewise, one proves $L^{*,k}(x) = L^*(x)$ for all $1 \le k \le J^*$, i.e., $\lim_{n\to\infty} T^n x - ng^*$ exists, or $x \in W$. □

Finally, the following theorem gives an abstract characterization of the function $L^*(x)$, but lacks utility until a better understanding of the set Γ (as defined below) becomes available.

For any infinite sequence of pure policies $\Phi=\{f^1,f^2,f^3,\ldots\}$, $f^n \in S_p$, where f^n is used when n periods remain before the termination of the planning period, define the n-period rewards and tpm's by:

$$q(\Phi,n) = q(f^n) + P(f^n)q(f^{n-1}) + \ldots + [P(f^n)\ldots P(f^2)]q(f^1)$$

$$P(\Phi,n) = P(f^n)\ldots P(f^1)$$

Let $\Gamma = \{(\alpha,\gamma) \mid \alpha \in E^N;\ \gamma$ is an N×N stochastic matrix;

(α,γ) is a limit point of $(q(\Phi, nJ^*) - nJ^*g^*, P(\Phi, nJ^*))$ for some infinite policy sequence $\Phi\}$.

<u>Theorem 3.2:</u> Characterization of $L^*(x)$

$$L^*(x)_i = \max_{(\alpha,\gamma)\in\Gamma} [\alpha + \gamma x_i]; \quad x \in E^N; \quad 1 \le i \le N. \qquad (3.12)$$

In addition, for each x there is a choice $(\alpha^*, \gamma^*) \in \Gamma$ which achieves all N maxima in (3.12).

Proof: Fix $(\alpha,\gamma) \in \Gamma$ and a policy sequence Φ such that $\lim_{k\to\infty}(q(\Phi, n_k J^*) - n_k J^* g^*, P(\Phi, n_k J^*)) = (\alpha,\gamma)$ for some sequence of increasing integers $\{n_k\}_{k=1}^{\infty}$. Next, note that for all $n \geq 1$:

$$v(nJ^*) \geq q(\Phi, nJ^*) + P(\Phi, nJ^*)x \qquad (3.13)$$

where $x = v(0)$. Replace n by n_k, subtract $n_k J^* g^*$ from both sides in (3.13) and let k tend to infinity, to conclude:

$$L^*(x) \geq \alpha + \gamma x \text{ for any } (\alpha,\gamma) \in \Gamma \qquad (3.14)$$

Finally, let Φ^* be such that $f^r \in S(r)$ for all $r \geq 1$, with $v(0) = x$. Then (3.13) with Φ replaced by Φ^* holds as a strict equality for all n; choose a subsequence $\{n_k\}_{k=1}^{\infty}$ such that the bounded sequence $P(\Phi^*, n_k J^*) \to$ (say) γ^* and consequently $q(\Phi^*, n_k J^*) - n_k J^* g^* = v(n_k J^*) - n_k J^* g^* - P(\Phi^*, n_k J^*)x$ $\to L^*(x) - \gamma^* x \equiv \alpha^*$. Thus, $(\alpha^*, \gamma^*) \in \Gamma$ and (3.14) holds as a strict equality when α and γ are replaced by α^* and γ^*. □

4. THE ASYMPTOTIC BEHAVIOUR OF S(n)

In this section we describe the properties of the sets of optimizing policies S(n), as n tends to infinity. The asymptotic behaviour differs sharply between the case where $v(0) \in W$ and the case where $\lim_{n\to\infty}[v(n) - ng^*]$ fails to exist. However, even in case $v(0) \in W$, contrived examples show a possibly irregular dependence of S(n) upon n. This suggests that convergence of S(n) cannot be safely relied upon as a termination criterion for value-iteration. This is in surprising contrast with

(i) the fact that for appropriate choices of the sequence $\{\varepsilon_n\}_{n=1}^{\infty} \downarrow 0$ the sequence of ε_n - optimal policies $\{S(n, \varepsilon_n)\}_{n=1}^{\infty}$ will converge whenever $v(0) \in W$.

(ii) the empirical fact that in "real-life" problems, the sets $\{S(n)\}_{n=1}^{\infty}$ converge invariably and unambiguously.

For a more detailed description of the following turnpike properties, cf. Federgruen and Schweitzer (1980a).

(i) if $v^* = \lim_{n\to\infty} [v(n) - ng^*]$ exists, then

$$S(n) \subseteq S^*(v^*) \subseteq S_{RMG} \tag{4.1}$$

for all n exceeding some n_0 which depends upon v(0). (The Cartesian product set $S^*(v^*)$ was defined by (3.5).) Thus S(n) will always settle upon maximal gain policies whenever $\{v(n) - ng^*\}_{n=1}^{\infty}$ has a limit.

An important unsolved problem is the estimation of n_0. Until upper bounds on n_0 are available, one cannot be sure that policies produced by a finite number of value-iteration steps are indeed maximal gain. Examples are known where $\sup_{v(0)\in W} n_0(v(0)) = \infty$, i.e., n_0 may exceed any given integer m by an appropriate bad choice of v(0).

(ii) If $v^* = \lim_{n\to\infty} [v(n) - ng^*]$ so that $S(n) \subseteq S^*(v^*)$ for large n, the asymptotic behaviour of S(n) may still be erratic [see, e.g., modifications of Shapiro (1968)]. S(n) could be a <u>strict</u> subset of $S^*(v^*)$ for every n, with some members of $S^*(v^*)$ never identified. Also, S(n) could <u>oscillate periodically</u> among members of $S^*(v^*)$ [see Brown (1965)] or could even oscillate in a non-periodic way among members of $S^*(v^*)$ (by modification of an example in Bather (1973)). This potential for irregular behaviour (when $S^*(v^*)$ is not a singleton) discourages use of convergence of S(n) as a termination criterion.

It is nevertheless possible to compute $S^*(v^*)$ correctly by means of ε-optimal policies, as follows. Let $\{\varepsilon_n\}_{n=1}^{\infty}$

denote any sequence of positive numbers which approaches 0 at a slower rate than the geometric rate of convergence of $v(n) - ng^* - v^*$ to $\underline{0}$, e.g., take $\varepsilon_n = n^{-1}$ (or the reciprocal of any positive polynomial in n). Then for all n exceeding some n_1,

$$S^*(v^*) = \{f \in S_p \mid v(n+1) \leq q(f) + P(f)v(n) + \varepsilon_n \underline{1}\}.$$
$$= S(n+1, \varepsilon_n)$$

Once again, no bounds are available for n_1.

Here again, all of the results mentioned under (i) and (ii) carry over to the non-stationary model where only geometric approximations for the parameters are available (cf. section 2).

(iii) If $\lim_{n\to\infty} [v(n) - ng^*]$ does not exist, then S(n) need not lie in S_{RMG} for all large n; it need not even intersect S_{RMG}. The first such example is given in [Lanery (1967)], where S(n) lies outside S_{RMG} for infinitely many n. Later the authors constructed an example [Federgruen and Schweitzer (1980a)] where S(n) is disjoint from S_{RMG} for every n. These examples contrast sharply with the behaviour in the discounted case, where S(n) can contain only optimal policies when n is sufficiently large.

For problems with many thousands of states, value-iteration is the most practical way to locate maximal gain policies. The importance of (i) and (iii) is that value-iteration can be relied on to identify maximal gain policies only if $\lim_{n\to\infty}[v(n) - ng^*]$ exists.

This motivates the search for conditions ensuring either $J^* = 1$ (so the limit exists for every v(0)) or $v(0) \in W$ (so the limit exists for this v(0)). Sufficient conditions for $J^* = 1$ were indicated above. If these conditions cannot be verified, and if there is concern about non-existence of $\lim_{n\to\infty}[v(n) - ng^*]$ due to the presence of periodic tpm's, it is suggested that a data-transformation be applied (see below) to ensure that $J^* = 1$.

5. BOUNDS AND LYAPUNOV FUNCTIONS

For discounted MDPs, where $\beta < 1$, the unique fixed point $v^* = Tv^*$ of the monotone contraction operator T is sought. The following bounds on v^* were given by MacQueen (1966) and later improved slightly by Porteus (1971).

$$x + \frac{(Tx - x)_{min}}{1 - \beta}\underline{1} \leq v^* \leq x + \frac{(Tx - x)_{max}}{1 - \beta}\underline{1} \text{ any } x \in E^N. \quad (5.1)$$

These bounds have the following three convenient properties:

invariance: they remain unchanged when x is replaced by $x + a\underline{1}$, $-\infty < a < +\infty$

sharpness: the bounds converge to v^* when x approaches v^*

monotonicity under value iteration: the bounds move monotonically inward when x is replaced by Tx.

The bounds in (5.1) are conveniently used during discounted value-iteration, with $x = v(n)$ and $Tx = v(n+1)$, since the bounds may be computed with almost no additional effort and converge monotonically and geometrically to v^*. The bounds are useful both as a termination criterion (stopping when $|v^* - v(n)|_\infty$ achieves a given precision) and for elimination of suboptimal actions [MacQueen (1967)]. For a recent survey on elimination tests, we refer to White (1978).

Consider now the construction of analogous bounds for undiscounted MDPs, where $\beta = 1$. Differences are immediately apparent because the discounted case has only one set of optimality equations for the N-vector v^* whereas the undiscounted case has a pair of coupled functional equations for two N-vectors g^* and v^*. Furthermore, in the discounted case

v^* is uniquely determined by these equations whereas in the undiscounted case only g^* is uniquely determined, while v^* is determined up to certain additive constants.

We describe first bounds on the gain rate vector g^*. Consider, initially, the special case where all components of g^* are equal: $g^* = \langle g^* \rangle \underline{1}$. This is the most important case in practice, arising in the so-called unichain case where each pure policy f has a tpm P(f) with a single closed irreducible set of states, plus possibly some transient states. Bounds on the scalar $\langle g^* \rangle$ were given first by Odoni (1969)

$$(Tx - x)_{min} \leq \langle g^* \rangle \leq (Tx - x)_{max} \quad \text{any } x \in E_N \tag{5.2}$$

with the properties

invariance: the bounds remain unchanged when x is replaced by $x + a\underline{1}$

sharpness: the bounds converge to $\langle g^* \rangle$ when x approaches $v^* \in V$

monotonicity under value iteration: the bounds move monotonically inward (but not necessarily strictly monotonically) when x is replaced by Tx.

These bounds are conveniently used during value-iteration, with $x = v(n)$ and $Tx = v(n+1)$, or (see eq. (2.3)) $x = y(n)$ and $Tx = Ty(n)$. If $\lim_{n\to\infty}(v(n) - ng^*)$ exists, then both upper and lower bounds converge to $\langle g^* \rangle$; however, if these limits do not exist, then a gap will occur between the asymptotic levels of the upper and lower bounds.

The bounds are useful as a termination criterion (stopping with an estimation of $\langle g^* \rangle$ which achieves a given precision) but so far have not been useful for elimination of suboptimal actions (see further discussion below) due to the absence of estimates of the deviation of $v(n) - n\langle g^* \rangle \underline{1}$ from v^*.

Consider next the general case where the components of g^* can be unequal. The authors obtained the following bounds on the maximal gain rate vector g^* [Schweitzer and Federgruen (1980a)]

$$g + \{\max_{f\in\Lambda(g)} [q(f) - g + P(f)x - x]\}_{\min}\underline{1} \leq$$

$$\leq g^* \leq g + \{\max_{f\in\Lambda(g)} [q(f) - g + P(f)x - x]\}_{\max}\underline{1}, \qquad (5.3)$$

$$g \in G,\ x \in E_N$$

where $G \equiv \{g \in E_N | g = \max_{f\in K} P(f)g\}$ and, for $g \in G$, $\Lambda(g) \equiv \{f\in K | g = P(f)g\}$ and where the expressions within braces in (5.3) are maximized component by component. Note that G is not empty, since $\underline{0}$, $\underline{1}$ and g^* are in G. Note also that when $g^* = \langle g^* \rangle \underline{1}$, the choice $g = \underline{0}$ reduces the above bounds to those of Odoni.

The above bounds have the properties:

<u>invariance</u>: they remain unchanged when replacing (g,x) by $(g + a\underline{1},\ x + b\underline{1})$ for any scalars a, b.

<u>sharpness</u>: the bounds converge to g^* when g approaches g^*, and x approaches some $v^* \in V$.

Unfortunately, the <u>monotonicity</u> property has eluded generalization to this case. We lack a value-iteration scheme for generating successive pairs of vectors {g,x} such that the bounds move monotonically inward. The main technical difficulty appears to be the absence of simple characterizations of G and $\Lambda(g)$, $g \in G$, or of simple ways to generate sequences of members of G. However, in the upper bound in (5.3), it is permitted to replace $\max_{f\in\Lambda(g)}$ by $\max_{f\in S_p}$, and the task of minimizing the upper bound is then related to the primal linear program for the gain rate vector [Denardo and Fox (1968)], [Hordijk and Kallenberg (1979)]:

$$\min_{g,x} \sum_i g_i$$

$$g_i \geq \sum_j P_{ij}^k g_j, \quad 1 \leq i \leq N, \; k \in K(i)$$

$$x_i \geq q_i^k - g_i + \sum_j P_{ij}^k x_j, \quad 1 \leq i \leq N, \; k \in K(i).$$

Finally, we describe bounds on the relative value vector $v^* \in V$. Consider, initially, the case $n^* = 1$, where $g^* = \langle g^* \rangle \underline{1}$ and v^* is unique up to an additive multiple of $\underline{1}$ (cf. (3.4)). It is convenient to measure the deviation of an N-vector x from v^* by $sp[x - v^*]$, which is invariant to the additive constant in v^*. Define

$$\phi(x) = sp[Tx - x]$$

which is non-negative and vanishes if and only if $x \in V$. The following theorem shows that $\phi(x)$ provides a computable measure for the distance between x and v^*:

<u>Theorem 5.1</u>: (cf. Federgruen and Schweitzer (1980b))

Assume $n^* = 1$ (see (3.4)). Then there exists a constant c_1, $1/2 \leq c_1 < \infty$ such that

$$\frac{\phi(x)}{2} \leq sp[x - v^*] \leq c_1 \phi(x) \text{ for all } x \in E^N \qquad (5.4)$$

if and only if

all states in $\hat{R} = \{i | i \in R(f), \text{ for some } f \in S_p\} \supseteq R^*$ can reach each other, i.e., if $i,j \in \hat{R}$ then there exists a policy $f \in S_R$ such that $P(f)_{ij}^r > 0$ for some $1 \leq r \leq N$. (5.5)

Remark: The condition (5.5) can be expressed in various equivalent ways. For example, (5.5) is equivalent to $\hat{R}$ being a communicating system (cf. Bather (1973)), or to the existence of a (randomized) policy that has $\hat{R}$ as its single subchain. Observe, in addition, that the combination of $n^* = 1$ and (5.5) certainly holds in the unichain case.

The bounds in (5.4) estimate the deviation of x from v^*, in terms of the computable quantity $\Phi(x)$. The upper bound in (5.4) is the direct analogue of (5.1) in the discounted case, which may be rewritten as

$$sp[x - v^*] \leq (1 - \beta)^{-1} sp[Tx - x]. \qquad (5.6)$$

The bounds in (5.4) again have the three desirable properties of (a) invariance (they remain unchanged when replacing x by $x + a\underline{1}$); (b) sharpness (they converge to 0 as x approaches $v^* + a\underline{1}$; (c) monotonicity of the upper bound under value-iteration ($\phi(Tx) \leq \phi(x)$).

In principle, the bounds in (5.4) permit both a termination criterion for achieving a given precision, and elimination of suboptimal actions (those not in $S^*(v^*)$, which set is uniquely determined if $n^* = 1$). Unfortunately, the bounds are not useful as written because c_1 can be enormously large for real problems. Our current upper bound for the unichain case, $c_1 = 4N/[\min\{P^k_{ij} | P^k_{ij} > 0\}]^N$, needs considerable refinement to make these bounds practical.

An alternative method of bounding v^* and eliminating suboptimal actions has been proposed (Federgruen, Schweitzer and Tijms (1978)) in the unichain case where every $P(f)$, $f \in S_p$, is unichained. The method uses the data-transformation (6.1), which converts the original undiscounted MDP into a new one, denoted by a tilde, with the same state and action spaces, S_{RMG} and v^{*rel} left intact, the scalar gain rate multiplied by a scalar $0 < \tau < 1$, and every $\tilde{P}(f)_{ii} \geq 1 - \tau > 0$. White's relative value scheme (2.3) for the transformed model has the form

$$\tilde{y}(n+1) = \tilde{Q}\tilde{y}(n), \quad \text{where} \tag{5.7}$$

$$\tilde{Q}x = \tilde{T}x - (\tilde{T}x)_N \underline{1}, \text{ and}$$

$$\tilde{T}x = (1-\tau)x + \tau Tx.$$

We recall that the scheme (5.7) has the property $\lim_{n\to\infty} \tilde{y}(n) = v^{*rel}$. If every $P(f)$, $f \in S_p$, is unichained, then $\tilde{Q}$ is an <u>N-step contraction operator</u> on $\tilde{E}^N = \{x \in E^N | x_N = 0\}$ with v^{*rel} as its unique fixed point. This permits the construction of monotonically and geometrically converging upper and lower bounds on v^{*rel}, in terms of $\tilde{y}(n)$, and to use these bounds to eliminate sub-optimal actions in complete analogy with MacQueen's procedures in the discounted case (cf. Federgruen, Schweitzer and Tijms (1978)).

This is believed to be the first published scheme employing value-iteration for monotonically and geometrically converging upper and lower bounds on v^{*rel}, and for permanent action elimination. Computational testing of the scheme is lacking, and it is unknown how to weaken the unichainedness assumption to merely $n^* = 1$. Finally, we refer to Hastings (1976) for a "temporary" elimination scheme of non-optimal actions.

For MDP's with $n^* = 1$, i.e., with $v^* \in V$ unique up to a multiple of $\underline{1}$, we define a continuous function $\phi: E^N \to E^1$ as a <u>Lyapunov function</u> if it satisfies the following conditions:

(a) $\phi(x) \geq 0$, $x \in E^N$; $\phi(x) = 0$ if and only if $x \in V$

(b) $\phi(Tx) \leq \phi(x)$; $x \in E^N$ (5.8)

(c) there exists an integer $M \geq 1$ such that $\phi(T^M x) < \phi(x)$ whenever $\phi(x) > 0$.

One such function is $\phi(x) = sp[Tx - x]$, as used above; if every $P(f)$ is unichained and has a positive diagonal (where the latter can be achieved via the above discussed data-transformation (6.1)), then (5.8c) holds with M = N.

Another possible Lyapunov function is $\phi(x) = sp[x - v^*]$ with (5.8c) following from (2.7) if $\lim_{n\to\infty}[T^n x - ng^*]$ exists. This Lyapunov function, however, cannot be numerically evaluated midway through the value-iteration computations because v^* is unknown. In the discounted case, the right hand side of (5.6) is a Lyapunov function with M = 1 and $\phi(Tx) \leq \beta\phi(x)$.

Lyapunov functions have two convenient uses, one theoretical and the other numerical. Theoretically, their existence ensures convergence of the value-iteration scheme $v(n) - ng^*$, i.e., construction of a Lyapunov function is a way of proving algorithm convergence [cf. Zangwill (1970)]. Computationally, their numerical value measures (or bounds) the deviation $v^* - x$ between the limit point and the current guess.

Lastly, we describe difficulties in constructing bounds on $v^* \in V$ in the general multichain case. If $n^* \geq 2$, $v^* \in V$ is not unique to a multiple of $\underline{1}$; instead, v^* is determined up to n^* additive constants (cf. section 3). Consequently, the expression $sp[x - v^*]$ is not uniquely defined until a particular choice of $v^* \in V$ is made. One natural measure of deviation, $\inf_{v\in V} sp[x-v]$, appears computationally intractable. Another natural choice, to measure the deviation of $x = v(n) - ng^* = T^n v(0) - ng^*$ from $v^* = L(v(0)) = \lim_{n\to\infty}[v(n) - ng^*] \in V$ via $sp[v(n) - ng^* - L(v)(0))]$ is again intractable because g^* and $L(v(0))$ are unknown while being midway through the calculations.

The contraction property (2.7) is not helpful because λ and n_0 are usually unknown.

A third choice is to compute bounds on the optimal bias w^* [Denardo (1970)] $\in V$. Here exact variational bounds are available [Schweitzer and Federgruen (1980a)]

$$v + \{ \max_{f \in S^*(v)} [-P(f)v + P(f)y - y] \}_{min} \underline{1} \le w^* \le$$

$$v + \{ \max_{f \in S^*(v)} [-P(f)v + P(f)y - y] \}_{max} \underline{1}, \; v \in V, \; y \in E_N \qquad (5.9)$$

where V and $S^*(v)$ were defined in (3.3) and (3.5), respectively. These bounds are both <u>invariant</u> and <u>sharp</u>.

In addition to the absence of a compelling reason to select w^* as the prototype member of V, the bounds in (5.9) are not computationally useful until simple ways are discovered to characterize V and $S^*(v)$, $v \in V$, and to generate sequences from V. These deficiencies in our computational procedures indicate that the multichain undiscounted case is still an open area for investigation.

6. DATA TRANSFORMATIONS

Data transformations of the parameters of our model are meant to convert one MDP into another (cf. Schweitzer (1971b), Schweitzer (1972), Federgruen and Schweitzer (1980c), Lippman (1975), Porteus (1975), and Porteus and Totten (1974)). Their main use is to create a transformed MDP which is easier either to analyze theoretically or to compute numerically. Data transformations generally destroy the interpretation of v(n) as the vector of maximal n-period rewards. However, they are useful for the infinite-horizon case, provided that the quantities of interest transform in a tractable manner.

A useful data-transformation for the undiscounted MDP, indicated by a tilde, is

$$\tilde{N} = N; \; \tilde{K}(i) = K(i), \; 1 \le i \le N$$

$$\tilde{q}_i^k = \tau q_i^k; \; \tilde{P}_{ij}^k = (1-\tau)\delta_{ij} + \tau P_{ij}^k; \qquad (6.1)$$

$$1 \le i,j \le N; \; k \in K(i)$$

where $0 < \tau < 1$. This may be interpreted as observing the process, and making decision, at intervals τ rather than at unit intervals. It has the properties

$$\tilde{\Pi}(f) = \Pi(f); \; \tilde{g}(f) = \tau g(f), \; f \in S_p$$
$$\tilde{g}^* = \tau g^*; \; \tilde{S}_{RMG} = S_{RMG}; \; \tilde{V} = V. \tag{6.2}$$

The important new feature is that every $\tilde{P}(f)_{ii} \geq 1 - \tau > 0$ so that every tpm $\tilde{P}(f)$ is aperiodic. It follows that value-iteration on the transformed problem will be guaranteed to converge for any choice of the scrap value vector, since $\tilde{J}^* = 1$. Thus value-iteration on the transformed problem will be sure to identify maximal gain policies, and is hence preferred over value-iteration in the original model, if the latter has periodic tpm's.

A second use of data transformations is to convert semi-Markovian decision processes (SMDP's) where the transitions are not equally spaced in time, into MDPs, where the transitions are one unit time apart. Consequently, techniques developed for the infinite-horizon MDP may be invoked for the infinite-horizon SMDP.

Consider first the undiscounted infinite-horizon semi-MDP. The functional equations to be solved are (cf. Jewell (1963)):

$$g_i^* = \max_{k \in K(i)} \sum_j P_{ij}^k g_j^*, \; 1 \leq i \leq N$$
$$v_i^* = \max_{k \in L(i)} \{q_i^k - g_i^* H_i^k + \sum_j P_{ij}^k v_j^*\}, \; 1 \leq i \leq N \tag{6.3}$$

where $H_i^k > 0$ is the mean holding time in state i, when action k is chosen, and where L(i) is defined as below (3.2).

The appropriate data transformation is (cf. Schweitzer (1971b))

$$\tilde{N} = N;\ \tilde{K}(i) = K(i),\ 1 \le i \le N$$

$$\tilde{q}_i^k = \tau q_i^k / H_i^k;\ \tilde{P}_{ij}^k = \left(1 - \frac{\tau}{H_i^k}\right)\delta_{ij} + \tau \frac{P_{ij}^k}{H_i^k}; \tag{6.4}$$

$$1 \le i,j \le N;\ k \in K(i)$$

$$\tilde{H}_i^k = \tau$$

where

$$0 < \tau \le \min\{H_i^k/(1 - P_{ii}^k) \mid (i,k) \text{ with } P_{ii}^k < 1\}. \tag{6.5}$$

The transformed problem is an undiscounted MDP with decisions at fixed intervals τ, with $\tilde{S}_{RMG} = S_{RMG}$, $\tilde{g}^* = \tau g^*$ and $\tilde{V} = V$. Moreover, by choosing τ strictly less than the upper bound in (6.5), we have $\tilde{P}(f)_{ii} > 0$ for all $1 \le i \le N$ and $f \in S_p$. As a consequence, value-iteration as applied to the transformed model will be guaranteed to converge and will yield maximal gain policies as well as a solution $v \in V$. (Schweitzer (1971b)).

In the special case where $g^* = \langle g^* \rangle \underline{1}$, Odoni's bounds (cf. (5.2)) for the scalar gain rate of the transformed problem are just the bounds given by Hastings (1971) and Schweitzer (1971a) for the gain rate of an SMDP with $g^* = \langle g^* \rangle \underline{1}$:

$$\min_i \max_{k \in K(i)} \left| \frac{q_i^k + \sum_j P_{ij}^k x_j - x_i}{H_i^k} \right| \le \langle g^* \rangle \le \tag{6.6}$$

$$\max_i \max_{k \in K(i)} \left| \frac{q_i^k + \sum_j P_{ij}^k x_j - x_i}{H_i^k} \right|,\ x \in E^N.$$

Consider next the discounted infinite horizon SMDP with functional equation:

$$v^* = Qv^*, \quad \text{where} \tag{6.7}$$

$$Qx_i = \max_{k \in K(i)} \left[q_i^k + \sum_{j=1}^{N} B_{ij}^k x_j \right], \quad 1 \le i \le N$$

and

$$B_{ij}^k \ge 0, \; B_{i,sum}^k = \sum_j B_{i,j}^k \le \delta < 1.$$

The operator Q is a monotone contraction operator with contraction modulus $\delta < 1$, hence it has a unique fixed point v^*, which can be approximated fast via the successive approximation scheme $v(n) = Qv(n-1) = Q^n v(0)$.

In addition, upper and lower bounds on $|v(n) - v^*|_\infty$ follow from standard contraction operator theory, and are based on the maximal and minimal magnitudes of $B_{i,sum}^k$ (cf. Porteus (1971), Hastings (1971)). Improvements on the latter can, however, be obtained via the following data transformation

$$\tilde{N} = N; \; \tilde{K}(i) = K(i) \text{ all } i$$

$$\tilde{q}_i^k = \frac{(1 - \tilde{\beta}) q_i^k}{1 - B_{i,sum}^k}; \; \tilde{B}_{ij}^k = \delta_{ij} + \frac{(1 - \tilde{\beta})(B_{ij}^k - \delta_{ij})}{1 - B_{i,sum}^k}; \tag{6.8}$$

$$1 \le i,j \le N; \; k \in K(i)$$

where $\tilde{\beta}$ is chosen to satisfy

$$0 \le \max_{\substack{1 \le i \le N \\ k \in K(i)}} \frac{B_{i,sum}^k - B_{ii}^k}{1 - B_{ii}^k} \le \tilde{\beta} < 1.$$

This ensures that $\tilde{B}^k_{ij} \geq 0$ and $\tilde{B}^k_{i,sum} = \tilde{\beta} < 1$. The transformed (tilde) process is a discounted MDP with discount factor $\tilde{\beta}$, and with the same fixed point v^* and the same set of optimal policies as the original SMDP. The tilde value-iteration scheme $\tilde{v}(n+1) = \tilde{Q}\tilde{v}(n)$ where $\tilde{Q}x_i = \max_{k \in K(i)} [\tilde{q}^k_i + \sum_j \tilde{B}^k_{ij} x_j]$, $1 \leq i \leq N$ may converge to v^* quicker than the original scheme $v(n+1) = Qv(n)$.

In addition, applying MacQueen's bounds to the operator $\tilde{Q}$, we obtain new bounds on the fixed point v^*:

$$x_i + \min_r \max_{k \in K(r)} \left[\frac{q^k_r + \sum_{j=1}^{N} B^k_{rj} x_j - x_r}{1 - B^k_{r,sum}} \right] \leq v^*_r$$

$$\leq x_i + \max_r \max_{k \in K(r)} \left[\frac{q^k_r + \sum_{j=1}^{N} B^k_{rj} x_j - x_r}{1 - B^k_{r,sum}} \right], \quad i \leq i \leq N \tag{6.9}$$

for any $x \in E^N$. These bounds are invariant when replacing x by $x + a\underline{1}$, sharp when x approaches v^*, and move monotonically inward when x is replaced by $\tilde{Q}x$. They reduce to MacQueens's bounds if every $B^k_{i,sum} = \beta$; e.g., if $B^k_{ij} = \beta P^k_{ij}$. If the row sums are unequal, the bounds in (6.9) appear to be tighter than those due to Porteus and Hastings.

These examples illustrate the usefulness of data-transformations in painlessly extending algorithms and bounds from one model to an "equivalent" one (especially from MDPs to SMDPs).

7. REFERENCES

1. Bather, J. "Optimal decision procedures for finite Markov Chains", *Adv. in Appl. Prob.*, 5, Parts, I, II and III,

328-339, 521-540, 541-553, (1973).

2. Bellman, R. "A Markovian decision process", *J. Math. Mech.*, **6**, 679-684, (1957).

3. Brown, B. "On the iterative method of dynamic programming on a finite space discrete time Markov Process", *Ann. Math. Stat.*, **36**, 1279-1285, (1965).

4. Denardo, E. "Contraction mappings in the theory underlying dynamic programming", *SIAM Rev.*, **9**, 165-177, (1967).

5. Denardo, E. "Computing a bias-optimal policy in a discrete-time Markov Decision Problem", *Oprns. Res.*, **18**, 279-289, (1970).

6. Denardo, E. and Fox, B. "Multichain Markov Renewal Programs", *SIAM J. Appl. Math.*, **16**, 468-487, (1968).

7. Derman, C. "Finite State Markovian Decision Processes", Academic Press, New York, (1970).

8. Federgruen, A. "The rate of convergence for backwards products of a convergent sequence of finite Markov matrices", Graduate School of Management Working Paper Series No. 7827, University of Rochester, (1978), (to appear in *Stoch. Proc. and their Appl.*).

9. Federgruen, A. and Schweitzer, P.J. "Discounted and undiscounted value iteration in Markov decision processes: A survey", in "Dynamic Programming and its Applications", M.L. Puterman ed., Academic Press, New York, 23-52, (1978a).

10. Federgruen, A. and Schweitzer, P.J. "Nonstationary Markov Decision Problems with converging parameters", Math. Center Report BW91/78, (1978b) (to appear in *J. Optimization Theory and Applications*).

11. Federgruen, A. and Schweitzer, P.J. "Turnpike properties in undiscounted Markov Decision Problems", (forthcoming), (1980a).

12. Federgruen, A. and Schweitzer, P.J. "A Lyapunov function for Markov Renewal Programming", (in preparation), (1980b).

13. Federgruen, A. and Schweitzer, P.J. "Data transformations for Markov Renewal Programming", (forthcoming), (1980c).

14. Federgruen, A., Schweitzer, P.J. and Tijms, H.C. "Contraction Mappings underlying undiscounted Markov Decision Problems", *J. Math. Anal. Appl.*, **65**, 711-730, (1978).

15. Hastings, N. "Bounds on the gain of a Markov Decision Process", *Oprns. Res.*, **19**, 240-244, (1971).

16. Hastings, N. "A test for nonoptimal actions in undiscounted finite Markov Decision Chains", *Man. Sci.*, **23**, 87-92, (1976).

17. Hastings, N. and Mello, J. "Tests for suboptimal actions in discounted Markov Programming", *Man. Sci.*, **19**, 1019-1022, (1973).

18. Hordijk, A. and Kallenberg, L. "Linear Programming and Markov Decision Chains", *Man. Sci.*, **25**, 352-362, (1979).

19. Howard, R. "Dynamic Programming and Markov Processes", John Wiley, New York, (1960).

20. Jewell, W. "Markov Renewal Programming", *Oprns. Res.*, **11**, 938-971, (1963).

21. Lanery, E. "Etude asymptotique des systèmes Markoviens à commande", *Rev. Inf. Rech. Op.*, **1**, 3-56, (1967).

22. Lippman, S. "Applying a new device in the optimization of exponential queuing systems", *Oprns. Res.*, **23**, 687-710, (1975).

23. MacQueen, J. "A Modified Dynamic Programming Method for Markovian Decision Problems", *J. Math. Anal. Appl.*, **14**, 38-43, (1966).

24. MacQueen, J. "A test for suboptimal actions in Markovian Decision Problems", *Oprns. Res.*, **15**, 559-561, (1967).

25. Morton, T. and Wecker, W. "Discounting, Ergodicity and Convergence for Markov Decision Processes", *Man. Sci.*, **23**, 890-900, (1977).

26. Odoni, A. "On finding the maximal gain for Markov Decision Processes", *Oprns. Res.*, **17**, 857-860, (1969).

27. Porteus, E. "Some bounds for discounted sequential decision processes", *Man. Sci.*, **18**, 7-11, (1971).

28. Porteus, E. "Bounds and transformations for discounted finite Markov decision chains", *Oprns. Res.*, **23**, 761-784, (1975).

29. Porteus, E. and Totten, J. "Extrapolations for iterative methods of solving M-matrix equations", GSB Report RT 209, Stanford University, Stanford, California, (1974).

30. Schweitzer, P.J. "Multiple Policy improvements in undiscounted Markov Renewal Programming", *Oprns. Res.*, **19**, 784-793, (1971a).

31. Schweitzer, P.J. "Iterative Solution of the functional equations for undiscounted Markov Renewal Programming", *J. Math. Anal. Appl.*, **34**, 495-501, (1971b).

32. Schweitzer, P.J. "Data Transformations for Markov Renewal Programming", ORSA National Meeting, Atlantic City, New Jersey, November 10, (1972).

33. Schweitzer, P.J. and Federgruen, A. "Functional Equations of undiscounted Markov Renewal Programming", Math. Center Report BW60/76, BW71/77, (1977a).

34. Schweitzer, P.J. and Federgruen, A. "Geometric convergence of value-iteration in multichain Markov Renewal Programming", *Adv. Appl. Prob.*, **11**, 188-217, (1979).

35. Schweitzer, P.J. and Federgruen, A. "The asymptotic behaviour of undiscounted value iteration in Markov Decision Problems", *Math. of O.R.*, **2**, 360-381, (1978a).

36. Schweitzer, P.J. and Federgruen, A. "Functional equations of undiscounted Markov Renewal Programming", *Math. of O.R.*, **3**, 308-322, (1978b).

37. Schweitzer, P.J. and Federgruen, A. "Variational Characteristics in Markov Renewal Programs", (forthcoming), (1980a).

38. Shapiro, J. "Turnpike planning horizons for a Markovian Decision Model", *Man. Sci.*, **14**, 292-300, (1968).

39. White, D. "Dynamic Programming, Markov Chains, and the method of successive approximations", *J. Math. Anal. Appl.*, **6**, 373-376, (1963).

40. White, D. "Elimination of non-optimal actions in Markov decision processes", in "Dynamic Programming and its Applications", M.L. Puterman ed., Academic Press, New York, 131-160, (1978).

41. Zangwill, W. "Nonlinear Programming; A Unified Approach", Englewood Cliffs, Prentice-Hall, Inc., (1969).

A SUFFICIENT CONDITION FOR THE EXISTENCE OF A STATIONARY 1-OPTIMAL PLAN IN COMPACT ACTION MARKOVIAN DECISION PROCESSES

S. S. Sheu and K.-J. Farn

(Institute of Applied Mathematics, National Tsing Hua University)

ABSTRACT

In this paper we shall propose a sufficient condition for the existence of a stationary optimal plan in a Markovian decision process with finite state space and compact action space using Blackwell's criterion of optimality. Our results extend a result of Hordijk who used the same condition to prove the existence of a stationary optimal plan using average cost as criterion.

1. INTRODUCTION

Consider a Markovian decision process of which the state spaces S is finite and the action space A is compact. On every period or day, we shall inspect the position of state s and then determine an action a. Then, we shall incur an immediate income or reward I(s,a), which may be negative, and the system moves to a new state s' according to the transition probability $q(s'|s,a)$. We shall assume $|I(s,a)| \leq M < \infty$ for all s and a. The problem is to control the process in the most effective way over an infinite future.

A plan or policy Π is a sequence $(\pi_1, \pi_2, \ldots)$ where π_n specifies a conditional probability distribution over A for every given history $h = (s_1, a_1, \ldots, s_n)$. A plan Π is (non-randomized) stationary if $\Pi = (f, f, \ldots)$, where f is a map from S to A. The stationary plan defined by f is denoted by f^∞.

Let $r_j(s,\Pi)$ denote the expected reward at the j-th period if the initial state is s and plan Π is used. The expected reward in the first n periods is denoted by $V^n(s,\Pi) = \sum_{j=0}^{n-1} r_j(s,\Pi)$. For a discounted factor β, $0 \leq \beta < 1$, the expected total discounted rewards is denoted by

$$V_\beta(s,\Pi) = \sum_{n=0}^{\infty} \beta^n r_n(s,\Pi).$$

A plan Π^* satisfying

$$V_\beta(s,\Pi^*) = \sup_{\Pi} V_\beta(s,\Pi)$$

for all s, is called a <u>β-optimal</u> plan.

In the discounted case, much of the interest is to prove existence theorems of β-optimal plans. Blackwell [3] and Maitra [11] showed that a β-optimal stationary plan exists when the action space is finite. Maitra [12] and Furukawa [7] extended their results to the case when the action space is compact.

Blackwell [2], Derman [5] and Veinott [14] respectively proposed three kinds of optimality for the non-discounted Markovian decision processes. Blackwell [2], Derman [5], [6], Denardo and Miller [4], Miller and Veinott [13] respectively proved the existence of a stationary optimal plan in various senses under the assumption that both state space and the action space are finite. Extensions to the compact action space case using Derman's criterion of optimality were discussed by Hordijk [8], [9].

In this paper, we shall consider the nondiscounted problem using Blackwell's criterion of optimality for the compact action space case. The paper is organized in this way. In Section 2, we sketch all the known results we need. In Section 3, we propose the conditions for the existence of a stationary optimal plan in the sense of Blackwell. Finally, examples are presented in Section 4.

2. PRELIMINARIES

In this section we shall develop the basic definitions and theorems to be used throughout the paper.

Blackwell [2] used the following criteria which depend on the case of small interest rates (i.e., values of β close to 1- or ρ close to 0+ where $\beta = (1+\rho)^{-1}$, $\rho > 0$).

2.1 Definition

A plan Π^* is <u>optimal</u> if there exists a $\beta_0 \in (0,1)$ such that

$$V_\beta(s,\Pi^*) \geq V_\beta(s,\Pi), \quad s \in S, \ \beta \in (\beta_0,1) \tag{2.1}$$

for every plan Π.

2.2 Definition

A plan Π^* is <u>1-optimal</u> if

$$\lim_{\beta \to 1-} \inf(V_\beta(s,\Pi^*) - V_\beta(s,\Pi)) \geq 0 \tag{2.2}$$

for all $s \in S$ and plan Π.

Another criterion of approaching the nondiscounted problem, which is due to Derman [5], depends on the average gain.

2.3 Definition

A plan Π^* is <u>average optimal</u> if

$$\lim_{n\to\infty} \inf V^n(s,\Pi^*)/n \geq \lim_{n\to\infty} \inf V^n(s,\Pi)/n \tag{2.3}$$

for all $s \in S$ and plan Π.

In the sequel (section 3), we shall discuss a process with special structure, the so-called completely ergodic process.

2.4 Definition

If, for a Markovian decision process, the Markov chain induced by the stationary plan f^∞ is ergodic for every f, then the process is called <u>completely ergodic</u>.

The following lemma is proved in Blackwell [2].

2.5 Lemma

Let P be a N × N Markov matrix.

(a) The sequence $\sum_{i=0}^{n-1} P^i/n$ converges as $n \to \infty$ to a markov matrix P* such that PP* = P*P = P*P* = P*.

(b) I - (P-P*) is nonsingular, and

$H(\beta) = \sum_{i=0}^{\infty} \beta^i (P^i - P^*) \to H = (I-P+P^*)^{-1} - P^*$ as $\beta \to 1-$.

$H(\beta) \cdot P^* = P^* \cdot H(\beta) = H \cdot P^* = P^* \cdot H = 0$ and

$(I-P) \cdot H = H \cdot (I-P) = I - P^*$.

(c) For every N × 1 column vector c, the system $P \cdot x = x$, $P^* \cdot x = P^* \cdot c$ has a unique solution x.

As in section 1, let $S = \{1,2,\ldots, N\}$ and let $F = A^N$. For $f \in F$, let I(f) be the N component column vector whose s-th component is I(s,f(s)). Let P(f) be the N × N Markov matrix whose s-s′th element is $q(s' \mid s,f(s))$. If $\Pi = (f_1, f_2, \ldots)$, $f_i \in F$, let $P^n(\Pi) = P(f_1)P(f_2)\ldots P(f_n)$ for $n > 0$ and $P^0(\Pi) = I$. The vector of expected total discounted rewards by using plan Π is

$$V_\beta(\Pi) = \sum_{n=0}^{\infty} \beta^n \cdot P^n(\Pi) \cdot I(f_{n+1})$$

where β is the discount factor.

The next theorem (see Miller and Veinott [13]) provides an expansion of V_β for β close to 1. To describe this it is convenient to define the norm of a (finite) matrix $C = (c_{ij})$ by $\|C\| = \max_i \sum_j |c_{ij}|$.

2.6 Theorem

If $f \in F$ and $0 < \rho < \|H(f)\|^{-1}$, then

$$V_\beta(f) = (1+\rho) \sum_{n=-1}^{\infty} \rho^n \, y_n(f) \tag{2.4}$$

where $\beta = (1+\rho)^{-1}$, $H(f) = (I-P(f)+P^*(f))^{-1} - P^*(f)$,
$y_{-1}(f) = P^*(f)\cdot I(f)$ and $y_n(f) = (-1)^n\cdot H(f)^{n+1}\cdot I(f)$, $n = 0,1,2,\ldots$ and $y_{-1}(f)$ is a unique solution of $(I-P(f))\cdot y_{-1}(f) = 0$, $P^*(f)\cdot y_{-1}(f) = P^*(f)\cdot I(f)$; $y_0(f)$ is a unique solution of

$$(I-P(f))\cdot y_0(f) = I(f) - y_{-1}(f),\ P^*(f)\cdot y_0(f) = 0.$$

Maitra [12] and Furukawa [7] established the following theorem for the discount problem when the action space is compact.

2.7 Theorem

Let A be a compact set. Suppose $I(s,a)$ and $q(s'|s,a)$ are continuous in a for each fixed $s \in S$ and $(s,s') \in S \times S$ respectively. Then, there exists a β-optimal stationary plan.

3. THE EXISTENCE OF A STATIONARY 1-OPTIMAL PLAN

From now on, the basic problem will be generalized in the following way. We suppose that, for each state $s = 1,2,\ldots,N$, we have a compact set A of actions. We shall establish a sufficient condition for the existence of a stationary 1-optimal plan. The condition can be weakened if the process is completely ergodic.

The following lemma is important in the later discussions.

3.1 Lemma

Let $f_1,f_2,\ldots,f_n$ belong to F. Then there exists a $f^* \in F$ such that for some $\beta^* \in (0,1)$,

$$V_\beta(s,f^*) \geq V_\beta(s,f_k)$$

for all $k = 1,2,\ldots,n$, and all $\beta \in [\beta^*,1)$ and all $s \in S$.

Proof. We only need to prove the case when $n = 2$, since the general case can be proved by induction.

Consider $V_\beta(f_1)$ and $V_\beta(f_2)$.

Note that for each s and f, the s-th coordinate of $V_\beta(f)$ is a rational function of β as the representation

$$V_\beta(f) = [I-\beta\cdot P(f)]^{-1} \cdot I(f)$$

shows. Therefore, there exists a β_s such that

$$V_\beta(s,f_1) \geq V_\beta(s,f_2) \text{ for all } \beta \in [\beta_s,1)$$

or $$V_\beta(s,f_1) < V_\beta(s,f_2) \text{ for all } \beta \in [\beta_s,1)$$

Let $$\beta^* = \max_s\{\beta_s\}$$

and $$m(s,\beta) = \max\{V_\beta(s,f_1),\ V_\beta(s,f_2)\}$$

for all $\beta \in [\beta^*,1)$ and

$$m(\beta) = \langle m(1,\beta),\ m(2,\beta),\ldots,m(N,\beta)\rangle^T.$$

We define f^* as follows

$$f^*(s) = f_1(s) \text{ when } V_{\beta^*}(s,f_1) \geq V_{\beta^*}(s,f_2)$$

$$f^*(s) = f_2(s) \text{ when } V_{\beta^*}(s,f_2) > V_{\beta^*}(s,f_1).$$

Then, $$L_\beta(f^*)m(\beta) = I(f^*) + \beta\cdot P(f^*)m(\beta) \geq m(\beta)$$

for all $\beta \in [\beta^*,1)$,
where $L_\beta(f)$ is the transformation which maps $N \times 1$ column vector ω into

$$L_\beta(f)\omega = I(f) + \beta\cdot P(f)\cdot\omega.$$

Thus, by the monotone property of $L_\beta(f)$, we have

$$V_\beta(s,f^*) \geq m(s,\beta) \text{ for all } \beta \in [\beta^*,1).$$

The proof is completed. ///

3.2 Remark

Lemma 3.1 implies that for any $f_1,\ldots,f_n \in F$, there exists a $f^* \in F$ such that $y_{-1}(s,f^*) \geq y_{-1}(s,f_k)$ for all $s \in S$ and

$k = 1,2,\ldots,n$.

Now we make the following assumptions:

Assumption I: For any $s \varepsilon S$, $I(s,\cdot)$ is a continuous function on A.

Assumption II: For any pair $(s,s') \varepsilon S \times S$, $q(s'|s,\cdot)$ is a continuous function on A.

Assumption III: $P^*(f)$ is continuous in f.

Our main result, described by the following theorem, concerns the existence of a stationary 1-optimal plan.

3.3 Theorem

If a Markovian decision process with compact action space and finite state space satisfies Assumptions I, II and III, then there exists a stationary 1-optimal plan.

Proof. For $f \varepsilon F$, $\rho < \|H(f)\|^{-1}$, we have

$$V_\beta(f) = (1+\rho) \sum_{n=-1}^{\infty} \rho^n y_n(f),$$

where
$$\rho = (1-\beta)\cdot\beta^{-1}$$

$$y_{-1}(f) = P^*(f)I(f)$$

$$y_n(f) = (-1)^n H^{n+1}(f),\ n = 1,2,\ldots .$$

Since F is compact, for each state s, there exists a $f_s \varepsilon F$ such that

$$y_{-1}(s,f_s) = \max_{f\varepsilon F} \{y_{-1}(s,f)\}.$$

By Remark 3.2, there exists a $f^* \varepsilon F$ satisfying

$$y_{-1}(s,f^*) \geq y_{-1}(s,f_s) \text{ for all } s \varepsilon S$$

i.e. $$y_{-1}(s,f^*) = \max_{f\varepsilon F} \{y_{-1}(s,f)\} \text{ for all } s \varepsilon S.$$

We define

$$F_0 = \{f | f \varepsilon F,\ y_{-1}(s,f) = y_{-1}(s,f^*),\ \text{for all } s \varepsilon S\}. \qquad (3.1)$$

That is, F_0 is the set of all $f \varepsilon F$ having maximal average return per unit time. It is evident that F_0 contains at least f^*.

Note that F_0 is a compact set. But,

$$\begin{aligned} y_0(f) &= H(f) I(f) \\ &= \left[I-P(f)+P^*(f)\right]^{-1} \cdot I(f) - P^*(f) \cdot I(f) \end{aligned}$$

which is also a continuous function in f.

Therefore, for each s, there exists a $f_s^* \varepsilon F_0$ such that

$$y_0(s,f_s^*) = \max_{f \varepsilon F_0} \{y_0(s,f)\}$$

By Lemma 3.1, there exists a $f^{**} \varepsilon F$ such that

$$V_\beta(s,f^{**}) \geq V_\beta(s,f_s^*), \quad \text{for all } s \text{ and all } \beta \text{ sufficient near to } 1.$$

which by (2.4) implies

$$y_{-1}(s,f^{**}) \geq y_{-1}(s,f_s^*) = y_{-1}(s,f_s^*),\ \forall\ s \ \varepsilon\ S.$$

and hence, $\quad y_{-1}(s,f^{**}) = y_{-1}(s,f^*)$, for all s.

That is $\quad f^{**} \varepsilon F_0$

Also, $\quad y_0(s,f^{**}) \geq y_0(s,f_s^*)$, for all s.

Therefore, $\quad y_0(s,f^{**}) = \max_{f \varepsilon F_0} \{y_0(s,f)\}$, for all s.

Using Theorem 2.6 and the fact that under Assumptions I, II, III, $\sum_{n=1}^{\infty} \rho^n y_n(f) \to 0$ uniformly in f as $\rho \to 0$, we obtain

$$\lim_{\beta\to 1^-} y_{-1}(s,f_\beta) = y_{-1}(s,f^{**}) \text{ for all } s \in S \tag{3.2}$$

where f_β is a β-optimal stationary plan, for each β.

Next, we claim

$$\lim_{\beta\to 1^-} \inf y_0(s,f_\beta) \le y_0(s,f^{**}),\ \forall\ s \in S. \tag{3.3}$$

For, there exists a convergent sequence f_{β_n} such that

$$f_{\beta_n} \to f_0 \in F \text{ as } \beta_n \to 1^-$$

Therefore,

$$y_{-1}(s,f_{\beta_n}) \to y_{-1}(s,f_0),\ \forall\ s \in S.$$

By (3.2), we conclude

$$y_{-1}(s,f_0) = y_{-1}(s,f^{**}),\ \forall\ s.$$

and hence,

$$f_0 \in F_0$$

Thus,

$$\begin{aligned}\lim_{\beta\to 1^-} \inf y_0(s,f_\beta) &\le \lim_{\beta_n\to 1^-} \inf y_0(s,f_{\beta_n}) \\ &= y_0(s,f_0) \\ &\le y_0(s,f^{**})\end{aligned}$$

by the definition of f^{**}.

By (3.2) and (3.3), for any plan Π we have

$$\begin{aligned}\lim_{\beta\to 1^-} \inf(V_\beta(s,f^{**})-V_\beta(s,\Pi)) &\ge \lim_{\beta\to 1^-} \inf(V_\beta(s,f^{**})-V_\beta(s,f_\beta)) + \\ &\quad \lim_{\beta\to 1^-} \inf(V_\beta(s,f_\beta)-V_\beta(s,\Pi)) \\ &\ge 0 \quad \text{for all } s \in S.\end{aligned}$$

The proof is completed. ///

3.4 Corollary

Suppose Assumptions I and II are satisfied and $\{P^n(f), n \geq 1\}$ is a family of equicontinuous functions on F. Then there exists a stationary 1-optimal plan.

Proof. Note that $\{\sum_{n=0}^{N-1} P^n(f)/N, N \geq 1\}$ is also equicontinuous. Assumption III, and hence the statement, follows from the fact that

$$\sum_{n=0}^{N-1} P^n(f)/N \longrightarrow P^*(f)$$

The proof is completed. ///

If the process is completely ergodic, Assumption III can be dropped.

3.5 Lemma

For a completely ergodic process, Assumption II implies Assumption III.

Proof. For any $f \in F$, let $P_N(f)$ be a $N \times N$ matrix whose N-th column is $1 = \{1,1,\ldots,1\}^T$ and the other elements are the same as those of $I - P(f)$.

Let $\pi(f) = \langle\pi(1,f), \pi(2,f),\ldots,\pi(N,f)\}$ be a stationary distribution of $P(f)$, then $\pi(f)P_N(f) = \langle 0,0,\ldots,0,1\rangle$. Since the process is completely ergodic, $\pi(f)$ is the unique solution of the system of linear equations

$$\pi(f)P_N(f) = \langle 0,0,\ldots,0,1\rangle \qquad (3.4)$$

Hence, $P_N(f)$ is nonsingular and

$$\pi(f) = \langle 0,0,\ldots,0,1\rangle \cdot P_N^{-1}(f)$$

which is also a continuous function of f. The proof is completed. ///

Combining Theorem 3.3 and Lemma 3.5, we immediately get the following:

3.6 Theorem

If a completely ergodic process with compact action space and finite state space satisfies Assumptions I, II, then there exists a stationary 1-optimal plan.

3.7 Remark

Under Assumptions I, II, III, each element in the set F_O defined as in (3.1) is average optimal by the following argument: for any plan Π, any state s,

$$\begin{aligned}\liminf_{n\to\infty} \frac{1}{n} V^n(s,\Pi) &\leq \liminf_{\beta\to 1^-} (1-\beta)V_\beta(s,\Pi) \\ &\leq \liminf_{\beta\to 1^-} (1-\beta)V_\beta(s,f_\beta) \\ &= y_{-1}(s,f^*), \text{ where } f^* \in F_O.\end{aligned}$$

3.8 Remark

Assumption III was also used by Hordijk to prove the existence of a stationary average optimal plan for compact action and denumerable state Markovian decision processes; see Hordijk [9].

3.9 Remark

Another criterion of optimality for the nondiscounted case was proposed by Veinott [14]. Veinott defined a plan Π^* to be average overtaking optimal if

$$\liminf_{N\to\infty} \frac{1}{N} \sum_{n=0}^{N-1} (V^n(s,\Pi^*) - V^n(s,\Pi)) \geq 0$$

for every $s \in S$ and every plan Π. By a theorem of Lippman [10], one can show that if the state space is finite, then a plan is average overtaking optimal if and only if it is 1-optimal.

4. EXAMPLES

This section contains all of our examples. The first example shows the condition of Corollary 3.3 can be satisfied. The second example shows that the condition of Corollary 3.3 is not necessary. The third example shows that there does not necessarily exist an average optimal stationary plan if Assumption III fails. In the fourth, we demonstrate that the existence of an average optimal stationary plan does not imply the existence of a stationary 1-optimal plan.

4.1 Example

Let $S = \{1,2\}$ and $A = [0,1/2]$

Consider the system defined by $q(1|1,a) = 0$, $q(2|1,a) = 1$ $q(1|2,a) = a$, $q(2|2,a) = 1-a$ where $0 \leq a \leq 1/2$ and f_a denotes the policy which always selects action a ($a \varepsilon [0,1/2]$). Hence, we obtain

$$q^*(1|1,f_a) = q^*(1|2,f_a) = \frac{a}{1+a}$$

$$q^*(2|1,f_a) = q^*(2|2,f_a) = \frac{1}{1+a}$$

The absolute difference between $q^n(s'|s,f_a)$ and $q^*(s'|s,f_a)$ is denoted by $d_a^n(s,s')$ where $(s,s') \varepsilon S \times S$, $n = 1,2,\ldots$. By computations,

$$d_a^n(1,1) = d_a^n(1,2) = \frac{a^n}{1+a}$$

$$d_a^n(2,1) = d_a^n(2,2) = \frac{a^{n+1}}{1+a}$$

for all $a \varepsilon [0,1/2]$ and $n = 1,2\ldots.$.

Therefore, $q^n(s'|s,f_a)$ converges to $q^*(s'|s,f_a)$ uniformly; but $q^*(s'|s,f_a)$ is continuous and A is compact, $q^n(s'|s,f_a)$ must be equicontinuous.

4.2 Example

Let the state space and the law of transitions be as in example 4.1. The only difference is that we extend the action space to $[0,1]$. Since $P(f_1)$ is periodic with period 2, where

$$q^{2n}(1|1,f_1) = q^{2n}(2|2,f_1) = q^{2n+1}(2|1,f_1)$$

$$= q^{2n+1}(1|2,f_1) = 1 \text{ for all } n = 1,2,\ldots .$$

On the other hand, $P^n(f_a)$ converges to $P^*(f_a)$, for all $a \in [0,1)$. Therefore, $q^n(s'|s,f)$, $n = 1,2,\ldots$ is not a family of equicontinuous functions despite that $q^*(s'|s,f)$ is continuous in f, for all $(s,s') \in S \times S$.

4.3 Example

Let $S = \{1,2,3\}$ and $A = [0,1]$.

Consider the system defined by $q(1|1,a) = (1-a)$, $q(2|1,a) = a(1-a)$, $q(3|1,a) = a^2$, $q(2|2,a) = q(3|3,a) = 1$, $q(1|2,a) = q(1|3,a) = q(2|3,a) = q(3|2,a) = 0$ and $I(1,a) = 1$, $I(2,a) = 7$, $I(3,a) = 4$ where $0 \le a \le 1$ and let f_a denote the policy which always selects action a $(a \in [0,1])$. Then, we obtain

$$q^*(2|1,f_a) = 1-a$$

$$q^*(3|1,f_a) = a$$

$$q^*(2|2,f_a) = q^*(3|3,f_a) = 1$$

and

$$q^*(1|1,f_a) = q^*(1|2,f_a) = q^*(1|3,f_a)$$

$$= q^*(2|3,f_a) = q^*(3|2,f_a) = 0$$

where $a > 0$, and $P^*(f_0)$ is an identity matrix.

We can easily evaluate the vector $y_{-1}(f_a) = \langle 7-3a,7,4\rangle^T$, $0 < a \le 1$. When $a = 0$, $y_{-1}(f_0) = \langle 1,7,4\rangle^T$. It follows that there is no stationary average optimal plan.

4.4 Example

Let $S = \{1,2\}$ and $A = [0,1]$.

Consider the system defined by $q(1|1,a) = 1-a$, $q(2|1,a) = a$, $q(1|2,a) = 0$, $q(2|2,a) = 1$ and $I(1,a) = 2a^{\frac{1}{2}}$, $I(2,a) = 0$ where $0 \leq a \leq 1$ and denote the policy which always selects action a ($a \in |\overline{0},1|$) by f_a. Then, we obtain

$$q^*(1|1,f_a) = q^*(1|2,f_a) = 0$$

$$q^*(2|1,f_a) = q^*(2|2,f_a) = 1$$

where $a > 0$, and $P^*(f_0)$ is an identity matrix.

We can easily evaluate the vector $y_{-1}(f_a) = \langle 0,0\rangle^T$, $0 \leq a \leq 1$, and $y_0(f_a) = \langle 2a^{-\frac{1}{2}},0\rangle^T$, when $a > 0$, and $y_0(f_0) = \langle 0,0\rangle^T$. The first component of $y_0(f_a)$ is unbounded as $a \downarrow 0$ so it is not possible to maximize $y_0(f_a)$; and therefore, there is no stationary 1-optimal plan.

On the other hand, solving

$$x = \max_{0 \leq a \leq 1} \{2a + \beta(1-a)x\}$$

we obtain

$$a = (1-\beta)\beta^{-1}, \quad x = 1/\sqrt{\beta(1-\beta)}$$

Therefore, for $\beta \geq \frac{1}{2}$, the plan f^∞ defined by $f_\beta(1) = 1-\beta/\beta$ is a β-optimal stationary plan and

$$V_\beta(s,f_\beta) = \begin{cases} 1/\sqrt{\beta(1-\beta)} & \text{if } s = 1 \\ 0 & \text{if } s = 2 \end{cases}$$

For any plan Π,

$$\begin{aligned} \liminf_{n\to\infty} \frac{1}{n} V^n(s,\Pi) &\leq \liminf_{\beta\to 1^-} (1-\beta)V_\beta(s,\Pi) \\ &\leq \liminf_{\beta\to 1^-} (1-\beta)V_\beta(s,f_\beta) \\ &= 0 \end{aligned}$$

Thus, every stationary plan is average optimal.

4.5 Remark

Example 4.4 also occurs in Bather [1] with different approaches and purposes.

5. ACKNOWLEDGEMENTS

Research supported by the National Science Council Grant NSC No. 66M-0204-03(02), Republic of China.

6. REFERENCES

1. Bather, J. "Optimal decision procedures for finite Markov chains Part I: Examples", *Adv. Appl. Prob.*, **5**, 328-339, (1973).

2. Blackwell, D. "Discrete dynamic programming", *Ann. Math. Statist.*, **33**, 719-726, (1962).

3. Blackwell, D. "Discounted dynamic programming", *Ann. Math. Statist.*, **36**, 226-235, (1965).

4. Denardo, E. V. and Miller, B. L. "An optimality condition for discrete dynamic programming with no discounting", *Ann. Math. Statist.*, **39**, 1220-1227, (1968).

5. Derman, C. "On sequential decisions and Markov chains", *Mgt. Sci.*, **9**, 16-24, (1962).

6. Derman, C. "On sequential control processes", *Ann. Math. Statist.*, **35**, 341-349, (1964).

7. Furukawa, N. "Markovian decision processes with compact action spaces", *Ann. Math. Statist.*, **43**, 1612-1622, (1972).

8. Hordijk, A. "A sufficient condition for the existence of an optimal policy with respect to the average cost criterion in Markovian decision processes", Transactions Sixth Prague Conf. on Information Theory, Statistical Decision Functions, Random Processes, Academia, Prague, 263-274, (1971)

9. Hordijk, A. "Dynamic programming and Markov potential theory", Mathematical Centre Tract 51, Amsterdam, (1974)

10. Lippman, S. "Criterion equivalence in discrete dynamic programming", *Operations Res.*, **17**, 920-923, (1969).

11. Maitra, A. "Dynamic programming for countable state systems", Sankhya Ser. A. 27, 241-248, (1965).

12. Maitra, A. "Discounted dynamic programming on compact metric spaces", Sankhya Ser. A. 30, 211-216, (1968).

13. Miller, B. L. and Veinott, Jr., A. F. "Discrete dynamic programming with small interest rate", *Ann. Math. Statist.*, **40**, 366-370, (1969).

14. Veinott, Jr., A. F. "On the finding optimal policies in discrete dynamic programming with no discounting", *Ann. Math. Statist.*, **37**, 1284-1294, (1966).

ON SOLVING MARKOV DECISION PROBLEMS BY LINEAR PROGRAMMING

A. Hordijk and L.C.M. Kallenberg

(University of Leiden, The Netherlands)

ABSTRACT

In this paper we show that optimal policies for Markov decision problems can be found by solving one linear programming problem. Furthermore, the relation between the feasible solutions of the linear program and the set of stationary policies is analysed.

1. INTRODUCTION

A process is observed at discrete time points $t = 1,2,\ldots$ to be in one of a finite set of possible states. The *state-space* is denoted by $E = \{1,2,\ldots,N\}$.

After observing the state of the process, an *action* must be chosen. Let $A(i)$ denote the set of all possible actions in state i. We assume that $A(i)$ has a finite number of elements.

If the system is in state i and action $a \in A(i)$ is chosen, then a *reward* r_{ia} is earned immediately and with probability p_{iaj} the system will be in state j at the next instant.

Let $\{X_t, t=1,2,\ldots\}$ respectively $\{Y_t, t=1,2,\ldots\}$ denote the sequences of observed states respectively chosen actions.

A *decision rule* π^t at time t is a function which assigns for each i the probability of taking action a at time t; in general, it may depend on all realized states up to and including time t and on all realized actions up to time t.

A *policy* R is a sequence of decision rules; $R = (\pi^1,\pi^2,\ldots,\pi^t,\ldots)$. For a *Markov* policy $R = (\pi^1,\pi^2,\ldots)$ we have that π^t, the decision rule at time t, is a function only of the state at time t. Policy $R = (\pi^1,\pi^2,\ldots)$ is a *stationary* policy if all decision rules are identical; we denote the stationary policy $R = (\pi, \pi, \ldots)$ also by π^∞. A stationary policy is said to be *pure* if its decision rule is nonrandomized. Consequently, a pure and stationary policy is completely described by a mapping $f : E \to \bigcup_{i \in E} A(i)$ such that $f(i) \in A(i)$, $i \in E$. We denote this policy by f^∞.

For any policy R and initial state i, we denote by $v_i^\alpha(R)$ respectively $\phi_i(R)$ the *total expected discounted reward* respectively the *average expected reward*, where α is the discount factor:

$$v_i^\alpha(R) = \sum_{t=1}^{\infty} \alpha^{t-1} \sum_j \sum_a \mathbb{P}_R(X_t=j,\ Y_t=a \mid X_1=i) \cdot r_{ja}$$

$$\phi_i(R) = \liminf_{T\to\infty} \frac{1}{T} \sum_{t=1}^{T} \sum_j \sum_a \mathbb{P}_R(X_t=j,\ Y_t=a \mid X_1=i) \cdot r_{ja}$$

We are interested in finding policies which are optimal with respect to these reward functions.

R^* is said to be α-*discounted optimal* if $v_i^\alpha(R^*) = v_i^\alpha$, $i \in E$, where $v_i^\alpha = \sup_R v_i^\alpha(R)$. R^* is *average optimal* if $\phi_i(R^*) = \phi_i$, $i \in E$, where $\phi_i = \sup_R \phi_i(R)$.

Although the use of a linear programming algorithm was proposed already in 1960 by d'Epenoux [5] for the discounted case, no satisfactory linear programming algorithm was available for the average reward case. De Ghellinck [6] as well as Manne [10] obtained linear programming formulations for the average reward criterion in the unichain case. The first analysis of linear programs for the multichain case is due to Denardo and Fox [2], [3]. Derman [4] streamlined and slightly improved their results. He shows that in order to find an optimal policy there have to be solved, in the worst case, two linear programming problems and one search problem. In section 3, we give a very simple rule to find a pure and stationary average optimal policy from an optimal solution of the linear programming problem introduced by Denardo and Fox [3]. Consequently, we can construct an average optimal policy by solving only *one* linear program.

In order to make a comparison between the discounted case and the average reward case we present a survey of the main results of the discounted case in section 2.

For the discounted case there is a complete characterization of the feasible solutions and of the extreme points of the linear program.

In section 4 we derive analogous results for the average reward case. As it turns out, we have to use equivalence classes of feasible solutions. We construct a one-to-one correspondence between the stationary policies and the representatives of the equivalence classes such that optimal policies correspond to optimal solutions. Pure policies are mapped on extreme points; however, in general, the converse is not true. Furthermore, we show by examples that the elements in an equivalence class are not easy to characterize.

2. DISCOUNTED REWARD CASE

In this section we summarize the main results for the discounted reward case. Where possible we use Derman's monograph [4] as reference.

Theorem 1: (Derman [4], pp. 22-23)

v^α is the unique solution of the optimality equation

$$v_i = \max_a\{r_{ia} + \alpha\sum_j p_{iaj}\, v_j\}, \quad i \in E.$$

Theorem 2: (Derman [4], pp. 22-23)

Let a_i, $i \in E$, be such that

$$r_{ia_i} + \alpha\sum_j p_{ia_ij}\, v_j^\alpha = \max_a\{r_{ia} + \alpha\sum_j p_{iaj}\, v_j^\alpha\}, \quad i \in E.$$

Then the pure and stationary policy f^∞, where $f(i) = a_i$, $i \in E$, is α-discounted optimal.

Definition: A function $\tilde{v} : E \to \mathbb{R}$ is α-*superharmonic* if

$$\tilde{v}_i \geq r_{ia} + \alpha \sum_j p_{iaj} \tilde{v}_j \qquad a \in A(i), \quad i \in E.$$

Theorem 3: (Hordijk [7], p. 25 and p. 55)

v^α is the (componentwise) smallest α-superharmonic function.

The theorems 2 and 3 lead us to consider the following linear programming problem

$$\min \left\{ \sum_j \beta_j \tilde{v}_j \;\middle|\; \tilde{v}_i \geq r_{ia} + \alpha \sum_j p_{iaj} \tilde{v}_j \quad a \in A(i), \quad i \in E \right\} \quad (1)$$

where $\beta_j > 0$, $j \in E$, are given numbers such that $\sum_j \beta_j = 1$.

From theorem 3 it follows that v^α is an optimal solution of (1).

The dual problem is

$$\max \left\{ \sum_i \sum_a r_{ia} x_{ia} \;\middle|\; \begin{array}{l} \sum_i \sum_a (\delta_{ij} - \alpha p_{iaj}) x_{ia} = \beta_j, \; j \in E \\ x_{ia} \geq 0 \;, \; a \in A(i), \; i \in E \end{array} \right\} \quad (2)$$

where δ_{ij} is Kronecker's delta.

Theorem 4: (Derman [4], p. 43)

If the simplex method is used to solve problem (2) and an optimal solution x is obtained, then the policy f^∞, where $f(i) = a_i$ such that $x_{ia_i} > 0$, $i \in E$, is an α-discounted optimal policy.

Theorem 5: (Derman [4], p.42)

The mapping $x_{ia}(\pi) = \left[\beta^T\left(I-\alpha P(\pi)\right)^{-1}\right]_i \cdot \pi_{ia}$, $a \in A(i)$,

$i \in E$, where $P(\pi) = \left(\sum_a p_{iaj}\ \pi_{ia}\right)$, is a one-to-one mapping of the stationary policies onto the set of feasible solutions of the dual problem (2).

Theorem 6: The mapping preserves the optimality property, i.e. π^∞ is an optimal policy if and only if $x(\pi)$ is an optimal solution of (2).

Proof: If π^∞ is an optimal policy, then

$$v^\alpha = v^\alpha(\pi^\infty) = \left(I-\alpha P(\pi)\right)^{-1} r(\pi), \quad \text{where } r(\pi) = \left(\sum_a r_{ia}\ \pi_{ia}\right).$$

$$\text{Hence, } \sum_i \sum_a r_{ia}\ x_{ia}(\pi) = \sum_i \left[\beta^T\left(I-\alpha P(\pi)\right)^{-1}\right]_i \cdot r_i(\pi) = \beta^T v^\alpha.$$

Since it follows from the theory of linear programming that the optimum of (1) is equal to the optimum of (2), we have that $x(\pi)$ is an optimal solution of the dual problem (2).

The proof that π^∞, where $\pi_{ia} = x_{ia} \big/ \sum_a x_{ia}$, $a \in A(i)$, $i \in E$, is an optimal policy if x is an optimal solution of the dual problem (2), can be found in Derman [4] p. 43.

Theorem 7: The pure and stationary policies are mapped onto the set of feasible extreme points of the dual problem.

Proof: Let f^∞ be a pure and stationary policy, where $f(i) = a_i$, $i \in E$. Suppose that the corresponding $x(f)$ is not an extreme point. Then, there exist feasible solutions x^1 and x^2 such that

$$x(f) = \lambda x^1 + (1-\lambda)x^2 \text{ for some } 0 < \lambda < 1, \text{ and } x^1 \neq x^2.$$

Since $x_{ia}(f) = 0$ $a \neq a_i$, $i \in E$, we have $x^1_{ia} = x^2_{ia} = x_{ia}(f) = 0$, $a \neq a_i$, $i \in E$.

Hence, x^1, x^2 and $x(f)$ are all mapped on the same policy f^∞. Consequently, $x^1 = x^2 = x(f)$ which implies a contradiction.

On the other hand, let x be a feasible extreme point of the dual problem. Since the dual problem has N constraints, an extreme point x has at most N positive components. The feasibility of x implies that

$$\sum_a x_{ja} = \beta_j + \alpha \sum_i \sum_a p_{iaj}\, x_{ia} \geq \beta_j > 0, \qquad j \in E.$$

Hence, for each $i \in E$ there is exactly one $a_i \in A(i)$ such that $x_{ia_i} > 0$. Consequently, the corresponding policy is pure.

3. CONSTRUCTION OF AN AVERAGE OPTIMAL PURE POLICY

Definition: A pair of functions $(\tilde{\phi},\tilde{u})$, where $\tilde{\phi} : E \to \mathbb{R}$ and $\tilde{u} : E \to \mathbb{R}$ is *superharmonic* if

$$\begin{cases} \tilde{\phi}_i \geq \sum_j p_{iaj}\, \tilde{\phi}_j & a \in A(i), \quad i \in E. \\ \tilde{\phi}_i + \tilde{u}_i \geq r_{ia} + \sum_j p_{iaj}\, \tilde{u}_j & a \in A(i), \quad i \in E. \end{cases}$$

Theorem 8: (Hordijk and Kallenberg [8])

ϕ is the (componentwise) smallest function for which there exists a function u such that (ϕ,u) is superharmonic.

In view of theorem 8 it is plausible to consider the following linear programming problem

$$\min \left\{ \sum_j \beta_j \tilde{\phi}_j \;\middle|\; \begin{array}{ll} \tilde{\phi}_i \geq \sum_j p_{iaj}\, \tilde{\phi}_j & a \in A(i), \quad i \in E \\ \tilde{\phi}_i + \tilde{u}_i \geq r_{ia} + \sum_j p_{iaj}\, \tilde{u}_j & a \in A(i), \quad i \in E \end{array} \right\} \qquad (3)$$

From theorem 8 it follows that ϕ is an optimal solution of (3).

The dual linear programming problem is

$$\max \left\{ \sum_i \sum_a r_{ia} x_{ia} \;\middle|\; \begin{array}{ll} \sum_i \sum_a (\delta_{ij} - p_{iaj}) x_{ia} = 0 & j \in E \\ \sum_a x_{ja} + \sum_i \sum_a (\delta_{ij} - p_{iaj}) y_{ia} = \beta_j & j \in E \\ x_{ia}, y_{ia} \geq 0 \quad a \in A(i),\ i \in E & \end{array} \right\} \tag{4}$$

<u>Theorem 9</u>: (Hordijk and Kallenberg [8])

If the simplex method is used to solve problem (4) and an optimal solution (x,y) is obtained, then the policy f^∞, where

$$f(i) = a_i \text{ such that } \begin{cases} x_{ia_i} > 0 & i \in E_x = \{i \mid \sum_a x_{ia} > 0\} \\ y_{ia_i} > 0 & i \notin E_x \end{cases} \tag{5}$$

is average optimal.

<u>Remark</u>: Rule (5) says that an optimal pure and stationary policy can directly be obtained from an optimal solution of the linear program by taking an arbitrary action $a \in \bar{A}(i)$, $i \in E$, where

$$\bar{A}(i) = \begin{cases} \{a \mid x_{ia} > 0\} & i \in E_x. \\ \{a \mid y_{ia} > 0\} & i \notin E_x. \end{cases}$$

4. STATIONARY POLICIES AND FEASIBLE SOLUTIONS

For a feasible solution (x,y) of the dual program (4) we define a stationary policy $\pi^\infty(x,y)$ by

$$\pi_{ia}(x,y) = \begin{cases} x_{ia} \Big/ \sum_a x_{ia} & a \in A(i),\ i \in E_x \\ y_{ia} \Big/ \sum_a y_{ia} & a \in A(i),\ i \notin E_x \end{cases} \tag{6}$$

Conversely, let π^∞ be a stationary policy and $P(\pi) = (\sum_a p_{iaj}\pi_{ia})$.

Then it holds that (see e.g. Kemeny and Snell [9])

a. $P^*(\pi) = \lim_{n\to\infty} \frac{1}{n} \sum_{k=1}^{n} P^{k-1}(\pi)$ exists and $P^*(\pi)\ P(\pi) =$

$P(\pi)\ P^*(\pi) = P^*(\pi)\ P^*(\pi) = P^*(\pi)$.

b. $D(\pi) = \left(I-P(\pi)+P^*(\pi)\right)^{-1} - P^*(\pi)$ exists and $D(\pi)\ P^*(\pi) =$

$P^*(\pi)\ D(\pi) = 0$.

The matrix $P(\pi)$ induces a Markov chain. Suppose that this Markov chain has m ergodic sets $E_1, E_2, \ldots, E_m$ and let T be the set of transient states. We define

$$\begin{aligned} x_{ia}(\pi) &= \left[\beta^T P^*(\pi)\right]_i \cdot \pi_{ia} & a \in A(i),\ i \in E \\ y_{ia}(\pi) &= \left[\beta^T D(\pi) + \gamma^T P^*(\pi)\right]_i \cdot \pi_{ia} & a \in A(i),\ i \in E \end{aligned} \tag{7}$$

where $$\gamma_l = \begin{cases} \max_{i \in E_j} \dfrac{-\sum_k \beta_k\, d_{ki}(\pi)}{\sum_k p^*_{ki}(\pi)} & l \in E_j \\ 0 & l \in T \end{cases}$$

Theorem 10: For any stationary policy π^∞, $\left(x(\pi), y(\pi)\right)$, defined by (7), is a feasible solution of the dual problem (4).

Proof:

1. $\sum_i \sum_a (\delta_{ij} - p_{iaj}) x_{ia}(\pi) = \left(\beta^T P^*(\pi)\right)_j - \left(\beta^T P^*(\pi)\, P(\pi)\right)_j = 0,$

   $j \in E.$

2. $\sum_a x_{ja}(\pi) + \sum_i \sum_a (\delta_{ij} - p_{iaj}) y_{ia}(\pi) = \left(\beta^T P^*(\pi)\right)_j + \left(\beta^T D(\pi) + \gamma^T P^*(\pi)\right)_j - \left(\beta^T D(\pi) P(\pi) + \gamma^T P^*(\pi) P(\pi)\right)_j = \left(\beta^T \left\{P^*(\pi) + D(\pi)\left(I - P(\pi) + P^*(\pi)\right)\right\}\right)_j = \beta_j,$

   $j \in E.$

3. $x_{ia}(\pi) \geq 0 \quad a \in A(i), \quad i \in E.$

4. $p^*_{ki}(\pi) = 0 \quad i \in T, \quad k \in E.$

Since $p_{ki}(\pi) = p^*_{ki}(\pi) = 0$ $i \in T$, $k \notin T$, we have $d_{ki}(\pi) = 0$ $i \in T$, $k \notin T$.

For $i \in T$, $k \in T$, we obtain

$$d_{ki}(\pi) = \left[\left(I - P(\pi) + P^*(\pi)\right)^{-1}\right]_{ki} - p^*_{ki}(\pi) = \left[\left(I - P(\pi)\right)^{-1}\right]_{ki}$$

$$= \sum_{j=0}^{\infty} p^j_{ki}(\pi) \geq 0.$$

Therefore,

for $i \in T$: $y_{ia}(\pi) = \sum_k \beta_k \, d_{ki}(\pi) \cdot \pi_{ia} \geq 0 \qquad a \in A(i)$

for $i \in E_j$: $y_{ia}(\pi) = \left\{ \sum_k \beta_k \, d_{ki}(\pi) + \gamma_i \sum_{k \in E_j} p^*_{ki}(\pi) \right\} \cdot \pi_{ia} \geq 0$

$a \in A(i)$

Hence, $y_{ia}(\pi) \geq 0 \quad a \in A(i), \quad i \in E.$

From the properties 1 until 4 it follows that $(x(\pi), y(\pi))$ is a feasible solution of (4).

Theorem 11: (Hordijk and Kallenberg [8])

a. If π^∞ is an optimal policy, then $(x(\pi), y(\pi))$ is an optimal solution of the dual problem (4).

b. If (x,y) is an optimal solution of (4), then the policy $\pi^\infty(x,y)$ is average optimal.

Theorem 12: (Hordijk and Kallenberg [8])

Let f^∞ be a pure and stationary policy.

Then, the corresponding feasible solution $(x(f), y(f))$ is an extreme point of the dual linear programming problem (4).

The feasible solutions of the dual program form an extremely complicated set. We will illustrate some characteristics by especially constructed examples. They show that:

(i) (Example 1) It is possible that - using the simplex method for solving problem (4) - we have $|\bar{A}(i)| \geq 2$ for some states i, which states can be transient and recurrent.
Hence, there may be several possibilities for choosing a pure optimal policy via rule (5), and rule (6) can assign a non pure optimal policy to a feasible extreme optimal point.

(ii) (Example 2) The mapping between the feasible solutions and the stationary policies is not one-to-one. However, we can define an equivalence relation on the set of feasible solutions such that there exists a one-to-one correspondence between the stationary policies and the equivalence classes.

(iii) (Example 3) The elements of an equivalence class are not easy to characterize.

<u>Example 1</u>:

$$E = \{1,2,3,4,5\};\ A(1)=A(3)=A(5)=\{1\},\ A(2)=A(4)=\{1,2\}$$

$$p_{113} = p_{213} = p_{224} = p_{314} = p_{415} = p_{423} = p_{514} = 1$$

$$\beta_1 = \frac{2}{10},\ \beta_2 = \frac{5}{10},\ \beta_3 = \frac{1}{10},\ \beta_4 = \frac{1}{10},\ \beta_5 = \frac{1}{10}.$$

$$r_{11} = r_{21} = r_{22} = r_{31} = r_{41} = r_{42} = r_{51} = 1.$$

Dual linear programming problem:

maximize $x_{11} + x_{21} + x_{22} + x_{31} + x_{41} + x_{42} + x_{51}$ subject to

$$\begin{array}{rcl}
x_{11} & = & 0 \\
x_{21} + x_{22} & = & 0 \\
-x_{11} - x_{21} + x_{31} - x_{42} & = & 0 \\
-x_{22} - x_{31} + x_{41} + x_{42} - x_{51} & = & 0 \\
-x_{41} + x_{51} & = & 0 \\
x_{11} + y_{11} & = & \frac{2}{10} \\
x_{21} + x_{22} + y_{21} + y_{22} & = & \frac{5}{10} \\
x_{31} - y_{11} - y_{21} + y_{31} - y_{42} & = & \frac{1}{10} \\
x_{41} + x_{42} - y_{22} - y_{31} + y_{41} + y_{42} - y_{51} & = & \frac{1}{10} \\
x_{51} - y_{41} + y_{51} & = & \frac{1}{10}
\end{array}$$

$$x_{ia}, y_{ia} \geq 0 \text{ for all } i,a.$$

The solution $x_{11} = 0$, $x_{21} = 0$, $x_{22} = 0$, $x_{31} = \frac{4}{10}$, $x_{41} = \frac{1}{10}$, $x_{42} = \frac{4}{10}$, $x_{51} = \frac{1}{10}$, $y_{11} = \frac{2}{10}$, $y_{21} = \frac{1}{10}$, $y_{22} = \frac{4}{10}$, $y_{31} = 0$, $y_{41} = 0$, $y_{42} = 0$, $y_{51} = 0$ is an extreme optimal point; hence, it can be obtained as final basic solution of the simplex method.

Also we see that $|A(2)| = |A(4)| = 2$.

Example 2:

$E = \{1,2,3,4\}$; $A(1) = A(4) = \{1\}$, $A(2) = A(3) = \{1,2\}$.

$p_{112} = p_{213} = p_{224} = p_{313} = p_{321} = p_{414} = 1$; $\beta_1 = \beta_2 = \beta_3 = \beta_4 = \frac{1}{4}$

$r_{11} = r_{21} = r_{22} = r_{31} = r_{32} = r_{41} = 1.$

Dual linear programming problem:

maximize $x_{11} + x_{21} + x_{22} + x_{31} + x_{32} + x_{41}$ subject to

$$
\begin{array}{lr}
x_{11} - x_{32} & = 0 \\
-x_{11} + x_{21} + x_{22} & = 0 \\
-x_{21} + (1-1)x_{31} + x_{32} & = 0 \\
-x_{22} + (1-1)x_{41} & = 0 \\
x_{11} + y_{11} - y_{32} & = \frac{1}{4} \\
x_{21} + x_{22} - y_{11} + y_{21} + y_{22} & = \frac{1}{4} \\
x_{31} + x_{32} - y_{21} + (1-1)y_{31} + y_{32} & = \frac{1}{4} \\
x_{41} - y_{22} + (1-1)y_{41} & = \frac{1}{4}
\end{array}
$$

$x_{ia}, y_{ia} \geq 0$ for all i,a.

The following two feasible solutions (x^1,y^1) and (x^2,y^2) are mapped on the same pure and stationary policy f^∞, where $f(i) = 1$ $i = 1,2,4$ and $f(3) = 2$.

$x^1_{11} = \frac{1}{4}$, $x^1_{21} = \frac{1}{4}$, $x^1_{22} = 0$, $x^1_{31} = 0$, $x^1_{32} = \frac{1}{4}$, $x^1_{41} = \frac{1}{4}$, $y^1_{11} = 0$,

$y^1_{21} = 0$, $y^1_{22} = 0$, $y^1_{31} = 0$, $y^1_{32} = 0$, $y^1_{41} = 0$ and $x^2_{11} = \frac{1}{6}$, $x^2_{21} = \frac{1}{6}$,

$x^2_{22} = 0$, $x^2_{31} = 0$, $x^2_{32} = \frac{1}{6}$, $x^2_{41} = \frac{1}{2}$, $y^2_{11} = \frac{1}{6}$, $y^2_{21} = 0$, $y^2_{22} = \frac{1}{4}$,

$y^2_{31} = 0$, $y^2_{32} = \frac{1}{12}$, $y^2_{41} = 0$.

Note that $(x^1,y^1) = (x(f),y(f))$ and hence, (x^1,y^1) is an extreme point.

We call two feasible solutions (x^1,y^1) and (x^2,y^2) *equivalent* if

$$\pi_{ia}(x^1,y^1) = \pi_{ia}(x^2,y^2) \text{ for all } a \in A(i),\ i \in E.$$

For a stationary policy π^∞, let $(X(\pi), Y(\pi))$ be the class of corresponding equivalent feasible solutions. We choose the point $(x(\pi),y(\pi))$ as the *representative* of this equivalence class.

Hence, the mapping defined by (7) is a one-to-one mapping of the stationary policies onto the set of representatives.

From example 2 it follows that $X(\pi)$ can have more than one element. Each $x \in X(\pi)$ is an invariant probability vector with regard to $P(\pi)$. However, the converse is not true. For example $x = (\frac{1}{3}, \frac{1}{3}, \frac{1}{3}, 0)^T$ is an invariant probability vector with regard to $P(f)$, where f is the policy used in example 2, but there is no feasible solution of the dual problem with $x_{41} = 0$.

In the unichain case (i.e. if there is only one ergodic set E_1), $x_i = \Sigma_a x_{ia}$ satisfies

$$\begin{cases} x_i = \sum_{j \in E_1} x_j \, p_{ji}(\pi) \\ \sum_{i \in E_1} x_i = 1 \end{cases}$$

It is well-known (e.g. Chung [1] p.33) that this system has a unique solution. Hence, $X(\pi)$ consists of one element: $X(\pi) = \{x(\pi)\}$.

As can be shown similarly to theorem 10 any (x,y), where $x = x(\pi)$ and $y \in Y^*(\pi) = \{y \mid y_{ia} = \left[\beta^T D(\pi) + c^T P^*(\pi)\right]_i \cdot \pi_{ia}$ for some $c \geq \gamma\}$, is a feasible solution. Hence, $Y^*(\pi) \subset Y(\pi)$.

The next example shows that even in the unichain case it is possible that $Y^*(\pi) \neq Y(\pi)$.

Example 3:

$E = \{1,2,3\}$; $A(1) = A(2) = \{1,2\}$, $A(3) = \{1\}$.

$p_{112} = p_{123} = p_{213} = p_{221} = p_{312} = 1$; $\beta_1 = \beta_3 = \frac{1}{4}$, $\beta_2 = \frac{1}{2}$.

$r_{11} = r_{12} = r_{21} = r_{22} = r_{31} = 1$.

Dual linear programming problem:

maximize $x_{11} + x_{12} + x_{21} + x_{22} + x_{31}$ subject to

$$\begin{array}{lllllllllll}
x_{11} & +x_{12} & & -x_{22} & & & & & & & = 0 \\
-x_{11} & & +x_{21} & +x_{22} & -x_{31} & & & & & & = 0 \\
 & -x_{12} & -x_{21} & & +x_{31} & & & & & & = 0 \\
x_{11} & +x_{12} & & & & +y_{11} & +y_{12} & & -y_{22} & & = \frac{1}{4} \\
 & & x_{21} & +x_{22} & & -y_{11} & & +y_{21} & +y_{22} & -y_{31} & = \frac{1}{2} \\
 & & & & x_{31} & & -y_{12} & -y_{21} & & +y_{31} & = \frac{1}{4}
\end{array}$$

$x_{ia}, y_{ia} \geq 0$ for all i,a.

Take the pure policy f^∞, where $f(i) = 1$, $i \in E$.

It can easily be verified that $E_1 = \{2,3\}$, $x_{11}(f) = 0$, $x_{12}(f) = 0$, $x_{21}(f) = \frac{1}{2}$, $x_{22}(f) = 0$, $x_{31}(f) = \frac{1}{2}$, $y_{11}(f) = \frac{1}{4}$, $y_{12}(f) = 0$, $y_{21}(f) = \frac{1}{4}$, $y_{22}(f) = 0$, $y_{31}(f) = 0$.

The vector y where $y_{11} = \frac{1}{2}$, $y_{12} = 0$, $y_{21} = \frac{1}{2}$, $y_{22} = \frac{1}{4}$, $y_{31} = \frac{1}{4}$ is also an element of $Y(\pi)$, since $(x(f),y)$ is also a feasible solution. Suppose that $y \in Y^*(\pi)$.

Since state 1 is transient, we obtain

$$y_{11} = \left[\beta^T D(f)\right]_1 = y_{11}(f) = \frac{1}{4} : \text{contradiction.}$$

Hence, $Y^*(\pi) \neq Y(\pi)$.

5. REFERENCES

1. Chung, K.L. "Markov chains with stationary transition probabilities", Springer, Berlin, (1960).

2. Denardo, E.V. "On linear programming in a Markov decision problem", *Management Science,* **16**, 281-288, (1970).

3. Denardo, E.V. and Fox, B.L. "Multichain Markov renewal programs", *SIAM Journal on Applied Mathematics,* **16**, 468-487, (1968).

4. Derman, C. "Finite state Markovian decision processes", Academic Press, New York, (1970).

5. d'Epenoux, F. "Sur un problème de production et de stockage dans l'aléatoire", Revue Francaise de Recherche Opérationelle, **14**, 3-16, (1960).

6. De Ghellinck, G.T. "Les problèmes de décisions sequentielles", *Cahiers du Centre d'Etudes de Recherche Opérationelle,* **2**, 161-179, (1960).

7. Hordijk, A. "Dynamic programming and Markov potential theory", Mathematical Centre Tract 51, Amsterdam, (1974).

8. Hordijk, A. and Kallenberg, L.C.M. "Linear programming and Markov decision chains", *Management Science,* **25**, 352-362, (1979).

9. Kemeny, J.G. and Snell, J.L. "Finite Markov chains", Van Nostrand, New York, (1960).

10. Manne, A.S. "Linear programming and sequential decisions", *Management Science,* **6**, 259-267, (1960).

AN ALGORITHM FOR AVERAGE COSTS DENUMERABLE STATE SEMI-MARKOV DECISION PROBLEMS WITH APPLICATIONS TO CONTROLLED PRODUCTION AND QUEUEING SYSTEMS

H.C. Tijms

(Vrije Universiteit, Amsterdam)

1. INTRODUCTION

In the last fifteen years there has been a considerable interest in the study of optimal design and control of production and queueing systems, cf. the bibliography by Crabill et al (1977) and the survey papers by Prabhu and Stidham (1974) and Sobel (1974). On the one hand the literature deals with steady-state analysis of an intuitively reasonable control rule having a simple form and on the other hand a large number of papers are concerned with verifying the optimality of such a simple control rule among a larger class of control rules. However, so far little attention has been paid to the development of computationally tractable algorithms for the numerical solution of the above control problems which usually involve an unbounded number of states.

In this paper we shall present for structured applications of average costs denumerable state semi-Markov decision problems a computational approach which appeared to be quite successful for the applications considered. This approach combines policy-iteration and embedding techniques in such a way that we need only to perform calculations for a finite number of states without having truncated the unbounded state space. For a specific application we exploit its structure and use an embedding technique to develop a tailor-made policy-iteration algorithm which deals only with a finite embedded set of states in any iteration and generates a sequence of improved policies having the desired simple form. In each of the applications considered the dimension of the embedded sets of states is considerably smaller than that of any appropriately chosen finite state space approximation and this dimensionality reduction is important in view of computations. It should be pointed out that the algorithm is designed for specially structured applications but need not work in general.

In section 2 we shall present the embedding approach for the average costs denumerable state semi-Markov decision model and in section 3 we shall give three applications. The first application considers an M/G/1 queueing system in which the queue size can be controlled by varying the service time distribution where fixed switching-costs are assumed. The second application deals with a production-inventory system in which discrete production occurs and the inventory is controlled by turning off or on the production facility where a start-up time is assumed. The third application considers an M/M/c queueing system with a variable number of servers where servers can be turned on or off with switching-costs.

We note that the embedding approach is also very useful in solving stochastic control problems in which the system can be continuously controlled and the decision processes are represented by controlled Markov drift processes involving compound Poisson processes. Typical applications arise in controlled dam-storage and production-inventory problems in which the inventory can be continuously controlled by varying the release and production rate respectively. These control problems in which the times and costs incurred between two successive decisions depend on the whole control rule used are not covered by the semi-Markov decision model. For this type of control problem a general Markov decision approach needing optimal stopping procedures was developed by De Leve (1964) and subsequently studied in De Leve et al, (1970, 1977), cf. also Tijms (1976a, 1977) and Tijms and Van Der Duyn Schouten (1978). We finally note that the above control problems could also be analysed by the technique of diffusion process approximations introduced by Bather (1966, 1968) and further studied by Chernoff and Petkau (1977), Faddy (1974), Puterman (1976), Rath (1977) and Whitt (1973).

2. THE EMBEDDING APPROACH

We are concerned with a dynamic system which at decision epochs beginning with epoch 0 is observed and classified into one of the states of a denumerable state space I. After observing the state of the system, a decision must be taken where for any state $i \in I$ a finite set of possible actions is available. If at a decision epoch action a is taken in state i, then the time until the next decision epoch and the state at the next decision epoch are random with a known joint probability distribution function which only depends on the last observed state i and the subsequently chosen action a. We further assume that a cost structure is imposed on the model in the following way. If action a is chosen in state i, then an immediate fixed cost is incurred and in addition until the next decision epoch the evolution of the system can be described by some stochastic process in which we incur costs (e.g. a cost rate and fixed costs) in a well-defined way where the cost evolution is only determined by the last observed state i and the subsequently chosen action a. For ease the costs are assumed to be non-negative. We note that such a detailed cost structure is typical for applications.

We now define the following familiar quantities. Given at epoch 0 the system is in state $i \in I$ and action $a \in A(i)$ is chosen, define

$p_{ij}(a)$ = probability that at the next decision epoch the state will be j.

$\tau(i,a)$ = unconditional expected transition time until the next decision epoch.

$c(i,a)$ = expected costs incurred until the next decision epoch.

We assume that $\inf_{i,a} \tau(i,a) > 0$. We take the long-run average expected costs per unit time as optimality criterion. A stationary policy to be denoted by f^∞ is a control rule which always prescribes the single action $f(i) \in A(i)$ whenever the system is observed in state i. We confine ourselves to a finite subclass F_0 of the class of all stationary policies such that the subset of states

$$I_0 = \left\{ i \in I \,\middle|\, f(i) \neq g(i) \text{ for some } f^\infty, g^\infty \in F_0 \right\}$$

is finite. In applications F_0 will typically consist of policies having a simple form so that we know or may reasonably expect that F_0 contains a policy which is average cost optimal within the class of all policies. Anyhow we shall not be concerned with the verification of such an optimality result but we shall only focus on the computation of the best policy within the class F_0. We make the following assumption.

<u>ASSUMPTION 1.</u> For any $f^\infty \in F_0$ there is a state $s_f \in I$ such that for any $i \in I$ the quantities $T(i,f)$ and $K(i,f)$ are finite where

$T(i,f)$ $[K(i,f)]$ = total expected time [total expected costs incurred] until the next decision epoch at which a transition occurs into state s_f given that the initial state is i and policy f^∞ is used.

Now we first derive some preliminary results before discussing the embedding approach. We first observe that, by assumption 1 and $\inf_{i,a}\tau(i,a) > 0$, the expected number of transitions until the first return to state s_f is finite for any initial state i when using $f^\infty \in F_0$. Consequently for any $f^\infty \in F_0$ there is a unique stationary probability distribution $\{\pi_j(f),\ j \in I\}$ such that for all $i,j \in I$

$$\lim_{n\to\infty} \frac{1}{n} \sum_{k=1}^{n} p_{ij}^{k}(f) = \pi_j(f), \qquad \pi_j(f) = \sum_{i\in I} p_{ij}(f(i))\,\pi_i(f), \tag{2.1}$$

where $P^n(f) = (p_{ij}^n(f))$ is the n-fold matrix product of the stochastic matrix $P(f) = (p_{ij}(f(i)))$ with itself, cf. Chung (1960). For any $f^\infty \in F_0$ define

$$g(f) = \sum_{j\in I} c(j,f,(j))\pi_j(f) \Big/ \sum_{j\in I} \tau(j,f(j))\pi_j(f). \tag{2.2}$$

Denote by $Z(t)$ the total costs incurred in $[0,t)$ and, for $n = 0,1,\ldots,$ let T_n be the nth decision epoch and let C_n be the total costs incurred in $[T_n, T_{n+1})$. Then, by the proof of Theorem 7.5 in Ross (1970) and the ergodic Theorem on p. 89 in Chung (1960) we find for any $f^\infty \in F_0$ that for initial state $i = s_f$

$$\lim_{t\to\infty} E_{i,f^\infty} \frac{[Z(t)]}{t} = \frac{K(s_f,f)}{T(s_f,f)} = \lim_{n\to\infty} \frac{E_{i,f^\infty}[\sum_{k=0}^{n} C_k]}{E_{i,f^\infty}[\sum_{k=0}^{n}(T_{k+1}-T_k)]} = g(f), \tag{2.3}$$

where E_{i,f^∞} denotes the expectation when the initial state is i and policy f^∞ is used. Further, using (2.1) and assumption 1, we have that (2.3) holds for any $i \in I$ where moreover, with probability 1, $Z(t)/t$ converges to $g(f)$ as $t\to\infty$. Hence under policy f^∞ the long-run average (expected) cost per unit time equals g(f) independent of the initial state. For any $f^\infty \in F_0$, define now the relative cost function w(i,f), i ∈ I by

$$w(i,f) = K(i,f)-g(f)T(i,f) \quad \text{for all } i \in I. \tag{2.4}$$

Observe that, by (2.3)-(2.4), for any $f^\infty \in F_0$

$$w(s_f,f) = 0. \tag{2.5}$$

We now introduce the following assumption.

ASSUMPTION 2. For any $f^\infty \in F_0$, $\Sigma_{j\in I}\, p_{ij}(a)w(j,f)$ and $\Sigma_{j\in I}\pi_j(\bar{f})w(j,f)$ converge absolutely for any $i \in I_0$, $a \in A(i)$ and $\bar{f}^\infty \in F_0$.

For any $f^\infty \in F_0$ define the "policy-improvement" quantity $T(i,a,f)$, $i \in I_0$ and $a \in A(i)$ by

$$T(i,a,f) = c(i,a)-g(f)\tau(i,a)+\sum_{j\in I} p_{ij}(a)w(j,f). \tag{2.6}$$

We have the following familiar results (cf. De Leve et al (1977) and Derman and Veinott (1967)).

THEOREM 2.1. (a) Suppose assumption 1 holds. Then, for any $f^\infty \in F_0$,

$$w(i,f) = c(i,f(i))-g(f)\tau(i,f(i))+\sum_{j\in I} p_{ij}(f(i))w(j,f), \quad i \in I \tag{2.7}$$

where $\Sigma_j p_{ij}(f(i))w(j,f)$ converges absolutely for any $i \in I$.
(b) Suppose assumptions 1-2 hold. If for some f^∞, $\bar{f}^\infty \in F_0$,

$$T(i,\bar{f}(i),f) \le w(i,f) \text{ for all } i \in I_0, \tag{2.8}$$

then $g(\bar{f}) \le g(f)$ where the strict inequality sign holds if in (2.8) the strict inequality sign holds for some state i which is positive recurrent under the stochastic matrix $P(\bar{f})$. The assertion remains true when the inequality signs are reversed.

(c) Suppose assumptions 1-2 hold. If for some $f_0^\infty \in F_0$,

$$\min_{a \in A(i)} T(i,a,f_0) = w(i,f_0) \text{ for all } i \in I_0, \qquad (2.9)$$

then $g(f_0) \leq g(f)$ for all $f^\infty \in F_0$, i.e. policy f_0^∞ is average cost optimal within the class F_0.

PROOF (a) Note that $K(i,f) = c(i,f(i)) + \Sigma_{j \neq s_f} p_{ij}(f(i))K(j,f)$ for all $i \in I$. A similar relation applies to $T(i,f)$. Together these relations, (2.4)- (2.5) and the nonnegativity of $K(i,f)$ and $T(i,f)$ imply part (a).

(b) Since $\bar{f}(i) = f(i)$ for all $i \notin I_0$, we have for all $i \in I$

$$c(i,\bar{f}(i)) - g(f)\tau(i,\bar{f}(i)) + \sum_{j \in I} p_{ij}(\bar{f}(i))w(j,f) \leq w(i,f).$$

Multiplying both sides of this inequality by $\pi_i(\bar{f})$, summing over i and using (2.1) and the fact that $\pi_i(\bar{f}) > 0$ for i positive recurrent under $P(\bar{f})$, we get part (b).

(c) This part is an immediate consequence of part (b)

Since $T(i,f(i),f) = w(i,f)$ for all $i \in I_0$ we can always construct a policy $\bar{f}^\infty \in F_0$ satisfying (2.8). Consequently, by part (b) of Theorem 1, we can always design a policy-iteration scheme which generates a sequence of improved policies within the class F_0 where the question of convergence to an optimal policy however remains to be settled. In practice we can only apply such an iteration scheme if for any policy $f^\infty \in F_0$ we can numerically evaluate the finite number of quantities

$$g(f),\ w(i,f) \text{ and } \sum_{j \in I} p_{ij}(a)w(j,f),\ i \in I_0 \text{ and } a \in A(i). \qquad (2.10)$$

In general we cannot numerically evaluate $g(f)$ and $w(i,f)$, $i \in I_0$ by directly solving the infinite system of linear equations (2.7). We shall now demonstrate how the quantities in (2.10) could be computed in specific applications by using an embedding technique and exploiting the structure of the problem considered.

For any policy $f^{\infty} \in F_0$, choose a finite set A_f with $s_f \in A_f$. Fix now policy $f^{\infty} \in F_0$. Consider the embedded Markov chain giving the state of the system at the decision epochs at which the system assumes a state in the embedded set A_f when policy f^{∞} is used. For this embedded Markov chain, define the following one-step transition probabilities, one-step expected transition times and one-step expected costs,

$\tilde{p}_{ij}(f)$ = the probability that the system will assume state j at the decision epoch at which the first return to the set A_f occurs when the initial state is $i \in I$, $j \in A_f$.

$\tilde{\tau}(i,f)$ = the total expected time until the decision epoch at which the first return to the set A_f occurs when the initial state is $i \in I$.

$\tilde{c}(i,f)$ = the total expected costs incurred until the decision epoch at which the first return to the set A_f occurs when the initial state is $i \in I$.

Observe that, by $s_f \in A_f$ and assumption 1, $\sum_{j \in A_f} \tilde{p}_{ij}(f) = 1$ and the quantities $\tilde{\tau}(i,f)$ and $\tilde{c}(i,f)$ are finite for any $i \in I$. We are now in a position to state the following key result.

THEOREM 2.2 Suppose that assumption 1 holds. Then for any $f^{\infty} \in F_0$ the finite system of linear equations in $\{g, v(i), i \in A_f\}$,

$$v(i) = \tilde{c}(i,f) - g\tilde{\tau}(i,f) + \sum_{j \in A_f} \tilde{p}_{ij}(f)v(j) \quad \text{for } i \in A_f \tag{2.11}$$

$$v(s_f) = 0 \tag{2.12}$$

has the unique solution $g = g(f)$, $v(i) = w(i,f)$, $i \in A_f$. Further,

$$w(i,f) = \tilde{c}(i,f) - g(f)\tilde{\tau}(i,f) + \sum_{j \in A_f} \tilde{p}_{ij}(f)w(j,f) \quad \text{for all } i \in I. \tag{2.13}$$

S with $\lambda ES < 1$ and $ES^2 < \infty$. Upon arrival a customer is immediately served if the server is idle and he waits in line if the server is busy. Given that at epoch 0 a service starts when $i \geq 1$ customers are present, define the following random variables

ζ_i = total amount of time spent by customers in the system during the first service.

τ_i = the first epoch at which the system becomes empty.

ν_i = the number of customers arriving in $(0,\tau_i)$.

W_i = total amount of time spent by customers in the system during $(0,\tau_i)$.

LEMMA 3.1 For any $i \geq 1$

$$E\zeta_i = iES+\tfrac{1}{2}\lambda ES^2,\ E\tau_i = \frac{iES}{1-\lambda ES},\ E\nu_i = \frac{i\lambda ES}{1-\lambda ES}, \tag{3.1}$$

$$EW_i = \tfrac{1}{2}i(i-1)E\tau_1+i\left\{\frac{ES}{1-\lambda ES} + \frac{\lambda ES^2}{2(1-\lambda ES)^2}\right\} \tag{3.2}$$

PROOF. Given n arrival epochs in $(0,s)$ the joint probability distribution of these arrival epochs is equal to that of the order statistics of n independent random variables uniformly distrbuted on $(0,s)$, cf. Ross (1970). Let S_1 be the length of the first service and let N_1 be the number of arrivals during the first service. Then $E(\zeta_i \mid S_1 = s, N_1 = n) = is+ns/2$ which gives $E\zeta_i$. To prove the other relations, note that the distributions of τ_i, ν_i and W_i are independent of the order in which the customers are served and note also that any customer in fact generates a busy period. By these standard arguments from queueing theory, we have for all $i \geq 1$

$$E\tau_i = iE\tau_1,\ E\nu_i = iE\nu_1 \text{ and } EW_i = \tfrac{1}{2}i(i-1)E\tau_1+iEW_1.$$

Hence it suffices to verify (3.1)-(3.2) for $i = 1$. We have

$$E(\tau_1|S_1 = s,\ N_1 = n) = s+nE\tau_1,\ E(W_1|S_1 = s,\ N_1 = n) = s+\frac{ns}{2}+$$
$$+\tfrac{1}{2}n(n-1)E\tau_1+nEW_1.$$

As already pointed out, by the flexibility of the policy-improvement step, the policy-iteration algorithm can always be designed in such a way that a sequence of improved policies within the class F_0 will be generated. It seems to be difficult to give conditions under which the policy-iteration algorithm will converge to a policy $f_0^\infty \in F_0$ for which the optimality condition (2.9) holds so that policy f_0^∞ is at least average cost optimal within the class F_0. In each of the applications considered the policy-iteration methods generated within the finite class F_0 a sequence of strictly improved policies so that after a finite number of iterations convergence happened to a policy f_0^∞ (say) for which we could numerically verify the optimality condition (2.9) in all examples tested. We found the well-known empirical phenomenon of the very fast convergence of the policy-iteration algorithm where no significant dependence on the size of I_0 appeared and the average cost in the iterations roughly decreased as an exponential function; in the examples tested the number of iterations varied between 3 and 15.

We conclude this section by remarking that in controlled production and queueing systems involving switching-costs the verification of the optimality of a simple policy within a larger class of policies is an extremely difficult theoretical problem for which still no satisfactory theory has been developed. Although we have not obtained any theoretical optimality result for the applications considered, it is our conjecture that for each of these applications the developed tailor-made policy-iteration algorithm may be a fruitful tool for the verification of the optimality of a simple policy within a larger class of policies.

3. APPLICATIONS TO CONTROLLED PRODUCTION AND QUEUEING SYSTEMS

Before discussing the applications, we first give for the standard M/G/1 queue some known results that will be frequently used hereafter. For completeness we include a simple and instructive derivation of these results.

3.1. *Some Busy Period Results for the M/G/1 Queue*

Consider a single server system where customers arrive in accordance with a Poisson process with rate λ and the service times of the customers are independent, non-negative random variables each being distributed as the random variable

PROOF. Fix $f^\infty \in F_0$. Since the stochastic matrix $P(f)$ has no two disjoint closed sets, the finite stochastic matrix $(\tilde{p}_{ij}(f))$, $i,j \in A_f$ has also no two disjoint closed sets, cf. Theorem 2.2 in Federgruen et al (1978). Now, by a well-known result in Markov decision theory, the system of linear equations (2.11)-(2.12) has a unique solution. Hence, by (2.5) and (2.12), it suffices to verify (2.13). To do this, observe that, by $s_f \in A_f$,

$$K(i,f) = \tilde{c}(i,f) + \sum_{\substack{j \in A_f \\ j \neq s_f}} \tilde{p}_{ij}(f) K(j,f) \quad \text{for all } i \in I.$$

A similar relation applies to $T(i,f)$, $i \in I$. Using these relations and (2.4)-(2.5), we get (2.13).

We now return to the problem of the numerical evaluation of the quantities in (2.10) for a given policy $f^\infty \in F_0$. In specific applications it is often possible to choose the set A_f in such a way that, by exploiting the structure of the application considered, analytical expressions (or simple recursion formulae) can be obtained for the quantities $\tilde{c}(i,f)$, $\tilde{\tau}(i,f)$ and $\tilde{p}_{ij}(f)$ for all $i \in I$ and $j \in A_f$. Then we first compute the numbers $g(f)$ and $w(i,f)$, $i \in A_f$ by solving the finite system of linear equations (2.11)-(2.12). Next, using (2.13), we can compute $w(i,f)$ and $\Sigma p_{ij}(a) w(j,f)$ for any $i \in I_0$ and $a \in A(i)$. Concerning the choice of A_f we remark that usually the quantities $K(s,f)$ and $T(s,f)$ referring to the single state s_f are easy to calculate for particular states but are hard to calculate for the other states and that by including certain of these latter states in the set A_f we may overcome the difficulties in computing $g(f)$ and $w(s,f)$. The order of the system of linear equations to be solved in the value-determination step of the policy-iteration algorithm is given by the dimension of set A_f when f^∞ is the current policy. In applications considered the dimension of the embedded set A_f turned out to be considerably smaller than that of any appropriately chosen approximating finite state space and this reduction is quite important from a computational point of view.

By unconditioning we get $E\tau_1$ and EW_1. The proof is completed by noting that, by Wald's equation, $(1+E\nu_1)ES = E\tau_1$.

<u>REMARK</u> 3.1.1. In the above we have assumed that the customers arrive one at a time. Consider now the case of batch arrivals at epochs generated by a Poisson process with rate λ where the batch sizes are independent positive random variables having a common discrete probability distribution with mean β. Assuming that $\lambda\beta ES < 1$ and customers are served one at a time, a trivial modification of the above proof shows that the relations (3.1)-(3.2) remain true provided that we replace λ by $\lambda\beta$.

<u>REMARK 3.1.2.</u> For the standard M/G/1 queue considered in Lemma 3.1, denote by L(t) the number of customers present at time t and for any i,j with $i \geq j \geq 0$ denote by τ_{ij} the first epoch at which the number of customers present equals j given that L(0) = i. In Lemma 3.1 we have found that for any j the function

$$w_{i,j}^{(k)} = E\left[\int_0^{\tau_{ij}} \{L(t)\}^k dt \,\middle|\, L(0)=i\right], \qquad i=j,j+1,\ldots$$

is a quadratic function in i when k=1. More generally, for any fixed j and k with ES^{k+1} finite, we have that the function $w_{i,j}^{(k)}$ is a (k+1)th order polynomial whose coefficients can be explicitly determined by using (after elaboration)

$$w_{i,j}^{(k)} = \int_0^\infty dF(s) \left[\sum_{n=0}^{\infty} e^{-\lambda s} \frac{(\lambda s)^n}{n!} \left\{ \sum_{p=0}^{n} (i+p)^k \frac{s}{n+1} + w_{i-1+n,j}^{(k)} \right\}\right]$$

for $i \geq j+1$,

where $w_{j,j}^{(k)} = 0$ and F is the probability distribution function of the service time S. Thus in the applications of this paper we may handle polynomial holding cost functions where for ease of presentation we only deal with linear holding costs.

We now discuss the first application

3.2. *An M/G/1 Queueing System with Controllable Service Time Distribution*

Consider a single server system where customers arrive in accordance with a Poisson process with rate λ. Each customer is served by using one of two available service types $k = 1,2$. At any service completion epoch the server has to decide which service type to use for the next service. The service time of a customer has probability distribution function F_k when service type k is used where $F_k(0) < 1$ for $k = 1,2$. It is assumed that $F_2(t) \geq F_1(t)$ for all $t > 0$ so that service type 2 is "faster" than service type 1. Denote by μ_k the first moment of F_k and for $j \geq 2$ denote by $\mu_k^{(j)}$ the jth moment of F_k. We assume that $\lambda\mu_2 < 1$, $\mu_1^{(2)} < \infty$ and $\mu_2^{(3)} < \infty$. The following costs are incurred. There is a holding cost at rate h.i when i customers are present and a service cost at rate r_k when the server is busy and uses service type k. Further, a fixed switching-cost of R_k is incurred when at a service completion epoch the server decides to switch from service type k to the other one. The cost parameters are assumed to be nonnegative.

This controlled queueing problem can be represented by a semi-Markov decision model in which the decision epochs are given by the service completion epochs and at any decision epoch the system can be classified into one of the states of the denumerable state space

$$I = \{i \mid i = 0,1,\ldots\} \cup \{i' \mid i = 0,1,\ldots\}$$

where state i(i') corresponds to the situation in which the number of customers present is i and service type 1(2) was used for the service just completed. For any state $s \in I$ the set of available actions is given by $A(s) = \{1,2\}$ where action k prescribes to use service type k for the next service. If action k is taken in state s, the time until the next decision epoch is distributed as the service time under service type k if $s \neq 0,0'$ and is distributed as the sum of the time between successive arrivals and the service time under service type k otherwise. Hence, for $k = 1,2$,

$$\tau(i,k) = \tau(i',k) = \mu_k \text{ for } i \geq 1 \text{ and } \tau(0,k) = \tau(0',k) = \frac{1}{\lambda} + \mu_k. \tag{3.3}$$

If action k is chosen in state s, we incur as immediate cost the appropriate switching-cost if any and until the next decision epoch the evolution of the system can be described by the queue length process given service type k where costs at a rate of $h.j+r_k\delta(j)$ are incurred when j customers are in the system with $\delta(0) = 0$ and $\delta(j) = 1$ for $j \geq 1$. Using (3.1), we have for $i \geq 0$

$$c(i,1) = h(i\vee 1)\mu_1 + \tfrac{1}{2}h\lambda\mu_1^{(2)} + r_1\mu_1,\ c(i',2)=h(i\vee 1)\mu_2 + +\tfrac{1}{2}h\lambda\mu_2^{(2)} + r_2\mu_2, \tag{3.4}$$

$$c(i,2) = R_1 + c(i',2),\ c(i',1) = R_2 + c(i,1). \tag{3.5}$$

where $i\vee 1 = \max(i,1)$. For k=1,2, define

$$p_k(j) = \int_0^\infty e^{-\lambda t} \frac{(\lambda t)^j}{j!} dF_k(t),\ j = 0,1,\ldots, \tag{3.6}$$

i.e. $p_k(j)$ is the probability that j customers arrive during a service time under service type k. The transition probabilities $p_{st}(k)$ can be directly expressed in terms of the probabilities $p_k(j)$ and for reasons of space we omit these obvious expressions.

For a given positive integer N, denote by $F_0 = F_0^{(N)}$ the class of stationary policies having the following simple form. Any policy $f^\infty \in F_0$ is characterized by two switch-over levels i_1 and i_2 with $0 \leq i_2 \leq i_1 < N$ and $i_1 \geq 1$. Under this policy to be denoted by $f^\infty = (i_1, i_2)$ the server switches from service type 1 to service type 2 only at the service completion epochs where the queue size is larger than i_1 and the server switches only from service type 2 to service 1 at the service completion epochs where the queue size is less than or equal to i_2. It is intuitively reasonable to expect that there exists an average cost optimal policy which belongs to the class F_0 with N sufficiently large. For the case of no switching-costs this optimality question was studied by Crabill (1972), Gallisch (1977) and Tijms (1976b). However, we wish only to compute the best policy within the class F_0. To do this we first note that, by choosing $s_f = i_2'$ for any policy $f^\infty = (i_1,i_2)$, assumption 1 of section 2 is satisfied. In fact we shall see below that the functions $T(s,f)$ and $K(s,f)$, $s \in I$ are bounded by a linear and quadratic function of s respectively. Without giving details, we note that by this result and the assumption of stochastically ordered service times with $\mu_2^{(3)} < \infty$ it can also be shown that assumption 2 of section 2 is satisfied. For any $f^\infty = (i_1,i_2) \in F_0$, we choose

$$A_f = \{i|i = 0,\ldots,i_1\} \cup \{i_2'\}.$$

We shall now demonstrate that for this choice of A_f analytical expressions can be given for the quantities $\tilde{c}(s,f)$, $\tilde{\tau}(s,f)$ and $\tilde{p}_{st}(f)$. Fix policy $f^\infty = (i_1,i_2)$. Observing that for initial state i' with $i > i_2$ the first entry state in A_f is state i_2' we find by Lemma 3.1

$$\tilde{c}(i',f) = h(i_2E\tau_{i-i_2}+EW_{i-i_2})+r_2E\tau_{i-i_2} \quad \text{for } i > i_2 \tag{3.7}$$

where $E\tau_j$ and EW_j are given by (3.1) and (3.2) in which $ES = \mu_2$ and $ES^2 = \mu_2^{(2)}$. Further

$$\tilde{c}(i,f) = R_1+\tilde{c}(i',f) \text{ for } i > i_1, \quad \tilde{c}(i',f) = R_2+\tilde{c}(i,f) \text{ for } 0\le i\le i_2$$

$$\tilde{c}(i,f) = c(i,1)+\sum_{j=i_1-i+2}^{\infty} \tilde{c}(i-1+j,f)p_1(j) \text{ for } 1 \le i \le i_1, \tag{3.8}$$

$$\tilde{c}(0,f) = \tilde{c}(1,f).$$

The formula for $\tilde{c}(i,f)$, $1 \le i \le i_1$ can be simplified. Therefore we introduce the following shorthand notation. For any $m \ge i_2-1$, $1 \le i \le m+2$ and $k = 1,2$, define

$$H(i,m,i_2,k,\alpha,\beta,\gamma) = \sum_{j=m-i+2}^{\infty} [\alpha E\tau_{i-1+j-i_2} + \beta EW_{i-1+j-i_2} + \gamma] p_k(j) \tag{3.9}$$

where α,β,γ are given constants. We find after some algebra

$$H(i,m,i_2,k,\alpha,\beta,\gamma) = [\bar{\alpha}.(i-1)^2 + \bar{\beta}.(i-1) + \bar{\gamma}][1 - \sum_{j=0}^{m-i+1} p_k(j)] +$$

$$+ [2\bar{\alpha}.(i-1) + \bar{\beta}.][\lambda\mu_k - \sum_{j=0}^{m-i+1} jp_k(j)] + \bar{\alpha}[\lambda^2\mu_k^{(2)} + \lambda\mu_k - \sum_{j=0}^{m-i+1} j^2 p_k(j)],$$

where

$$\bar{\alpha} = \frac{\frac{1}{2}\beta\mu_2}{1-\lambda\mu_2}, \quad \bar{\beta} = \frac{1}{1-\lambda\mu_2}\left\{\alpha\mu_2 + \tfrac{1}{2}\beta\mu_2 - \beta\mu_2 i_2 + \frac{\beta\lambda\mu_2^{(2)}}{2(1-\lambda\mu_2)}\right\}, \quad \bar{\gamma} = -\bar{\alpha}i_2^2 - \bar{\beta}i_2 + \gamma.$$

Using (3.1), (3.2), (3.4) and (3.7)-(3.9), we find for $1 \le i \le i_1$,

$$\tilde{c}(i,f) = hi\mu_1 + \tfrac{1}{2}h\lambda\mu_1^{(2)} + r_1\mu_1 + H(i,i_1,i_2,1, r_2+hi_2,h,R_1). \tag{3.10}$$

The formulae for $\tilde{\tau}(i,f)$ and $\tilde{\tau}(i',f)$ for $i \ge 1$ follow by putting $r_1 = r_2 = 1$, $h = R_1 = R_2 = 0$ in the corresponding formulae for $\tilde{c}(i,f)$ and $\tilde{c}(i',f)$, $i \ge 1$. Clearly

$$\tilde{\tau}(0,f) = \tilde{\tau}(0',f) = \frac{1}{\lambda} + \tilde{\tau}(1,f). \tag{3.11}$$

Further, it follows directly that

$$\tilde{p}_{i'i_2'}(f) = 1 \text{ for } i > i_2, \; \tilde{p}_{ii_2'}(f) = 1 \text{ for } i > i_1,$$

$$\tilde{p}_{it}(f) = p_1(t-i+1) \text{ for } 1 \le i \le i_1, \; i-1 \le t \le i_1,$$

$$\tilde{p}_{ii_2'}(f) = 1 - \sum_{j=0}^{i_1-i+1} p_1(j) \text{ for } 1 \le i \le i_1,$$

$$\tilde{p}_{i't}(f) = \tilde{p}_{it}(f) \text{ for } 0 \le i \le i_2, \; \tilde{p}_{0t}(f) = \tilde{p}_{1t}(f).$$

Hence we see that the functions $\tilde{c}(s,f)$, $\tilde{\tau}(s,f)$ and $\tilde{p}_{st}(f)$ depend on f in a simple way and can be easily computed. We next specify the system of linear equations (2.11)-(2.12). Therefore we first observe that, by (2.7) and (3.8),

$$w(i',f) = c(i',1)-g(f)\tau(i',1)+\sum_{j=0}^{\infty} p_1(j)w(i-\delta(i)+j,f) =$$
$$= R_2+w(i,f) \quad \text{for } 0 \le i \le i_2. \tag{3.12}$$

Further, by (2.7), (3.8) and (3.11), we find

$$w(0,f) = \frac{-g(f)}{\lambda}+w(1,f). \tag{3.13}$$

By (3.10), (3.12)-(3.13) and $s_f = i_2'$, the linear equations (2.11)-(2.12) reduce to

$$v(0)+\frac{g}{\lambda}-v(1) = 0 \tag{3.14}$$

$$v(i)+g\left\{\mu_1+H(i,i_1,i_2,1,1,0,0)\right\} - \sum_{j=0}^{i_1-i+1} p_1(j)v(i-1+j) =$$
$$= hi\mu_1+\tfrac{1}{2}h\lambda\mu_1^{(2)}+r_1\mu_1+H(i,i_1,i_2,1,r_2+hi_2,h,R_1) \quad \text{for } 1 \le i \le i_1 \tag{3.15}$$

$$v(i_2) = -R_2. \tag{3.16}$$

This system of linear equations can be very efficiently solved. Successively for $i = i_1,\ldots,1$ we can express $v(i-1)$ as a linear combination of g and $v(i_1)$ by using the equation (3.15) for v(i). Next, by using the two equations (3.14) and (3.16), we can solve for g and $v(i_1)$. Hence, by (3.15), $v(i) = \alpha(i)v(i_1)+\beta(i)g+\gamma(i)$ for $0 \le i \le i_1$ where $\alpha(i)$, $\beta(i)$, $\gamma(i)$ can be successively computed for $i = i_1,\ldots,0$. We have $\alpha(i_1) = 1$ and

$$\alpha(i-1) = p_1(0)^{-1}\{\alpha(i) - \sum_{j=1}^{i_1-i+1} p_1(j)\alpha(i-1+j)\} \text{ for } i = i_1,\ldots,1.$$

Similar recursions apply to $\beta(i)$ and $\gamma(i)$. We next specify $w(s,f)$ for $s \notin A_f$. For $0 \le i \le i_2$ we have that $w(i',f)$ is given by (3.12). Using (2.13), (3.7) and $w(i_2',f) = 0$, we find

$$w(i',f) = hEW_{i-i_2} + \left\{r_2+hi_2-g(f)\right\}E\tau_{i-i_2} \quad \text{for } i > i_2 \tag{3.17}$$

where $E\tau_j$ and EW_j are given by (3.1)-(3.2) with $ES = \mu_2$ and $ES^2 = \mu_2^{(2)}$. Finally, by (2.7),

$$w(i,f) = R_1+w(i',f) \text{ for } i > i_1. \tag{3.18}$$

We next specify the test quantity $T(s,a,f)$. By (2.6)-(2.7) and (3.5),

$$T(i',1,f) = c(i',1)-g(f)\tau(i',1) + \sum_{j=0}^{\infty} p_1(j)w(i-1+j,f)$$

$$= \begin{cases} R_2+w(i,f) & \text{for } i_2 < i \le i_1 \\ R_2+T(i,1,f) & \text{for } i > i_1. \end{cases}$$

Using (3.4), (3.9), (3.12) and (3.17), we get from (2.6) that

$$T(i',2,f) = hi\mu_2+\tfrac{1}{2}h\lambda\mu_2^{(2)}+r_2\mu_2-g(f)\mu_2 +$$

$$+ \sum_{j=0}^{i_2-i+1} p_2(j)\{R_2+w(i-1+j,f)\} +$$

$$+H(i,i_2,i_2,2,r_2+hi_2-g(f),h,0) \text{ for } 1 \le i < i_2.$$

Using (3.4), (3.9) and (3.17)-(3.18), we get from (2.6) that

$$T(i,1,f) = hi\mu_1+\tfrac{1}{2}h\lambda\mu_1^{(2)}+r_1\mu_1-g(f)\mu_1+\{1-\delta(i-i_1-1)\}p_1(0)w(i_1,f)+$$

$$+H(i,i-1-\delta(i-i_1-1),i_2,1,r_2+hi_2-g(f),h,R_1) \text{ for } i > i_1,$$

where $\delta(j) = 1$ for $j \geq 1$ and $\delta(0) = 0$. Finally, by (2.6)-(2.7) and (3.5),

$$T(i,2,f) = \begin{cases} R_1 + T(i',2,f) & \text{for } 1 \leq i \leq i_2 \\ R_1 + w(i',f) & \text{for } i_2 < i \leq i_1 \end{cases}$$

We shall now describe the algorithm. Choose an integer N and a policy $f^\infty = (i_1, i_2)$ with $0 \leq i_2 \leq i_1 < N$ and $i_1 \geq 1$.

ALGORITHM

STEP 1 (value-determination step). Let $f^\infty = (i_1, i_2)$ be current policy. Compute from (3.14)-(3.16) the numbers $g(f)$ and $w(i,f)$, $i = 0,\ldots,i_1$.

STEP 2 (policy-improvement step). (a) First determine an integer $\bar{i}_2$ with $0 \leq \bar{i}_2 \leq i_1$. Define $\bar{i}_2$ as the largest integer k such that $i_2 < k \leq i_1$ and $T(i',1,f) < w(i',f)$ for all $i_2 < i \leq k$ if such an integer k exists, otherwise let $\bar{i}_2$ be equal to $\ell-1$ with ℓ the smallest integer such that $1 \leq \ell \leq i_2$ and $T(i',2,f) < w(i',f)$ for all $\ell \leq i \leq i_2$ if such an integer ℓ exists, and otherwise let $\bar{i}_2 = i_2$.

(b) Next determine an integer $\bar{i}_1$ with $\bar{i}_2 \leq \bar{i}_1 < N$ and $\bar{i}_1 \geq 1$. Define $\bar{i}_1$ as the largest integer k such that $i_1 + 1 \leq k < N$ and $T(i,1,f) < w(i,f)$ for all $i_1+1 \leq i \leq k$ if such an integer k exists, otherwise let $\bar{i}_1$ be equal to $\ell-1$ with ℓ the smallest integer such that $\bar{i}_2 < \ell \leq i_1$, $\ell \geq 2$ and $T(i,2,f) < w(i,f)$ for all $\ell \leq i \leq i_1$ if such an integer ℓ exists, and otherwise let $\bar{i}_1 = i_1$.

STEP 3 Let $\bar{f}^\infty = (\bar{i}_1, \bar{i}_2)$. If $\bar{f}^\infty = f^\infty$, then stop, otherwise go to step 1 with the previous policy $f^\infty = (i_1, i_2)$ replaced by the new policy $\bar{f}^\infty = (\bar{i}_1, \bar{i}_2)$.

The algorithm stops after a finite number of iterations since the class F_0 is finite and $g(\bar{f}) < g(f)$ if the new policy

$\bar{f}^{\infty} = (\bar{i}_1, \bar{i}_2)$ is not equal to the previous policy $f^{\infty} = (i_1, i_2)$. This follows from Theorem 2.1(b) by observing that the states $\bar{i}_1$ and $\bar{i}_2'$ are positive recurrent under policy $\bar{f}^{\infty}$. In all examples tested we could numerically verify the condition (2.9) for the finally obtained policy f_0^{∞} (say) so that this policy is at least average cost optimal within the class $F_0 = F_0^{(N)}$. We note that the chosen integer N only appears in step 2(b) of the algorithm. It gives no difficulties to enlarge N during the algorithm if desired. It needs hardly to be said that the embedding approach results in a policy iteration algorithm which compares quite favourably with any computational approach for an approximating finite state space model since the dimension of an approximating finite state space will usually be much larger than that of the embedded set A_f for which besides the system of linear equations (3.14)-(3.16) can be very efficiently solved.

Consider the following numerical example with constant service times.

$$\lambda = 1, \ \mu_1 = 1, \ \mu_2 = 0.8, \ h = 0.02, \ r_1 = 2, \ r_2 = 50.$$

In Table I we give for various values of the switching-cost the (i_1, i_2) policies generated by the algorithm where n and $g(i_1, i_2)$ denote the iteration number and the average costs. We have chosen N = 200 and $(i_1=100, \ i_2=0)$ as starting policy. In the two examples in Table I an optimal policy was found after 8 and 6 iterations respectively.

REMARK 3.2

(a) As a special case of the above results we can compute the average expected queue size under a given (i_1, i_2) rule by solving a finite system of linear equations. In Federgruen and Tijms (1979) a computational method has been given which recursively computes the steady-state probabilities of the queue size under a given (i_1, i_2) rule, cf. also Hordijk and Tijms (1976) for the standard M/G/1 queue.

(b) The analysis can be rather straightforwardly extended to the case where more than two service types are available. However, in view of results in Faddy (1974) and Sobel (1974), we might expect that in many situations the fairly complicated control rules involving more than two service types are not significantly better than the bang-bang control rules involving only the slowest and the fastest service type.

(c) The analysis applies also to the case where customers arrive in batches, cf. remark 3.1. Further, the holding cost rate may be allowed to depend on the service type used where in fact a nonlinear holding cost rate may be assumed when the fastest service type is not used. The analysis also carries on if the arrival rate depends on the service type used. However, for a controllable arrival rate it might be desirable to include the epochs of arrivals occurring when service type 1 is used as extra decision epochs. If the service time distribution F_1 is exponential, this can be done with only obvious modifications of the analysis.

TABLE I

The Iterations

	$R_1 = R_2 = 0$		$R_1 = R_2 = 50$	
n	(i_1,i_2)	; $g(i_1,i_2)$	(i_1,i_2)	; $g(i_1,i_2)$
1	(100,0)	; 4.49718	(100,0)	; 4.50654
2	(122,100)	; 3.98023	(122,100)	; 3.99908
3	(82,82)	; 3.97213	(114,78)	; 3.97869
4	(96,82)	; 3.95903	(109,84)	; 3.97847
5	(97,96)	; 3.95357	(110,82)	; 3.97789
6	(94,94)	; 3.95328	(111,81)	; 3.97781
7	(95,94)	; 3.95327		
8	(95,95)	; 3.95325		

3.3 *A Discrete Production-inventory Problem with Start-up Times*

Consider a production system which operates only intermittently to manufacture a single product. Production is stopped if the inventory is sufficiently high and production is restarted if the inventory has dropped sufficiently low. Demands for the product occur at epochs generated by a Poisson process with rate λ where demand in excess of stock is backordered. For ease of presentation we will assume that the demand size is equal to one unit (cf. remark 3.3 below for the case of a general discrete demand distribution). The units of the product are manufactured one at a time where the time to manufacture one unit is distributed as the positive random variable T_p with $\lambda ET_p < 1$ and $ET_p^2 < \infty$. After the completion of the production of one unit, either a new production is started or the production facility is shut-down. At any demand epoch occurring when the system is shut-down, the production facility can be reactivated where it takes a start-up time distributed as the nonnegative random variable T_s and $ET_s < \infty$ before the next production actually starts. We assume an upper bound U ($\leq \infty$) for the number of units that can be kept in stock. The following costs are considered. There is holding cost of $h > 0$ per unit kept in stock per unit time and for any unit backordered there is a fixed backorder cost $\pi_1 \geq 0$ and a linear backorder cost $\pi_2 > 0$ per unit time the backorder exists. There is an operating cost at rate $r_1 \geq 0$, $r_2 \geq 0$ and $r_3 \geq 0$ when the production facility is producing, shut-down and being reactivated respectively. Finally, a fixed set-up cost of $R \geq 0$ is incurred when the production system is reactivated.

This production problem can be represented by a semi-Markov decision model in which the decision epochs are given by the production completion epochs and the demand epochs occurring when the system is shut-down. At any decision epoch the system can be classified into one of the states of the denumerable state space.

$$I = \{i \mid i \leq U\} \cup \{i' \mid i \leq U\},$$

where state i corresponds to the situation in which the stock on hand minus stock on backorder equals i and a unit has been just made and state i' corresponds to the situation in which the stock on hand minus stock on backorder equals i and a demand has just occurred when the system is shut-down. For any state $s \in I$, two possible actions $a = 0$ and $a = 1$ are available. In state i the action $a = 0$ ($a = 1$) means to stop production (to continue production) and in state i' the action $a = 0$ ($a = 1$) means to keep the system shut-down (to reactivate the system). We clearly have

$$c(i,0) = c(i',0) = \begin{cases} (hi+r_2)/\lambda & \text{for } i \geq 1 \\ -\pi_2 i/\lambda+\pi_1+r_2/\lambda & \text{for } i \leq 0. \end{cases} \tag{3.19}$$

To give the formula for $c(i,1)$, denote by F_p and F_s the probability distribution functions of T_p and T_s respectively and let the random variable A_k denote the epoch at which the demand for the kth unit occurs. Observe that A_k has a gamma distribution with parameters (k,λ). For $k = 0,1,\ldots,$ let

$$p_k = \int_0^\infty e^{-\lambda t}\frac{(\lambda t)^k}{k!}\, dF_p(t) \text{ and } q_k = \int_0^\infty e^{-\lambda t}\frac{(\lambda t)^k}{k!}\, dF_s(t),$$

i.e. $\{p_k\}$ and $\{q_k\}$ are the probability distributions of the cumulative demand in a production time and start-up time respectively. Using the well-known fact that given n demand epochs have occurred in $(0,t)$ these n demand epochs have expectation $\frac{1}{2}t$ on the average, we find

$$c(i,1) = \begin{cases} -\pi_2 iET_p+\pi_2\lambda ET_p^2/2+\pi_1\lambda ET_p+r_1ET_p, & i < 0 \\ h\sum_{k=1}^{i} E\min(T_p,A_k)+r_1ET_p+ \\ +\int_0^\infty dF_p(t) \sum_{n=i+1}^{\infty} e^{-\lambda t}\frac{(\lambda t)^n}{n!}(n-i)(\pi_1+\pi_2 t/2), & i \geq 0. \end{cases} \tag{3.20}$$

After some routine algebra, we find for $i \geq 0$

$$c(i,1) = \frac{h}{\lambda}\sum_{k=1}^{i}\{k-\sum_{j=0}^{k}(k-j)p_j\}+\pi_1\{\lambda ET_p-\sum_{n=0}^{i}np_n-i+i\sum_{n=0}^{i}p_n\}+$$

$$+\frac{\pi_2}{2\lambda}\{\lambda^2ET_p^2-\sum_{n=0}^{i+1}n(n-1)p_n-i\lambda ET_p+i\sum_{n=0}^{i+1}np_n\}+r_1ET_p. \tag{3.21}$$

In view of computations it is important to note that c(i,1) for $i \geq 0$ can be recursively computed. For any i denote by $c_q(i)$ the expression obtained from c(i,1) by replacing T_p, $p_n(n \geq 1)$ and r_1 by T_s, $q_n(n \geq 1)$ and r_3 respectively. Then we have

$$c(i',1) = R+c_q(i)+\sum_{k=0}^{\infty}c(i-k,1)q_k \text{ for all } i, \tag{3.22}$$

which expression can be further simplified by substituting the formula for c(i,1), i < 0. The formula for $\tau(s,a)$ clearly follows by putting $r_1 = r_2=r_3 = 1$ and $h = \pi_1 = \pi_2 = R = 0$ in the corresponding formula for c(s,a). The transition probabilities $p_{st}(a)$ can be easily expressed in the probability distributions $\{p_k\}$ and $\{q_k\}$. For reasons of space we omit these obvious expressions.

Before defining the class F_o, we introduce the following shorthand notations. Let

$$a = \frac{\pi_2 ET_p}{2(1-\lambda ET_p)}, \quad b = -a-\frac{1}{1-\lambda ET_p}\left\{\frac{\pi_2\lambda ET_p^2}{2(1-\lambda ET_p)}+\pi_1\lambda ET_p+r_1ET_p\right\}$$

$$c = \frac{-ET_p}{1-\lambda ET_p},$$

and for i = 0,1,..., let

$$P_1(i) = \sum_{k=i+2}^{\infty}\{a(i+1-k)^2+b(i+1-k)\}p_k, \quad P_2(i) = c\sum_{k=i+2}^{\infty}(i+1-k)p_k \tag{3.23}$$

Next, using (3.23) and (3.26), we find

$$\tilde{c}(i,f) = c(i,1) + \sum_{k=i+2}^{\infty} p_k\, \tilde{c}(i+1-k,f) + \delta_{iM} p_0\, \tilde{c}(M+1,f) =$$

$$= c(i,1) + P_1(i) + \delta_{iM} p_0\, \tilde{c}(M+1,f) \quad \text{for } 0 \le i \le M,$$

where $\delta_{ii} = 1$ and $\delta_{ij} = 0$ for $j \neq i$. Further, we find

$$\tilde{c}(i',f) = \begin{cases} h\{i(i+1)-m(m+1)\}/2\lambda + r_2(i-m)/\lambda \\ \text{for } i > m \text{ if } m \ge 0 \\ \{hi(i+1)+\pi_2 m(m+1)\}/2\lambda - \pi_1 m + r_2(i-m)/\lambda \\ \text{for } i \ge 1 \text{ if } m < 0 \\ \pi_2\{m(m+1)-i(i+1)\}/2\lambda - \pi_1(m-i) + r_2(i-m)/\lambda \\ \text{for } m < i \le 0 \text{ if } m < 0. \end{cases}$$

$$\tilde{c}(i',f) = c(i',1) + \sum_{k=0}^{\infty} q_k\, \tilde{c}(i-k,f) =$$

$$= c(i',1) + \sum_{k=0}^{i} q_k\, \tilde{c}(i-k,f) + Q_1(i) \quad \text{for } i \le m,$$

$$\tilde{c}(i,f) = \tilde{c}(i',f) \quad \text{for } i > M.$$

The formula for $\tilde{\tau}(s,f)$ follows by putting $r_1 = r_2 = r_3 = 1$ and $h = \pi_1 = \pi_2 = R=0$ in the corresponding one for $\tilde{c}(s,f)$.

Explicit expressions for the transition probabilities $\tilde{p}_{st}(f)$ can also be easily derived. We shall not give these expressions apart since they will appear in the system of linear equations to be specified below. Before specifying (2.11)-(2.12), we first relate $w(i,f)$ and $w(i',f)$ for $i \le m$. By (2.7) and (3.22),

$$w(i',f) = c(i',1) - g(f)\tau(i',1) + \sum_{k=0}^{\infty} q_k \sum_{j=0}^{\infty} p_j w(i-k+1-j,f) =$$

$$Q_1(i) = \sum_{k=i+1}^{\infty} \{a(i-k)^2+b(i-k)\}q_k,\quad Q_2(i) = c\sum_{k=i+1}^{\infty}(i-k)q_k. \tag{3.24}$$

We have

$$P_1(i) = \{a(i+1)^2+b(i+1)\}\{1-\sum_{k=0}^{i+1} p_k\}-\{2a(i+1)+b\}\{\lambda ET_p - \sum_{k=0}^{i+1} kp_k\}+ \tag{3.25}$$

$$+a\{\lambda^2 ET_p^2+\lambda ET_p - \sum_{k=0}^{i+1} k^2 p_k\}.$$

Similar expressions apply to $P_2(i)$, $Q_1(i)$ and $Q_2(i)$.

For a given integer N with $0 \leq N \leq U$, denote by $F_0 = F_0^{(N)}$ the finite class of stationary policies having the following simple form. Any policy $f^\infty \in F_0$ is characterized by two integers m and M with $-N < m \leq M < N$ and under this policy to be denoted by $f^\infty = (m,M)$ the inventory level (= stock on hand minus stock on backorder) is controlled as follows. If the inventory level is larger than M, do not produce until the inventory level drops to or below m and at that time reactivate the production facility and continue production until the inventory level becomes M+1. The optimality of such a simple control rule within a larger class of control rules has been studied by Sobel (1970). However, we only focus on the computation of the best policy within the class F_0 of simple policies. To avoid an overburdened notation, we consider only policies $f^\infty = (m,M) \in F_0$ with $M \geq 0$. For any policy $f^\infty = (m,M) \in F_0$, choose $s_f = m'$ and

$$A_f = \{i \mid i = 0,\ldots,M\} \cup \{m'\}.$$

Fix now $f^\infty = (m,M)$. To give an explicit expression for $\tilde{c}(i,f)$, $i < 0$, we observe that under the (m,M) policy the inventory process when the system is producing can be described by the queue length process in the M/G/1 queue where the service time is distributed as the production time T_p. Together this observation and the relations (3.1)-(3.2) with S replaced by T_p imply

$$\tilde{c}(i,f) = \pi_2 EW_{-i}+\pi_1 E\nu_{-i}+r_1 E\tau_{-i} = ai^2+bi \text{ for } i < 0. \tag{3.26}$$

$$= R+c_q(i)-g(f)ET_s+\sum_{k=0}^{\infty} q_k\{c(i-k,1)-g(f)\tau(i-k,1)+$$

$$\sum_{j=0}^{\infty} p_j w(i-k+1-j,f)\}= \tag{3.27}$$

$$R + c_q(i)-g(f)ET_s+\sum_{k=0}^{\infty} q_k w(i-k,f) \text{ for } i \le m.$$

We further note that, by (2.13),

$$w(i,f)=\tilde{c}(i,f)-g(f)\tilde{\tau}(i,f)+w(0,f) \text{ for } i < 0. \tag{3.28}$$

Hence, using (3.24), (3.26) and (3.28), we have

$$w(i',f) = R+c_q(i)-g(f)ET_s+\sum_{k=0}^{i} q_k w(i-k,f)+ \tag{3.29}$$

$$Q_1(i)-g(f)Q_2(i)+(1-\sum_{k=0}^{i} q_k)w(0,f) \text{ for } i \le m.$$

By $w(m',f) = 0$ and (3.29), the linear equations (2.11)-(2.12) become

$$v(i)+g\tilde{\tau}(i,f)-\sum_{k=\delta_{iM}}^{i} p_k v(i+1-k)-(1-\sum_{k=0}^{i} p_k)v(0) = \tag{3.30}$$

$$\tilde{c}(i,f),\ 0 \le i \le M$$

$$g\{ET_s+Q_2(m)\}-\sum_{k=0}^{m} q_k v(m-k)-(1-\sum_{k=0}^{m} q_k)v(0) = R+c_q(m)+Q_1(m). \tag{3.31}$$

This system of linear equations can also be very efficiently solved. Successively for $i = 0,\dots, M-1$ we can express $v(i+1)$ as a linear combination of $v(0)$ and g by using equation (3.30) for $v(i)$. Then, using the equation (3.30) for $v(M)$ and equation (3.31), we can solve for g and $v(0)$ and next compute $v(1),\dots,v(M)$. We now specify the other $w(s,f)$. Next to the relations (3.28)-(3.29), we have

$$w(i',f) = \tilde{c}(i',f)-g(f)\tilde{\tau}(i',f),\ i > m \text{ and } w(i,f) = w(i',f),\ i > M.$$

We next specify the test quantity $T(s,a,f)$. Similarly to (3.27), we derive from (2.6), (2.7), (3.22), (3.24), (3.26) and (3.28) that, for all $m < i \leq M$

$$T(i',1,f) = c(i',1)-g(f)\tau(i',1)+\sum_{k=0}^{\infty} q_k \sum_{j=0}^{\infty} p_j w(i-k+1-j,f) =$$

$$= R+c_q(i)-g(f)ET_s+\sum_{k=0}^{\infty} q_k w(i-k,f) =$$

$$= R+c_q(i)+Q_1(i)-g(f)\{ET_s+Q_2(i)\}+\sum_{k=0}^{i} q_k w(i-k,f)+(1-\sum_{k=0}^{i} q_k)w(0,f)$$

and, for all $i > M$,

$$T(i',1,f) = R+c_q(i)-g(f)ET_s+\sum_{k=0}^{i-M-1} q_k T(i-k,1,f)+\sum_{k=i-M}^{\infty} q_k w(i-k,f)$$

which expression can be further simplified by using (3.26) and (3.28). Further by (2.6)-(2.7), (3.28), (3.26) and (3.23), we find

$$T(i',0,f) = c(i',0)-g(f)\tau(i',0)+w((i-1)',f) \text{ for } i \leq m,$$

$$T(i,1,f) = c(i,1)-g(f)\tau(i,1)+\sum_{j=0}^{i+1} p_j w(i+1-j,f)+P_1(i)-g(f)P_2(i)+$$

$$+(1-\sum_{j=0}^{i+1} p_j)w(0,f) \text{ for } i > M,$$

$$T(i,0,f) = \begin{cases} w(i',f) & \text{for } m < i \leq M \\ T(i',0,f) & \text{for } i \leq m. \end{cases}$$

We can now describe the algorithm. Choose an integer N and a policy $f^{\infty} = (m,M)$ with $-N < m \leq M < N$ and $M \geq 0$.

ALGORITHM

STEP 1 (value-determination step). Let $f^{\infty} = (m,M)$ be the current policy. Compute from (3.30)-(3.31) the numbers $g(f)$ and $w(i,f)$, $i = 0,\ldots,M$.

STEP 2 (policy-improvement step). (a) First determine an integer $\underline{m}$ with $-N < \underline{m} \leq M$. Define $\overline{m}$ as the largest integer k such that $m < k \leq M$ and $T(i',1,f) < w(i',f)$ for all $m < i \leq k$ if such an integer k exists, otherwise let $\overline{m}$ be equal to $\ell-1$ with ℓ the smallest integer such that $-N+1 < \ell \leq m$ and $T(i',0,f) < w(i',f)$ for all $\ell \leq i \leq m$ if such an integer ℓ exists, and otherwise let $\overline{m} = m$.

(b) Next determine an integer M with $\overline{m} \leq \overline{M} < N$ and $\overline{M} \geq 0$. Define $\overline{M}$ as the largest integer k such that $M+1 \leq k < N$ and $T(i,1,f) < w(i,f)$ for all $M+1 \leq i \leq k$ if such an integer k exists, otherwise let $\overline{M}$ be equal to $\ell-1$ with ℓ the smallest integer such that $\overline{m} < \ell \leq M$, $\ell \geq 1$ and $T(i,0,f) < w(i,f)$ for all $\ell \leq i \leq M$ if such an integer ℓ exists, and otherwise let $\overline{M} = M$.

STEP 3 Let $\overline{f}^{\infty} = (\overline{m},\overline{M})$. If $\overline{f}^{\infty} = f^{\infty}$ then stop, otherwise go to step 1 with the previous policy f^{∞} replaced by the new policy $\overline{f}^{\infty}$.

This algorithm stops after a finite number of iterations since the class F_0 is finite and $g(\overline{f}) < g(f)$ if the new policy $\overline{f}^{\infty}$ is not equal to the previous policy f^{∞}. In all examples tested we found that the finally obtained policy was average cost optimal within the class F_0.

Consider the following numerical example in which both T_p and T_s are deterministic.

$$T_p = 0.1,\ h = 0.05,\ \pi_1 = 25,\ \pi_2 = 2.5,\ r_1 = r_2 = 0,$$

$$r_3 = 100,\ R = 0.$$

In Table II we give for various values of λ and ET_s an optimal policy (m^*,M^*) with g^* its average costs and n^* the number of iterations after which this policy was found. We have chosen $N = 300$ and $(m = 150,\ M = 150)$ as starting policy for the algorithm. Observe from these examples that the minimal average cost is not a monotonic function in λ.

TABLE II

Optimal Policies

	$T_s = 0$			$T_s = 2$		
λ	(m^*,M^*)	g^*	(n^*)	(m^*,M^*)	g^*	(n^*)
8.5	(23,23)	1.1726	(6)	(32,118)	5.8151	(10)
9	(33,33)	1.6893	(9)	(39,110)	5.3281	(10)
9.5	(61,61)	3.1113	(10)	(62,114)	5.2948	(9)
9.75	(112,112)	5.6669	(11)	(109,153)	6.8447	(10)
9.9	(249,249)	12.5070	(13)	(242,283)	12.9911	(9)

REMARK 3.3. Consider now the case in which any demand has a discrete probability distribution $\{r_k, k \geq 1\}$ where we assume that $\lambda\beta ET_p < 1$ where β is the average demand size. For any policy $f^\infty = (m,M)$ with $M \geq 0$, choose now $s_f = M$ and $A_f = \{i \mid i = 0,\dots,M\}$. Using remark 3.1, an examination of the above analysis shows that only some technical modifications are required. In particular the formula for $w(i',f)$, $i > m$ will now involve the renewal function associated with the demand distribution.

3.3 *An M/M/c Queueing System with a Variable Number of Servers*

We consider an M/M/c queueing system in which the number of servers operating can be adjusted both at arrival and service completion epochs. The customers arrive in accordance with a Poisson process with rate λ and there are c independent servers available each having an exponentially distributed service time with mean $1/\mu$ where $\lambda/c\mu < 1$. The cost structure includes a holding cost of $h > 0$ per customer in the system per unit time, an operating cost of $w > 0$ per server turned on per unit time and a switch-over cost of $K(a,b)$ when the number of servers turned on is adjusted from a to b where $K(a,a) = 0$ and

$$K(a,b) = \begin{cases} K^+ + k^+.(b-a) & \text{for } b > a \\ K^- + k^-.(a-b) & \text{for } b < a \end{cases}$$

with K^+, K^-, k^+, $k^- \geq 0$.

This control problem can be represented by a semi-Markov decision model in which the decision epochs are given by the arrival and service completion epochs and at any decision epoch the system can be classified into one of the states of the denumerable state space

$$I = \{(i,s) \mid i = 0,1,\ldots;\ s = 0,\ldots,c\}$$

where state (i,s) corresponds to the situation in which i customers are present and s servers are turned on. For any state the set of possible actions consists of the actions a = 0,...,c where action a prescribes to adjust the number of servers turned on to a. The formulae for the one-step transition probabilities and one-step expected transition times and costs are obvious and omitted for reasons of space.

For a given integer N, denote by $F_0 = F_0^{(N)}$ the finite class of stationary policies having the following simple form. Any policy $f^\infty \in F_0$ is characterized by integers s(i), S(i), t(i) and T(i) for i = 0,1,... such that

(a) $-1 \leq s(i) < S(i) \leq T(i) < t(i) \leq c+1$ for $i \geq 0$
where $s(N) = c-1$ and $t(N) = c+1$

(b) $s(i) \leq s(i+1)$ and $t(i) \leq t(i+1)$ for $i \geq 0$.

If at a decision epoch there are i customers present and s servers turned on then under this policy $f^\infty = (s(i), S(i), T(i), t(i))$ the number of servers on is adjusted upward to S(i) when $s \leq s(i)$, is kept unaltered when $s(i) < s < t(i)$ and is adjusted downward to T(i) when $s \geq t(i)$. Observe that under any policy in the class F_0 all c servers are turned on or left on when N or more customers are present. It is still an unproven conjecture that for N sufficiently large the class $F_0^{(N)}$ contains a policy which is average cost optimal within the class of all possible policies, cf. Robin (1976) and Sobel (1974). However, we will only deal with the computation of a policy with minimal average cost within the class F_0. We note that as in Yadin and Naor (1976) an expression for the average costs under a given set of parameters s(i), S(i), T(i) and t(i) can be derived by steady-state analysis but the numerical evaluation of the optimal set of parameters from this expression is extremely difficult. We shall now briefly outline how to develop by the embedding approach a computationally tractable algorithm for computing the best policy within the class F_O.

For any policy $f^\infty = (s(i), S(i), T(i), t(i)) \in F_O$, define i_f as the smallest $i \leq N$ with $s(i) = c-1$ and $t(i) = c+1$, i.e. under policy f^∞ all c servers are turned on or left on when i_f or more customers are present. We choose now $s_f = (i_f, c)$ and

$$A_f = \{(i,s) \mid s = 0,\dots,s(i) \text{ and } s = t(i),\dots,c;\ i = 1,\dots, i_f-1\} \cup$$

$$\cup \{(0,s) \mid s = 0,\dots,c\} \cup \{(i_f,s) \mid s = 0,\dots,c\}.$$

For this choice of A_f we can easily derive analytical expressions for the quantities $\tilde{c}(t,f)$, $\tilde{\tau}(t,f)$ and $\tilde{p}_{tu}(f)$ for $t \in I$ and $u \in A_f$. These analytical expressions are obtained by solving second-order linear difference equations. To illustrate this, choose any (i,s) with $0 < i < i_f$ and $s(i) < s < t(i)$ and let L be the largest integer such that $0 \leq L < i$ and $(L,s) \in A_f$ and let R be the smallest integer such that $i < R \leq i_f$ and $(R,s) \in A_f$. Then $\tilde{c}((j,s),f)$ for $L < j < R$ is given by the solution of the second-order linear difference equation

$$x(j) = \frac{1}{\lambda+\mu\min(j,s)} \{hj+ws+\lambda x(j+1)+\mu\min(j,s)x(j-1)\},\quad L < j < R$$

where $x(L) = x(R) = 0$. We omit here further details and refer to the similar equations (3.1), (3.2), (3.7), (3.8) and (3.10) in De Leve et al (1977) where this queueing problem has been analysed by using a related but more complex embedding approach. For given policy $f^\infty = (s(i), S(i), T(i), t(i)) \in F_O$ we have, by the choice $s_f = (i_f, c)$ and the following relation for $i = 0,\dots, i_f$

$$w((i,s),f) = \begin{cases} K^{+}+k^{+}.(S(i)-s)+w((i,S(i)),f), & s \leq s(i) \\ K^{-}+k^{-}.(s-T(i))+w((i,T(i)),f), & s \geq t(i), \end{cases}$$

that the system of linear equations (2.11)-(2.12) can be reduced to a system of $2i_f+t(0)-s(0)-3$ linear equations in the unknown g, $v((i,S(i)))$, $v((i,T(i)))$ for $0 < i < i_f$ and $v((0,s))$ for $s(0) < s < t(0)$. We omit the obvious details. Once these linear equations have been solved, we can compute any $w((i,s),f)$ and $T((i,s),a,f)$.

To explain the policy-improvement step of the algorithm to be stated below, we note that for any $0 \leq i < i_f$ and $s = 0,\ldots,c$

$$T((i,s),a,f) = K(s,a)+w((i,a),f) \text{ for } s(i) < a < t(i)$$

Now, by the structure of the switch-over costs, we have for any fixed i with $0 \leq i < i_f$ that the smallest integer $\overline{S}(i)$ (say) which minimizes $K(0,a)+ w((i,a),f)$ for $s(i) < a < t(i)$ is less than or equal to the largest integer $\overline{T}(i)$ (say) which minimizes $K(c,a)+w((i,a),f)$ for $s(i) < a < t(i)$. Further, for any $s \leq s(i)$ the integer $\overline{S}(i)$ minimizes $T((i,s),a,f)$ for $s(i) < a < t(i)$ and for any $s \geq t(i)$ the integer $\overline{T}(i)$ minimizes $T((i,s),a,f)$ for $s(i) < a < t(i)$.

We now describe the algorithm. Let N be given integer and choose some policy $f^{\infty} \in F_0$.

ALGORITHM

STEP 1 (value-determination step). Let $f^{\infty} = (s(i), S(i), T(i), t(i))$ be the current policy. Compute the numbers $g(f)$, $w((i,S(i)),f)$, $w((i,T(i)),f)$ for $0 < i < i_f$ and $w((0,s),f)$ for $s(0) < s < t(0)$ by solving the above described system of linear equations.

STEP 2 (policy-improvement step). (a) For any $i = 0,\ldots, i_f-1$ determine $\overline{S}(i)$ as the smallest integer which minimizes $K(0,a)+w((i,a),f)$ for $s(i) < a < t(i)$ and determine $\overline{T}(i)$ as the largest integer which minimizes $K(c,a)+ w((i,a),f)$ for $s(i) < a < t(i)$.

(b) Successively for $i=0,\ldots,N-1$, determine $\overline{s}(i)$ and $\overline{t}(i)$

TABLE III

Optimal Policies

	$K^+=K^-=0$, $g^*=$ 1240.14				$K^+=K^-=75$, $g^*=$ 1247.67			
i	s*(i)	S*(i)	T*(i)	t*(i)	s*(i)	S*(i)	T*(i)	t*(i)
0	-1	0	6	7	-1	0	7	9
1	0	1	6	7	-1	0	7	9
2	1	2	6	7	-1	0	7	9
3	1	2	7	8	-1	0	7	9
4	2	3	7	8	-1	0	8	10
5	3	4	8	9	-1	0	8	10
6	4	5	8	9	-1	0	10	11
7	4	5	9	10	2	7	10	11
8	5	6	10	11	2	7	10	11
9	6	7	10	11	4	8	10	11
10	6	7	10	11	6	8	10	11
11	7	8	10	11	6	8	10	11
12	7	8	10	11	6	10	10	11
13	8	9	10	11	6	10	10	11
14	9	10	10	11	7	10	10	11
15					8	10	10	11
16					9	10	10	11

as follows where we put $s(-1) = -1$ and $t(-1) = 0$. Define $\underline{s}(i)$ as the largest integer k such that $s(i)+1 \leq k \leq \min[\overline{S}(i)-1, s(i+1)]$ and $K(s,\overline{S}(i))+w((i,\overline{S}(i)),f) < w(i,s),f)$ for all $s(i) < s \leq k$ if such an integer k exists, otherwise let $\underline{s}(i)$ be equal to $\ell-1$ with ℓ the smallest integer such that $\min[0, s(i-1)+1] \leq \ell \leq s(i)$ and $T((i,s),s,f) \leq K(s,\overline{S}(i))+w((i,\overline{S}(i)),f)$ for all $\ell \leq s \leq s(i)$ if such an integer ℓ exists, otherwise let $\overline{s}(i) = s(i)$. Define $\overline{t}(i)$ as the smallest integer k such that $\max[\overline{T}(i)+1, t(i-1)] \leq k < t(i)$ and $K(s,\overline{T}(i))+w((i,\overline{T}(i)),f) < w((i,s),f)$ for all $k \leq s < t(i)$ if such an integer k exists, otherwise let $\overline{t}(i)$ be equal $\ell+1$ with ℓ the largest integer such that $t(i) \leq \ell \leq \min[t(i+1)-1, c]$ and $T((i,s),s,f) \leq K(s,\overline{T}(i))+w((i,\overline{T}(i)),f)$ for all $t(i) \leq s \leq \ell$ if such an integer ℓ exists, otherwise let $\overline{t}(i) = t(i)$.

STEP 3. Let $\overline{f}^{\infty} = (\overline{s}(i), \overline{S}(i), \overline{T}(i), \overline{t}(i))$. If $\overline{f}^{\infty} = f^{\infty}$ then stop, otherwise go to step 1 with f^{∞} replaced by $\overline{f}^{\infty}$.

We conclude with a numerical example in which

$$\lambda = 9.5,\ c = 10,\ \mu = 1,\ h = 10,\ w = 100,\ k^{+} = k^{-} = 50$$

In Table III we give for various values of K^{+} and K^{-} an optimal policy $(s^*(i), S^*(i), T^*(i)\ t^*(i))$ and the minimal average cost g^*.

4. ACKNOWLEDGEMENT

I am very much indebted to Rob van der Horst and Leander Jansen for writing the computer programs.

5. REFERENCES

1. Bather, J. "A Continuous Time Inventory Model", *J. Appl. Prob.* **3**, 538-549. (1966).

2. Bather, J. "A Diffusion Model for the Control of a Dam", *J. Appl. Prob.* **5**, 55-71, (1968).

3. Chernoff, H. and Petkau, A.J. "Optimal Control of a Brownian Motion", *Siam. J. Appl. Math.* **34** 717-731. (1978).

4. Chung, K.L. "Markov Chains with Stationary Transition Probabilities", Springer-Verlag, Berlin. (1960).

5. Crabill, T.B. "Optimal Control of a Service Facility with Variable Exponential Service Time and Constant Arrival Rate, *Manag. Sci.* **18**, 560-566. (1972).

6. Crabill, T.B., Gross, D. and Magazine, M.J. "A Classified Bibliography of Research on Optimal Design and Control of Queues", *Oper. Res.* **25**, 219-232. (1977)

7. De Leve, G. "Generalized Markovian Decision Process, Part I: Model and Method Part II: Probabilistic Background", Mathematical Centre Tracts No. 3 and 4, Mathematisch Centrum, Amsterdam. (1964)

8. De Leve, G., Tijms, H.C. and Weeda, P.J. "Generalized Markovian Decision Processes, Applications", Mathematical Centre Tract No. 5, Mathematisch Centrum, Amsterdam. (1970)

9. De Leve, G., Federgruen, A. and Tijms, H.C. "A General Markov Decision Method, I: Model and Techniques, II: Applications", *Adv. Appl. Prob.* **9**, 297-315, 316-335. (1977)

10. Derman, C. and Veinott Jr., A.F. "A Solution to a Countable System of Equations Arising in Markov Decision Processes" *Ann. Math. Statist.* **38**, 582-584, (1967).

11. Faddy, M.J. "Optimal Control of Finite Dams: Continuous Output Procedure" *Adv. Appl. Prob.* **6** 689-710.

12. Federgruen, A., Hordijk, A. and Tijms, H.C. "Denumerable State Semi-Markov Decision Processes with Unbounded Costs, Average Cost Criterion", Report BW 92/78, Mathematisch Centrum, Amsterdam. (1978).

13. Federgruen, A. and Tijms, H.C. "Computation of the Stationary Distribution of the Queue Size in an M/G/1 queueing System with Variable Service Rate", Report BW 96/79, Mathematisch Centrum, Amsterdam. (1979).

14. Gallisch, E. "On Monotone Optimal Policies in a Queueing Model of M/G/1 Type with Controllable Service Time Distribution", Institute for Applied Mathematics, University of Bonn, (1977).

15. Hordijk, A. and Tijms, H.C. "A Simple Proof of the Equivalence of the Limiting Distributions of the Continuous-Time and Embedded Process of the Queue Size in the M/G/1 queue", Statist. Neerl. 30, 97-100. (1976).

16. Prabhu, N.U. and Stidham, S. "Optimal Control of Queueing Systems", in: Mathematical Methods in Queueing Theory (A.B. Clarke ed.), 263-294, Lecture Notes in Economics and Mathematical Systems 98, Springer-Verlag, Berlin, (1974).

17. Puterman, M.L. "A Diffusion Process Model for a Storage System", in: Logistics (M.A. Geisler ed.) North-Holland, Amsterdam. 143-159, (1975).

18. Rath, J. "The Optimal Policy for a Controlled Brownian Motion Process", *Siam J. Appl. Math.* **35**, 115-125, (1977).

19. Robin, M. "Some Optimal Control Problems for Queueing Systems", in: Stochastic Systems: Modeling, Identification and Optimization (R.J. Wets ed.), 154-169, Mathematical Programming Study 6, North-Holland, Amerstam, (1976).

20. Ross, S.M. "Applied Probability Models with Optimization Applications", Holden-Day, San Francisco, (1970).

21. Sobel, M.J. "Optimal Average-Cost Policy for a Queue with Start-up and Shut-down Costs", *Oper. Res.* **18**, 145-162, (1970).

22. Sobel, M.J. "Optimal Operation of Queues", in: Mathematical Methods in Queueing Theory, 231-261, Lecture Notes in Economics and Mathematical Systems 98, Springer-Verlag, Berlin, (1974).

23. Tijms, H.C. "Optimal Control of the Workload in an M/G/1 Queueing System with Removable Server", *Math. Operationsforsch. u. Statist.* **7**, 933-943. (1976a)

24. Tijms, H.C. "On the Optimality of a Switch-over Policy for Controlling the Queue Size in an M/G/1 Queue with Variable Service Rate", in: Optimization Techniques I (J. Cea ed.), 112-120, Lecture Notes on Computer Science 40, Springer-Verlag, Berlin, (1976b).

25. Tijms, H.C. "On a Switch-over Policy for Controlling the Workload in a Queueing System with Two Constant Service Rates and Fixed Switch-over Costs, *Zeitschrift fur Oper. Res.* **21**, 19-32, (1977).

26. Tijms, H.C. and Van Der Duyn Schouten, F.A. "Inventory Control with Two Switch-over Levels for a Class of M/G/1 Queueing Systems with Variable Arrival and Service Rate", *Stoch. Proc. and Appl.* **6**, 213-222, (1978).

27. Whitt, W. "Diffusion Models for Production and Inventory Systems", Dep. of Administrative Sciences, Yale University, (1973).

28. Yadin, M. and Naor, P. "On Queueing Systems with Variable Service Capacities", *Naval Res. Log. Quart.* **14**, 43-53, (1967).

CONNECTEDNESS CONDITIONS FOR DENUMERABLE STATE MARKOV DECISION PROCESSES

Lyn C. Thomas

(Department of Decision Theory, University of Manchester)

ABSTRACT

In this paper we describe some of the conditions put on the set of transition matrices of denumerable state Markov Decision Processes, in order to ensure results like the existence of average optimal policies. We display the equivalences and implications between these conditions.

1. INTRODUCTION

Consider a Markov Decision Process, where the state space, S, is countable. When in state i, one can choose an action, k, from the set of possible actions K_i. Thus at each step we can choose an overall action which is an element of $\bigotimes_{i \in S} K_i$, and a policy is a sequence of such elements, ordered by the number of steps already performed. Corresponding to each overall action, there is a transition matrix, P, of the probabilities of which state the process will move to in the next step. To ease notation, we will sometimes identify each overall action with its corresponding transition matrix. Let $\mathcal{P}$ be the set of all such transition matrices. One can think of them as describing the stationary Markov policies. Thus when in state i we choose action k (or overall action P), there is an immediate cost of r_i^k (or $r(i,P)$), and a probability p_{ij}^k (or P_{ij}) of moving to state j for the next step. The action sets, K_i, can be arbitrary, but unless otherwise stated we will assume they are compact with respect to some topology. Thus by Tychynov's theorem $\bigotimes_{i \in S} K_i$ will also be compact. The costs are always finite, and unless otherwise stated will be considered to be bounded as well, so that $|r_i^k| \leq M$, $\forall\, i \in S$,

$k \in K_i$.

The object of this paper is to examine some of the conditions that have been required of the set, P, of transition matrices in order that certain results hold for the corresponding Markov decision process. We concentrate in particular on the conditions that guarantee the existence of an optimal stationary policy in the average cost case. We then derive the implications and equivalences amongst these conditions, and give counter-examples where implications do not hold. A similar exercise was performed in [23] for the finite state case.

In Sections 2 and 3, we give a very brief sketch of the results we require from the discounted cost and the average expected cost types of Markov decision process - as much to establish notation as anything else. Section 4 describes the conditions that have been imposed on P, and also what results hold, if they are satisfied. In Section 5 we state the theorem, describing how these conditions are connected, and prove the implications, most of which are trivial. Finally, in Section 6 we give counter-examples to show which implications among the conditions do not hold.

2. DISCOUNTED COST CASE

If one tries to maximise total expected costs, one has to impose conditions that ensure the total cost under any policy is finite. The two usual ways of doing this are to discount future costs by a discount factor β, or to ensure that there is a positive probability of the process leaving the system. We shall concentrate on the discounted case terminology, and so if π is the policy consisting of the overall actions $(P^1, P^2, \ldots, P^n)$ then the discounted cost is

$$v_\beta^\pi(i) = \sum_{t=1}^{\infty} \sum_{j \in S} \beta^{t-1} (P^1 . P^2 . \ldots . P^{t-1})_{ij} r(j, P^t)$$

The matrices P will be superscripted by the corresponding overall action. If powers of them are required then brackets will be used i.e. $(P^1)^2$ is the product of the transition matrix corresponding to action 1, with itself.

In the case when the costs are uniformly bounded, the

operator T, with $(Tv)_i = \inf_{k \in K_i} \{r_i^k + \beta \sum_{j \in S} p_{ij}^k v_j\}$ is shown by Denardo [3] to be a contraction operator. The fixed point of this contraction, v_β, satisfies the optimality equation

$$v_\beta(i) = \inf_{k \in K_i} \{r_i^k + \beta \sum_{j \in S} p_{ij}^k v_\beta(j)\}$$

It then follows that $v_\beta(i)$ is the optimal value, since $v_\beta(i) = \inf_\pi v_\beta^\pi(i)$, as is shown in Ross [21, Chapter 7]. Moreover, since $T^n u$ converges to v_β for all starting values u, this implies value iteration converges. It is also an easy exercise to show that policy iteration converges, see [16].

If the costs are unbounded, the approach as it stands does not work. If we follow the approach of Lippman [18] and Wessels [27] and introduce a weighted supremum norm i.e. $||v|| = \sup_{i \in S} |v(i)/w(i)|$ for some positive finite function w, the results go through provided the transition matrices P are contracting operators in this new norm. Harrison [12] overcame the difficulties by introducing the idea of a shifted space where the costs do not vary too much from one another, and these ideas have been generalised by van Nunen [24] and van Nunen and Wessels [25,26].

3. AVERAGE COST CASE

For a general policy π with the sequence of transition matrices $P^1, P^2, \ldots, P^n, \ldots$ the average cost per step is defined by

$$\phi^\pi(i) = \lim_{n \to \infty} \frac{1}{n} \sum_{s=1}^{n} \sum_{j \in S} (P^1 . P^2 . \ldots P^{s-1})_{ij} r(j, P^s) \quad (3.1)$$

Howard [16] showed that under certain weak conditions

$$v_n(i,\pi) = \sum_{t=1}^{n} \sum_{j \in S} (P^1 . P^2 . \ldots . P^{t-1})_{ij} r(j, P^t) \to ng + v_i \quad (3.2)$$

where g is the "gain" and v_i the bias term. Thus one could identify the average cost with g. The limit in 3.1 need not exist, and in that case the usual criterion is to take the lim sup, but as Flynn [11] pointed out there is a host of possible choices. One says that a policy π^* is average cost optimal in the strong sense if

$$\limsup \frac{v_n(i,\pi^*)}{n} \leq \liminf_{n\to\infty} \frac{v_n(i,\pi)}{n}, \ \forall\ i\in S, \text{ any } \pi$$

This implies the other optimality criteria as is shown in [11].

One immediate question to ask is: "Are there always optimal policies that minimise the average cost?". The answer is no, even in the case when the sets K_i are finite and the costs are bounded, as the following example from Ross [21] shows.

Let the state space be $\{i\}$, $1 \leq i < \infty$, with $K_{2i+1} = 2$, $K_{2i} = 1$, where with abuse of notation from now on we let K_i be the number of actions in the action set corresponding to state i. Let $r^k_{2i+1} = 1$, $k = 1$ or 2, $r_{2i} = 1/i$, $P^1_{2i+1,2i+3} = 1 = P^2_{2i+1,2i+2}$, $P_{2i,2i} = 1$. Then, using action 1 in 2i+1 always gives $\phi^\pi(1) = 1$. If action 2 is chosen with positive probability in some state, say in state 2n-1 with probability p_n, then $\phi^\pi(1) \geq p_n/n$. However, by choosing action 1 for enough steps and then action 2 we can make the optimal value as near 0 as we like. Fisher and Ross [10] pointed out that a similar example can be constructed where there is an optimal policy which is not stationary.

The reason these examples fail to produce optimal stationary policies is that the states are not connected to one another. The process passes through each state just once until it gets absorbed in one of the even states. If the process kept returning to some fixed set of states, then the decisions made there would have an obvious effect on the average cost, rather than the indirect one of the examples above. Thus to ensure the existence of an optimal policy we need some connectedness conditions.

In the average cost case, the optimality equation is

$$g + v_i = \inf_{k\in K_i} \{r^k_i + \sum_{j\in S} p^k_{ij} v_j\}, \ \forall\ i\in S \tag{3.3}$$

Federgruen and Tijms [8] pointed out that the proof of Theorem 7.6 in [21] implies that if there is a bounded solution to the optimality equation, the stationary policy that minimises the right hand side of 3.3 is average cost optimal in the strong sense. (Weak continuity conditions on the r_i^k and p_{ij}^k together with the compactness of K_i ensure the existence of a minimiser.) So it is enough to show the optimality equation has a bounded solution to ensure existence of an optimal policy. This can either be proved directly, where conditions C4, C5 are used, or by taking the limit of discounted optimality equations as the discount factor tends to one, which is the way the proofs using C11, C12 go. If discounted optimality policies exist and have a limit policy as β tends to one, one can prove the existence of an average cost optimal policy directly, by showing it is this limiting policy, see Hordijk [13].

4. CONNECTEDNESS CONDITIONS

In this section we state the twenty-one conditions that we will look at, and note briefly the context in which they were used.

Let $\mu_{iF}(P)$ be the expected number of steps until first arriving in the set F under policy P and starting at state i. Recall that a closed communicating class of states A for a matrix P has the properties

(i) $\sum_{j \in A} P_{ij} = 1, \ \forall \ i \in A$

(ii) For each $i, j \in A$, there is an $n, m \geq 1$ such that $P_{ij}^n > 0$ and $P_{ji}^m > 0$.

A state i of a matrix P is recurrent if $\sum_{n=0}^{\infty} (P)_{ii}^n = \infty$ and if it is of period d then it is a positive recurrent if $\lim_{n \to \infty} (P)_{ii}^{nd}$ is positive.

The connectedness conditions are as follows.

<u>C1</u> There is a state $0 \in S$ such that $P_{i0} \geq \alpha > 0, \ \forall \ i \in S, \ P \in \mathcal{P}$.

Under this condition, Ross [20] showed that the average cost Markov Decision Process can be transformed into a discounted Markov Decision Process. You keep the action spaces and the rewards the same, and let $\tilde{P}_{ij} = P_{ij}/1-\alpha$ and $\tilde{P}_{io} = (P_{io}-\alpha)/(1-\alpha)$. Then the value v for the discounted process with discount factor $1-\alpha$, and transition probabilities $\tilde{P}$ satisfies $\alpha v^{\pi}_{1-\alpha}(0) = \phi^{\pi}(0)$, for all policies π. So all the results in discounted problems still hold, i.e. an optimal stationary Markov policy exists amongst all policies, value iteration and policy iteration converge to the optimal solution.

C2 (Strong scrambling). There is a $\alpha > 0$, such that $\sum_{j\in S} \min[P_{i_1 j}, P_{i_2 j}] \geq \alpha$, $\forall\ i_1, i_2 \in S$, $\forall\ P \in \mathcal{P}$.

Federgruen and Tijms [8] show that this condition implies that the operator T, defined by

$$(Tv)(i) = \inf_{k\in K_i} \{r^k_i + \sum_{j\in S} p^k_{ij} v(j)\},\ \forall\ i\ S$$

is a contraction operator in the span norm, $\|\ \ \|$, where $\|u\| = \sup_{i\in S} u(i) - \inf_{i\in S} u(i)$. This in turn proves the existence of a bounded solution to the optimality equation and also gives the bound $|v_n(i)-ng-v_i| \leq (1-p)^n \|v_o - v\|$, $\forall\ i\in S$, $n \geq 1$, where $(1-p)$ is the contraction factor of T. Thus the convergence of $v_n(i)$ to $ng + v_i$ in (3.2) is exponential and uniform in i.

C3 There is a state $0\in S$, $\alpha > 0$, $n \geq 1$, such that $(P_1\ P_2 \cdots P_n)_{io} \geq \alpha$, $\forall\ i\in S$, $P_k \in \mathcal{P}$, $k = 1,\ldots,n$.

White [28] proved the convergence of the relative value algorithm in the finite state case under this condition. Kushner [17] showed that the result also holds in the infinite state case. The existence of an optimal stationary policy then follows as a trivial corollary.

C4 (Scrambling). There is an $n \geq 1$, $\alpha > 0$ such that $\sum_{j\in S} \min[P^n_{i_1 j}, P^n_{i_2 j}] \geq \alpha$, $\forall\ i_1, i_2 \in S$, $P \in \mathcal{P}$.

Federgruen, Hordijk and Tijms [6] showed that the optimal equation has a bounded solution under this condition, and so using Ross's result there is an average optimal policy in the strong sense. Value iteration also converges under it. The same three authors in a different paper [7] showed that this condition was equivalent to four others, namely

C4a There is an $n \geq 1$, $\rho > 0$ such that for $P \in \mathcal{P}$, there is a probability distribution $\pi(P)$ for which

$$\left| \sum_{j \in A} P^m_{ij} - \sum_{j \in A} \pi_j(P) \right| \leq (1-\rho)^{([m/n])}, \forall\ i \in S,\ A \subset S,\ m \geq 1$$

and each $P \in \mathcal{P}$ has only one closed communicating class.

C4b (Strong ergodicity). For any $P \in \mathcal{P}$, there is a probability distribution $\pi(P)$ such that $P^n_{ij} \to \pi(P)_j$ uniformly in i and P as $n \to \infty$, for any $j \in S$, and each $P \in \mathcal{P}$ has only one closed communicating class.

C4c There is an $n \geq 1$, $\alpha > 0$ such that for any $P \in \mathcal{P}$, there is a state $s(P)$ for which

$$P^n_{is(P)} \geq \alpha,\ \forall\ i \in S.$$

C4d There is a finite set F, $n \geq 1$, $\alpha > 0$ such that for any $P \in \mathcal{P}$, there is a state $s(P) \in F$ with $P^n_{is(P)} \geq \alpha$, $\forall\ i \in S$.

C5 (Simultaneous Doeblin). There is a finite set F, $n \geq 1$, $\alpha > 0$ such that $\sum_{j \in F} P^n_{ij} \geq \alpha$, $\forall\ i \in S$, $P \in \mathcal{P}$, and each $P \in \mathcal{P}$ has only one closed communicating class.

Hordijk [13] proved the existence of an optimal policy under the lim sup criterion among all policies in $\mathcal{P}$ under this condition. Federgruen, Hordijk and Tijms [6] showed that it implied that the optimality equation has a bounded solution, and that value iteration converged. In the note [7] on recurrence condition on stochastic matrices, C5 is shown equivalent to the following three conditions.

C5a There is a finite set F, $B < \infty$ such that $\mu_{iF}(P) \leq B$, $\forall\ i \in S$, $P \in \mathcal{P}$, and each $P \in \mathcal{P}$ has only one closed communicating class.

C5b There is a $B < \infty$ such that for any $P \in \mathcal{P}$ there is an $s(P) \in S$ with $\mu_{is(P)}(P) \leq B$, $\forall\, i \in S$.

C5c There is a finite set F, $B < \infty$ such that for any $P \in \mathcal{P}$ there is an $s(P) \in F$ with $\mu_{is(P)}(P) \leq B$, $\forall\, i \in S$.

$\widetilde{C5}$ C5 plus P is aperiodic for all $P \in \mathcal{P}$.

Federgruen, Hordijk and Tijms [7] also proved that C5 is equivalent to C4.

C6 Let $\pi(P)_{ij} = \lim_{n\to\infty} \frac{1}{n} \sum_{s=0}^{n-1} P^s_{ij}$. For any $\varepsilon > 0$, there is a finite set F such that $\sum_{j\in F} \pi(P)_{ij} \geq 1-\varepsilon$, $\forall\, P \in \mathcal{P}$, $\forall\, i \in S$, and each P has only one closed communicating class. The former condition can be written by saying that the set $\{\pi(P)_i \cdot$ $i \in S$, $P \in \mathcal{P}\}$ is a tight collection of probability measures for each $i \in S$.

C7 $\sum_{j\in S} \pi(P)_{ij} = 1$, $\forall\, i \in S$, $P \in \mathcal{P}$, and $\pi(P)_{ij}$ is continuous on $\mathcal{P}$ for some topology.

Both C6 and C7 were used by Hordijk [13] to show the existence of a lim sup average optimal policy, by looking directly at the limit of the discounted optimal policies as the discount factor tends to 1. C6 is easier to verify than C7.

C8 There is a $B < \infty$, a state $0 \in S$, such that $\mu_{io}(P) \leq B$, $\forall\, i \in S$, $P \in \mathcal{P}$.

Derman [4], Derman and Veinott [5], and Federgruen and Tijms [8] all used this - the oldest of these conditions - to prove that in the average cost case there is a bounded solution to the optimality equation. [8] also proved that value and policy iteration converge under this condition.

C9 (Liapunov function). There is a state $0 \in S$, and finite non-negative numbers y_i, $i \in S$, such that

(i) $|r(i,P)| + 1 + \sum_{j\in S} \tilde{P}_{ij} y_j \leq y_i$, $\forall\, i \in S$, $P \in \mathcal{P}$

(ii) $\sum_{j\in S} \tilde{P}_{ij}y_j$ is continuous for $P\in\mathcal{P}$.

(iii) $\lim_{n\to\infty} \sum_{j\in S} \tilde{P}^n_{ij}y_j = 0$ where $\tilde{P}_{ij} = \begin{cases} P_{ij} & j \neq 0 \\ 0 & j = 0 \end{cases}$

This condition,introduced by Hordijk [13], allows unbounded costs, which none of the previous conditions allowed. It guarantees that the optimality equation has a finite solution, and so that an average cost optimal policy in the strong sense exists. A slight extension of it was used by Hordijk and Sladky [14] to show that there exists n-discount optimal and n-average optimal policies. A slight modification of this condition is the following.

$\underline{\tilde{C9}}$ This is C9 except that the y_i's must be bounded non-negative numbers.

If the y_i's are bounded, then the costs are also bounded, but not necessarily vice versa. It is then an easy exercise to show that $\tilde{C9}$ implies C8.

<u>C10</u> For any $\varepsilon > 0$, there is a finite set F, such that $\sum_{j\in F} P_{ij} \geq 1-\varepsilon$, $\forall$ $i\in S$, $P\in\mathcal{P}$ (i.e. P_i, $P\in\mathcal{P}$ is a tight set of probability measures), and for each $P\in\mathcal{P}$ there is only one closed communicating class.

This was used by Hordijk [13] to show the existence of a stationary optimal policy among all $P\in\mathcal{P}$. (If $\mathcal{P}$ contains all randomised policies, then this policy is optimal among all policies.) It is an easy condition to check.

<u>C11</u> Let $v_\beta(i)$ be the optimal cost starting in state i, under β-discounting criterion, then $v_\beta(i)-v_\beta(j)$ is uniformly bounded for $i,j\in S$, $0 \leq \beta \leq 1$. Taylor [22] used this in a replacement model with finite action spaces, to prove the existence of a stationary deterministic optimal policy by showing the optimality equation has a bounded solution - the limit of the discounted equation's solutions. It works in general problems.

<u>C12</u> There is a sequence $\beta_r \uparrow 1$, such that $v_{\beta_r}(i)-v_{\beta_r}(j)$ is uniformly bounded for $i,j\in S$, $r \geq 1$.

Ross [20] rewrote Taylor's condition, and showed that this condition also implied that the optimal equation has a bounded solution if the action spaces are finite. He also showed that the β_r-discounted optimal policies are ε-optimal for large enough r.

<u>C13</u> $\inf_{P \in \mathcal{P}} \mu_{ij}(P) \leqslant B, \ \forall\ i,j \in S.$

This was suggested by Hordijk [13] as an easy condition to check which implies Taylor's condition and so the existence of an average cost optimal policy.

<u>C14</u> There is a state $0 \in S$ and a sequence $\beta_r \uparrow 1$ such that $\mu_{io}(P_{\beta_r}) \leqslant B, \ \forall\ i \in S,\ r \geqslant 1$, where P_{β_r} is the optimal policy in the β_r-discounted case.

Ross [20] used this as an easy check which implies C12.

<u>C15</u> For any average optimal policy P, there is only one closed communicating class, which is also aperiodic, and for all $P \in \mathcal{P}$, $\sum_{j \in S} \pi_{ij}(P) = 1, \ \forall\ i \in S$.

For finite action spaces, Hordijk, Schweitzer, Tijms [14] proved that if there is a bounded solution to the optimality equations, then $v_n(i)-ng-v(i)$ converges to a finite limit independent of i.

The next five conditions include the ideas of communicating systems and recurrent systems. These were introduced by Bather [1] in the context of finite state spaces, and convex decision spaces, and used for denumerable state systems by Hordijk [13]. $(S,\mathcal{P})$ is a communicating system if for any $i,j \in S$ there is a $P \in \mathcal{P}$ such that $P^n_{ij} > 0$ for some $n \geqslant 1$. It is a recurrent system if in addition for any i,j there is a $P \in \mathcal{P}$ under which we are certain of reaching j from i.

<u>C16</u> $(S,\mathcal{P})$ is recurrent and for any $\varepsilon > 0$ P_i. $P \in \mathcal{P}$, $i \in S$ is a tight collection of probability measures.

C17 (S,P) is recurrent and there is a finite set of states F, $\alpha > 0$, $n \geq 1$ such that $\sum_{j \in F} P^n_{ij} \geq \alpha$, $\forall\ i \in S$, $P \in P$.

C18 (S,P) is recurrent and $\{\pi(P)_{i\cdot}, P \in P\}$ is a tight collection of probability measures.

These three conditions were used by Hordijk [13] to prove the existence of an optimal stationary policy among all the policies in P.

C19 (S,P) is a communicating system.

Bather [1] used this in the finite state case to prove the existence of an optimal stationary policy. Hordijk [13] pointed out that under C5,the simultaneous Doeblin condition, this condition is equivalent to (S,P) being a recurrent system. Thus in the finite state case C16, C17, C18 all collapse to this condition.

C20 $\forall\ P \in P$, the states are all positive recurrent and in one communicating class.

This condition was used by Derman [4] to ensure that if there is a bounded solution to the optimality equation, the optimal policy among pure stationary policies is optimal over all policies, for the average cost case. It also ensures that policy iteration converges.

C21 $\forall\ P \in P$, there is an M,N such that $P_{ij} = 0$, $i > M$, $j > i+N$.

If the costs are unbounded, Robinson [19] showed that the solution of the optimality equation will be average cost optimal if this condition holds.

5. RELATIONSHIPS AMONG THE CONDITIONS

We now state the main result of this paper, which describes the relations among C1 to C21. We assume that these conditions hold together with the rewards being bounded, except in the case of C9, and the action spaces being compact.

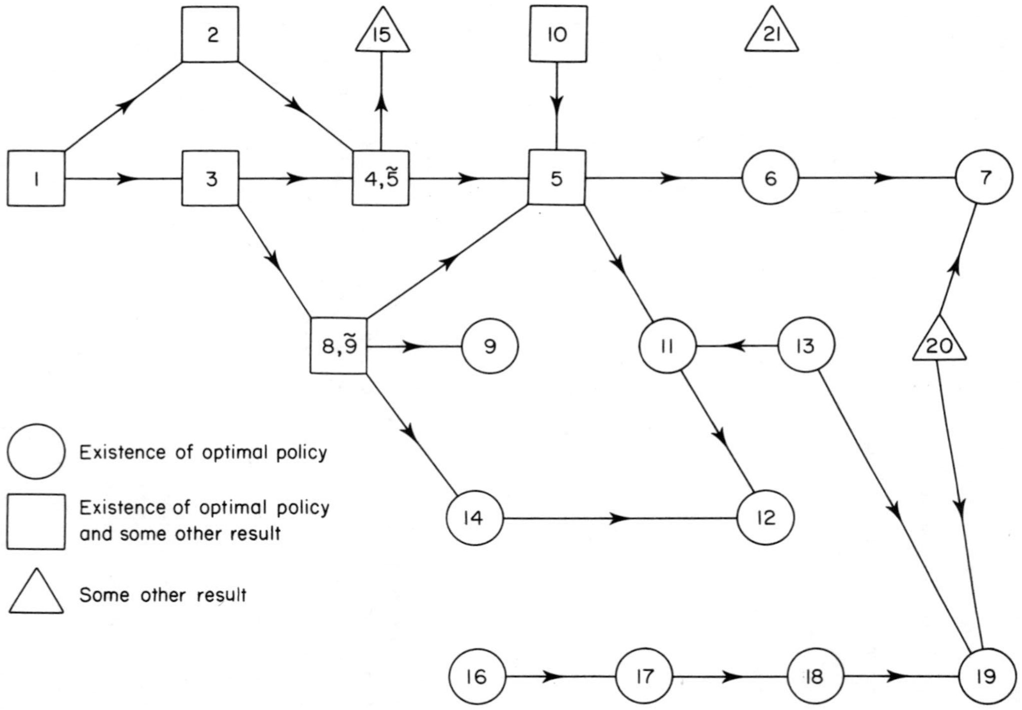
1
2
3
4,$\tilde{5}$
15
5
10
6
7
21
8,$\tilde{9}$
9
11
13
20
14
12
16
17
18
19
Existence of optimal policy
Existence of optimal policy
and some other result
Some other result

Fig. 1

Theorem

(a) The following implications hold between the conditions C1 $\Rightarrow$C2 $\Rightarrow$C4$\Leftrightarrow\tilde{C}5$ $\Rightarrow$C5 $\Rightarrow$C6 $\Rightarrow$C7; C1 $\Rightarrow$C3 $\Rightarrow$C8 $\Leftrightarrow\tilde{C}9$ $\Rightarrow$C5; C3 $\Rightarrow$C4; C8 $\Rightarrow$C14 $\Rightarrow$C12; C8 $\Rightarrow$C9; C4 $\Rightarrow$C15; C10 $\Rightarrow$C5; C5 $\Rightarrow$C11 $\Rightarrow$C12; C13 $\Rightarrow$C11; C16 $\Rightarrow$C17 $\Rightarrow$C18 $\Rightarrow$C19; C13 $\Rightarrow$C19; C20 $\Rightarrow$C19; C20 $\Rightarrow$C7.

(b) There are counter-examples to all possible implications which are not in the transitive closure of the implications of (a) except that we have no examples to show. C12 $\not\Rightarrow$C11, C2 $\not\Rightarrow$C14, C14 $\not\Rightarrow$C11, C16,17 $\not\Rightarrow$C11, C12, C14.

It is easiest to get a feeling for the result by examining figure 1, where the implications between the conditions are represented by arrowed lines. We shall prove the implications in this section and give the counter-examples in the next.

Proof of (a) C1 $\Rightarrow$C2 $\Rightarrow$C4, and C4d $\Rightarrow\tilde{C}5$ trivially, while C4 $\Rightarrow$C5 follows from Federgruen, Hordijk, Tijms [7] as mentioned in the text. C5 $\Rightarrow$C6 is theorem 11.4 of [13, p.96] and C6 $\Rightarrow$C7 is a corollary of lemma 10.2 of [13, p.83]. C1 is C3 with n = 1.

To prove that C3 $\Rightarrow$C8, assume that X_n is the state of the process at the nth step and $N_i = \min\{n | X_n = 0, X_m \neq 0, 2 \leqslant m \leqslant n-1, X_1 = i\}$. Then C3 implies that $P(N_i > n) \leqslant 1-\alpha$, and $P(N_i > 2n | N_i > n) \leqslant 1-\alpha$. So $P(N_i > 2n) \leqslant (1-\alpha)P(N_i > n) \leqslant (1-\alpha)^2$. Similarly $P(N_i > k_n) \leqslant (1-\alpha)^k$. Thus

$$\text{Exp}(N_i) = 1 + \sum_{t=1}^{\infty} P(N_i > t) \leqslant 1 + n \sum_{k=0}^{\infty} (k+1)(1-\alpha)^k$$

$$\leqslant 1 + n/\alpha + n(1-\alpha)\alpha^2$$

Thus C8 holds.

The equivalence of C8 and $\tilde{C}9$ follows from the discussion in Hordijk's book [13, pp.104-105]. C8 $\Rightarrow$C5a and C3 $\Rightarrow$C4d trivially with F = {0}. Similarly C8 $\Rightarrow$C14 is immediate, while Ross [20, theorem 1.4] proves that C14 $\Rightarrow$ C12. $\tilde{C}9$ $\Rightarrow$C9 is obvious from the definition of $\tilde{C}9$.

To prove C4,$\tilde{C}5 \Rightarrow$C15, we note that $\tilde{C}5$ requires aperiodicity, while C4a says that there is only one closed communicating class. Since $\tilde{C}5$ holds, then the state s(P) of C5b must be in the closed communicating class. Since s(P) is positive recurrent, then the class is also positive recurrent, as this is a class property.

C10 $\Rightarrow$C5a, since if $\sum_{j\in F} P_{ij} \geq 1-\varepsilon$, then $\mu_{iF}(P) \leq (1-\varepsilon)^{-1}$. C5b $\Rightarrow$C11 holds, since theorem 1.4 of Ross [20] still holds if the state, with bounded first passage time from every other state, varies with β, which in turn certainly is the case if C5b holds.

C11 $\Rightarrow$C12 is trivial and C13 $\Rightarrow$C11 is theorem 12.7 of Hordijk's book [13]. Similarly, C16 $\Rightarrow$C17 is proved by lemma 11.2 of [13] and C17 $\Rightarrow$C18 is theorem 11.4 of the same text. C18 $\Rightarrow$C19, C13 $\Rightarrow$C19 and C20 $\Rightarrow$C19 are all trivially true. C20 $\Rightarrow$C7 because the positive recurrence of the states implies that $\sum_{j\in S} \pi(P)_{ij} = 1$, while lemma 10.2 of [13] says that $\pi(P)_{ij}$ is continuous, provided the limit points have no disjoint closed sets. C20 ensures that this occurs.

6. COUNTER-EXAMPLES

To work out which are the pairs of implications that one has to produce counter-examples to is a non-trivial logical exercise in itself. Our strategy is first to produce counter-examples showing that all the implications in (a) are not equivalences. Then we shall endeavour to produce counter-examples showing that there are no other equivalences. In doing this, we use the fact that if we have two separate strings of implications, it is sufficient to show that the most restrictive condition in one string does not imply the least restrictive in the other, for there to be no implications between the two strings. In the examples, assume that there is only one action in each action set K_i, unless it is stated to the contrary. Recall we let K_i represent the action set and the number in it. Also, $S = \{1,2,3,\dots\}$.

<u>Proof of (b)</u> <u>C2 $\not\Rightarrow$ C1, C3 $\not\Rightarrow$ C1</u> Take $P_{11} = P_{12} = \frac{1}{2}$, $P_{22} = P_{23} = \frac{1}{2}$, $P_{i1} = P_{i3} = \frac{1}{2}$, $i \geq 3$, which does not satisfy C1 but satisfies C2 and, since $(P_{i1})^2 \geq \frac{1}{4}$, C3.

<u>C4 $\not\Rightarrow$ C2, C4 $\not\Rightarrow$ C3</u> $K_2 = 2$ with $P_{12} = 1 = P_{i1}$, $i \geq 3$, $P^1_{22} = 1$, $P^2_{21} = P^2_{23} = \frac{1}{2}$. This satisfies C4, since $(P_1)^2_{i2} = 1$, $(P^2)^3_{i1} \geq \frac{1}{4}$, but C2 is not satisfied, and as $((P^1)^2P^2)^n_{i1} = ((P^1)^2P^2)^n_{i3} = \frac{1}{2}$, while $(P^1)^n_{i2} = 1$ neither is C3.

<u>C5 $\not\Rightarrow$ C4</u> $P_{2i,1} = P_{2i+1,2} = 1$, $\forall$ i. C5 is satisfied with $K = \{1,2\}$, but there is no state s such that $P^n_{is} \geq \alpha$, $\forall$ i, for any possible n.

<u>C6 $\not\Rightarrow$ C5</u> $P_{1i} = 2^{-i}$, $i \geq 1$, $P_{i,i-1} = 1$, $i \geq 2$. $\pi_{ij}(P)$ is such that $\pi_{i1}(P) = \sum_{s=1}^{\infty} \frac{1}{s}(\frac{1}{2})^s = \log 2$, $\pi_{ij}(P) = \sum_{s=j}^{\infty} \frac{1}{s}(\frac{1}{2})^s \leq \frac{2}{j2^j}$. Thus for any $\varepsilon > 0$, choose $F = \{1,2,\ldots,N\}$ where $\sum_{j=N}^{\infty} 2/j2^j \leq 4/N2^N \leq \varepsilon$ and so $\sum_{j\in F} \pi_{ij}(P) \geq 1-\varepsilon$. However, for any finite set F, let $k^* = \max\{k | k \in F\}$, then for $i > n+k^*$ $P^n_{ij} = 0$, $\forall$ j F, so C5 is not satisfied.

<u>C7 $\not\Rightarrow$ C6</u> $P_{ii} = 1$, so $\pi_{ii}(P) = 1$ and C7 is satisfied, since *P* consists of only one element, but there are an infinite number of closed communicating classes, each consisting of one state only.

<u>C8 $\not\Rightarrow$ C15</u> (<u>so C8 $\not\Rightarrow$ C3</u>). Let $P_{12} = 1 = P_{i1}$, $i \geq 2$, $\mu_{i1} = 2$, $\forall$ i, but P is not periodic.

<u>C2 $\not\Rightarrow$ C9</u> (<u>so C4 $\not\Rightarrow$ C3</u>) $K_1 = K_2 = K_3 = 2$, with $P^1_{11} = P^1_{13} = \frac{1}{2}$, $P^2_{11} = P^2_{12} = \frac{1}{2}$, $P^1_{21} = P^1_{22} = P^2_{22} = P^2_{23} = \frac{1}{2}$, $P^1_{31} = P^1_{33} = P^2_{32} = P^2_{33} = \frac{1}{2}$, $P_{i1} = P_{i2} = P_{i3} = \frac{1}{3}$, $i > 3$. C2 holds with

$\alpha = \frac{1}{3}$. Suppose $r_i^k = 0$, $\forall$ i,k, then in C9 it does not matter whether we choose 1,2 or 3 as the special state, as they are symmetric. If we choose 1, we require the y_i to satisfy $y_2 \geq 2+y_3$, and $y_3 \geq 2+y_2$, which is not possible. The same inequalities hold if we take the state to be $\{i\}$, $i \geq 4$.

<u>C14 $\not\Rightarrow$C7</u> <u>(so C14 $\not\Rightarrow$C8)</u> $P_{i1}^1 = 1$, $r_i^1 = 0$, $P_{ii+1}^2 = 1$, $r_i^2 = 1$, satisfies C14 but not C7.

<u>C12 $\not\Rightarrow$C14</u> Take $r_i = 0$, $\forall$ i, $P_{ii+1} = 1$, $\forall$ i.

<u>C9 $\not\Rightarrow\tilde{C}$9</u> (even if costs are bounded) Let $r_i^k = 0$, $P_{io} = 1/i$, $P_{ii} = 1-1/i$, so Liapunov function equations become $1 \leq y_i/i$. So $y_i \geq i$ and these are finite but not bounded.

<u>C15 $\not\Rightarrow$C4</u> is trivial, since C15 only depends on the transition matrix of the optimal policy, while C4 is a condition on all transition matrices.

<u>C5 $\not\Rightarrow$C10</u> Let $P_{i1} = P_{ii} = \frac{1}{2}$, which satisfies C5 with $\alpha = \frac{1}{2}$, $F = \{1\}$, but not C10 for $\varepsilon < \frac{1}{2}$.

<u>C11 $\not\Rightarrow$C5</u>, <u>C11 $\not\Rightarrow$C13</u> C11 (and C12) cannot imply anything which does not depend on the costs, because if we put r_i^k = 0, $\forall$ $i\in S$, $k\in K_i$, they hold no matter how strange are the transition matrices.

<u>C17 $\not\Rightarrow$C16</u> $P_{2i,2i-1} = 1 = P_{2i+1,1}$, $P_{1i} = 2^{-i}$, $i \geq 1$. Thus $(P_{i1})^2 \geq \frac{1}{4}$, and it is a communicating system, which is recurrent, but if $\sum_{j\in K} P_{2i,j} > 1-\varepsilon$ for all i, then K must contain all odd-numbered integers, so C16 fails.

<u>C18 $\not\Rightarrow$C17</u> Let $P_{1i} = 2^{-i}$, $i \geq 1$, $P_{i,i-1} = 1$, $i \geq 2$ (as in C6 $\not\Rightarrow$C5). This is recurrent and a communicating system, and the rest follows as in the proof that C6 $\not\Rightarrow$C5.

<u>C19 $\not\Rightarrow$C18, C19 $\not\Rightarrow$C20, C19 $\not\Rightarrow$C13</u> are all trivial. Let K_i = 2, $\forall$ $i\in S$, with $P_{ii+1}^1 = 1$, $\forall$ i, $P_{ii-1}^2 = 1$, $i \geq 2$, $P_{11}^2 = 1$.

This is a communicating system, but $\pi(P^1)_{ij} = 0$, $\forall$ i,j and $\inf_{P \in P} \mu_{ij}(P) = |i-j|$ which is not bounded for all i and j.

C7 $\not\Rightarrow$ C20 $P_{i1} = 1$, $\forall$ i is trivially a counter-example.

We now look at over forty more counter-examples which show there can be no more equivalences in the diagram.

C1 $\not\Rightarrow$ C10 and C1 $\not\Rightarrow$ C19 $P_{i1} = \alpha$, $P_{ii} = 1-\alpha$, $i \geqslant 2$, $P_{11} = 1$ cannot satisfy C10 if we take $\varepsilon < 1-\alpha$. It is not a communicating system. So C19 fails.

C1 $\not\Rightarrow$ C21 Choose $P_{i1} = \alpha$, $P_{ii^2} = 1-\alpha$, $\forall$ $i \in S$, again satisfies C1 trivially but not C21.

C2 $\not\Rightarrow$ C9 was already proved by the example that showed C4 $\not\Rightarrow$ C3.

C3 $\not\Rightarrow$ C2   Let $P_{12} = P_{13} = \frac{1}{2}$, $P_{21} = 1$, $P_{i1} = P_{i2} = \frac{1}{2}$, $i \geqslant 3$, this does not satisfy C2 but as $(P_{i3})^2 \geqslant \frac{1}{4}$, $\forall$ i, it satisfies C3.

C6 $\not\Rightarrow$ C12   Let $P_{ii-1} = 1$, $i > 1$, $P_{11} = 1$, $r_1 = 0$, $r_i = 1$, $i > 1$, then $\pi(P)_{i1} = 1$, $\forall$ i, so C6 holds. However, $v_\beta(i) = \frac{1-\beta^i}{1-\beta}$, $v_\beta(0) = 0$, and the difference is bounded by $(1-\beta)^{-1}$. As $\beta \to \infty$, the difference is not uniformly bounded. So C12 does not hold.

C9 $\not\Rightarrow$ C12, C15 $\not\Rightarrow$ C12 The above example also satisfies C9 if we take {1} to be the special state. The equations become $y_1 \geqslant 2$, $y_2 \geqslant 1$, $y_i \geqslant 1+y_{i-1}$, $i \geqslant 3$ which has a solution $y_1 = 2$, $y_i = i-1$, $i \geqslant 2$. This also satisfies (b) and (c) or C9. Trivially, C15 is satisfied as there is only one closed communicating class and $\pi(P)_{i1} = 1$, $\forall$ i.

C9 $\not\Rightarrow$ C7   Since C9 allows unbounded costs, while the C7 has implicit in it the requirement that costs be bounded. However, if we restrict C9 to bounded costs, then it implies C7 as follows. C9 with bounded costs means that for each P there is only one closed communication class and that is positive recurrent (see Lemmas 2.7, 5.3 of [13]). This means that $\sum_{j \in S} \pi_{ij}(P) = 1$, while Lemma 10.2 of [13] with the fact that each P only has one closed class give the continuity condition of C7.

C10 $\not\Rightarrow$C15 $P_{12} = 1$, $P_{i1} = 1$, $i \geq 2$, satisfies C10 with $F = \{1,2\}$, which is also the only closed class, but it is periodic, so C15 fails.

C10 $\not\Rightarrow$C9, C10 $\not\Rightarrow$C19 The example which proved that C2 $\not\Rightarrow$ C9 works for these two cases. Let $K_1 = K_2 = K_3 = 2$, $P^1_{11} = P^1_{13} = \frac{1}{2}$, $P^2_{11} = P^2_{12} = \frac{1}{2}$, $P^1_{21} = P^1_{22} = P^2_{22} = P^2_{23} = \frac{1}{2}$, $P^1_{31} = P^1_{33} = P^2_{32} = P^2_{33} = \frac{1}{2}$, $P_{i1} = P_{i2} = P_{i3} = \frac{1}{3}$, $i \geq 3$. This satisfies C10 with $F = \{1,2,3\}$, and there is only one closed communicating class for each policy, but as in C2 $\not\Rightarrow$ C9, there is no Liapunov function. Also, as we cannot get to any state i, $i > 3$, C19 is not satisfied.

C10 $\not\Rightarrow$C14 Let $K_2 = \{k|k \geq 2, k \text{ integer}\}$. This is the only counter-example where the action space is not compact. Let $P_{i2} = 1$, $\forall\, i > 3$, $P^k_{22} = 1-1/k$, $P^k_{21} = 1/k$, $P_{11} = 1$, $r_i = 0$, $i > 3$, $r_1 = 1$, $r^2_2 = -1$ and $r^k_2 = (2k-2/2k-1)r^{k-1}_2$. Then C10 holds for all k with $F = \{1,2\}$, and there is only one closed communicating class {1}. However, the values of the rewards are chosen, so that as $\beta \to 1$ the optimal policy is increasing in k towards infinity. Since $\mu_{21}(P_k) = k$, this cannot be bounded and so C14 fails.

C10 $\not\Rightarrow$C21 Let $P_{i1} = 1-1/i^2$, $P_{ii^2} = 1/i^2$, $i \geq 2$, $P_{11} = 1$. Then C10 holds where, if $n = [1/\varepsilon]+ 1$, $F = \{1,2,\ldots,n\}$ and the only closed class is {1}. However, C21 fails.

C13 $\not\Rightarrow$C7, C13 $\not\Rightarrow$C5, C13 $\not\Rightarrow$C15, C13 $\not\Rightarrow$C9 Let K_i be countable for all i with $P^o_{ii+1} = 1$, $\forall\, i\in S$, $P^j_{ij} = 1$, $\forall\, i,j\in S$, $j\in K_i$. Then $\inf_{P\in P} \mu_{ij}(P) = 1$, $\forall\, i,j$, but $\pi(P^o)_{ij} = 0$, $\forall\, i,j$. So C7 fails. C5 also fails, since $\sum_{s\in F} (P^{\underline{j}})^n_{is} \neq 0$, where $P^{\underline{j}}$ means that j is choice in all K_i, only if $j\in F$. If it is to be positive for all $P^{\underline{j}}$, then F must be infinite. Moreover, the fact that $\pi(P^o)_{ij} = 0$, $\forall\, i,j$ means C15 is also not satisfied. Lastly, look at C9. Whichever state we choose as special, say $\{n\}$, for $i,j > n$, the equations become, using $P^{\underline{i}}$ and $P^{\underline{j}}$, that $y_i \geq y_j+1$ and $y_j \geq y_i+1$, which cannot be satisfied.

<u>C13 $\not\Rightarrow$C18, C13 $\not\Rightarrow$C20, C13 $\not\Rightarrow$C21, C13 $\not\Rightarrow$C14</u> We use the same example as above. C18 does not hold, since $\pi(P^j)_{ij} = 1$, and for $\sum_{j\in F} \pi(P)_{ij} \geq 1-\varepsilon$ for all P means F must contain all states. C20 fails because, for the policy P^j where j is chosen in each K_i, the communicating class is just $\{j\}$. Trivially also C21 is violated, and if we choose $r_i^i = 1$, $r_i^j = 0$, $j \neq i$, then the optimal policy to maximise reward is to choose action i in state i for all β. Thus $\mu_{ii}(P_{\beta_r}) = 1$, $\mu_{ij}(P_{\beta_r}) = \infty$ and C14 fails.

<u>C14 $\not\Rightarrow$C7</u> is shown by the same example that implied C14 $\not\Rightarrow$ C8

<u>C16 $\not\Rightarrow$C7</u>, C15 $\not\Rightarrow$C7, . Let $K_1 = K_2 = \{\theta | 0 \leq \theta \leq \frac{1}{2}\}$, where $P_{11}^\theta = 1-\theta$, $P_{12}^\theta = \theta$, $P_{21}^\theta = \theta$, $P_{2i}^\theta = (1-\theta)/2^{i-1}$, $i \geq 2$, $P_{i2}^\theta = 1$, $i \geq 3$. The system is a recurrent one in Bather's sense and if we chose N so that $(\frac{1}{2})^{N+1} < \varepsilon$, and let $F = \{1,..,N\}$, then $\sum_{j\in F} P_{ij}^\theta \geq 1-\varepsilon$ for all θ. So C16 holds. However, if we solve the equations for the invariant measure, we find $\pi(P^\theta)_{11} = \frac{2}{5}-\theta$, $\pi(P^o)_{11} = 1$, so C7 fails as $\pi(P)_{ij}$ is not continuous. If we put in rewards $r_i^\theta = -\theta$, $\forall$ $i\in S$, $\theta\in K_i$, the optimal policy will be $\theta = \frac{1}{2}$ for all states. This is aperiodic and has one closed communicating class, but C7 does not hold.

<u>C16 $\not\Rightarrow$C15</u> Let $K_1=K_2=3$, $K_i = 2$, $i \geq 3$ for all i with $P_{i1}^1 = 1-\frac{1}{i} = P_{i1}^2$, $P_{ii+1}^1 = P_{ii-1}^2 = 1/i$ for $i \geq 2$, $P_{12}^1 = P_{11}^2 = P_{11}^3 = 1$, $P_{11}^3 = P_{22}^3 = 1$. This is recurrent and satisfies $\sum_{j\in F} P_{ij} \geq 1-\varepsilon$ if $F = \{1,...,[1/\varepsilon] + 1\}$. If we define $r_i^1 = r_i^2 = 0$, $r_1^3 = r_2^3 = 1$, then optimal policy is action 3 in states 1 and 2, which has two closed communicating classes and so does not satisfy C15.

C16 $\not\Rightarrow$C20, C16 $\not\Rightarrow$C21 Let $K_i = 2$, $\forall$ i, and $P^1_{i1} = 1-2/i$, $P^1_{ii+1} = 1/i$, $P^1_{ii2} = 1/i$ for $i \geq 2$, $P^1_{11} = \frac{1}{2} = P^1_{21}$, $P^2_{ii-1} = 1/i$, $P^2_{i1} = 1-1/i$, $i \geq 2$, $P^2_{11} = 1$. This is recurrent and if $n > (2/\varepsilon)^2$ then if $F = \{1,\ldots,n\}$ $\sum_{j\in F} P_{ij} \geq 1-\varepsilon$ for all P. As $P_{ii2} \neq 0$, it does not satisfy C21, and as P^2 has a closed set consisting of {1} only it does not satisfy C20.

C16 $\not\Rightarrow$C9 We add some extra actions to the example which showed that C2 $\not\Rightarrow$C9. $K_1 = K_2 = 2$, $K_3 = 3$, $K_i = 3$, $i \geq 4$, so $P^1_{i1} = P^1_{i2} = P^1_{i3} = \frac{1}{3}$, $P^2_{ii+1} = 1/i = P^3_{i,i-1}$;$P^2_{i1} = P^3_{i1} = 1-1/i$, $i \geq 4$, $P^1_{11} = P^1_{13} = \frac{1}{2}$, $P^2_{11} = P^2_{12} = \frac{1}{2}$, $P^1_{21} = P^1_{22} = P^2_{22} = P^2_{23} = \frac{1}{2}$, $P^1_{31} = P^1_{33} = P^2_{32} = P^2_{33} = \frac{1}{2}$, $P^3_{34} = 1$. The extra actions make the system recurrent. Also, if $n > 1/\varepsilon$, then for $F = \{1,2,\ldots,n\}$, $\sum_{j\in F} P_{ij} \geq 1-\varepsilon$, $\forall$ $i\in S$, $P\in \mathcal{P}$ so C16 is satisfied. The argument in C2 $\not\Rightarrow$C9 shows that the example does not satisfy C9.

C20 $\not\Rightarrow$C18, C20 $\not\Rightarrow$C6, C20 $\not\Rightarrow$C9 Fisher [9] showed that if $\mathcal{P}$ is compact and each $P\in\mathcal{P}$ satisfies C20, then C9 holds if the action spaces are finite. However, an example due to Hordijk [13, p.53] shows that this is not the case if the action spaces are infinite. Let $K_i = \{1,2,3,\ldots,\infty\}$, $S = \{0,1,2\ldots\}$ with $P^k_{i+1i} = 1$, $i \geq 1$, $\forall$ $k\in K_i$, $P^k_{ij} = 2^{-j}$, $1 \leq j \geq k$, $P^k_{ij} = 2^{-k}(4^k-k)^{-1}$, $k+1 \leq n \leq 4^k$, $P^\infty_{ij} = 2^{-j}$. Then this satisfies C20 and Hordijk showed that $\mu(P^k)_{\theta,o} = 1 + \sum_{n=1}^{k} 2^{-n}n + 2^{-k}(4^k-k)^{-1} \sum_{n=k+1}^{4^k} n$. Similarly $\mu(P^k)_{ii} = i+1 + 2^i\mu(P\ )_{oo} + 1$ if $k > i$. Since $\mu(P^k)_{oo}$ tends to infinity as $k \to\infty$ and $\pi(P^k)_{ij} = 1/\mu(P^k)_{jj}$, it follows that we cannot find a finite set F such that $\varepsilon > 0$, $\sum_{j\in F} \pi(P)_{ij} \geq 1-\varepsilon$ for all P. Thus C18 and C6 fail. Following Hordijk [13] we can show that with $r^k_i = 0$ the solutions y_k of C9 must satisfy $y_i \geqslant \mu(P^k)_{oo}$, $\forall$ k. Since the right hand side tends to infinity, there are no finite y_i's satisfying C9.

C20 $\not\Rightarrow$C12 $P_{ii-1} = 1$, $i \geqslant 2$, $P_{ij} = 2^{-j}$, $j \geqslant 1$. This has all the states in one closed communicating class and $\mu(P)_{jj} \leqslant j+3(2^j)+1$ so is positive recurrent. Take $r_i = 0$, $i \leqslant N$ where $N \geqslant 3$ and $r_i = 1$, $i > N$, then for any discount factor $\beta, v_\beta(0) \leqslant \frac{1}{2}(1-\beta)^{-1}$ while as i tends to infinity $v_\beta(i)$ tends to $(1-\beta)^{-1}$. Thus $v_\beta(0)-v_\beta(i)$ is not uniformly bounded for any sequence $\beta_r \uparrow 1$.

C20 $\not\Rightarrow$C15 $P_{12} = 1$, $P_{ii-1} = \frac{3}{4}$, $P_{ii+1} = \frac{1}{4}$. By the argument used in Chung [2, p.25] this is positive recurrent with all states in one communicating class, but is periodic with period two, so C15 is not satisfied.

C20 $\not\Rightarrow$C21 Let $P_{ij} = 2^{-j}$, $\forall$ i,j$\in$S, then $\mu(P)_{ij} = 2^j$, so the system is positive recurrent, and all in one communicating class, but C21 does not hold.

C21 $\not\Rightarrow$C15, C21 $\not\Rightarrow$C7, C21 $\not\Rightarrow$C19 Let $P_{ii+1} = 1$, $\forall$ i$\in$S, so C21 holds with N = M = 1, but $\pi(P)_{ij} = 0$, $\forall$ i,j, so C7 and C15 fail. It is also not communicating.

C21 $\not\Rightarrow$C12, C21 $\not\Rightarrow$C9 Let $P_{ii} = 1$, $\forall$ i, $r_i = 1/i$. Then C12 fails because $v_\beta(i)-v_\beta(j) = (1/i-1/j)(1-\beta)^{-1}$ which is not uniformly bounded as β tends to 1. For C9 we require $1 + 1/i + y_i \leqslant y_i$, which is impossible.

7. CONCLUSIONS

(1) We have concentrated on the conditions put on the transition matrices. It should not be thought therefore that if there are two conditions which imply the same result, the more restrictive one is made redundant by the less restrictive one. These conditions can only be taken in conjunction with the conditions on the rewards, the action spaces and the various continuity conditions, and some of our conditions work with much weaker corresponding conditions on the rewards, etc., than others.

(2) For a particular example, it is a non-trivial exercise to check whether or not it satisfies some of the conditions. It would appear that conditions that can be checked easily (i.e. there exists a good computational algorithm) might be of more use than weaker, but more complicated conditions.

(3) We have not incorporated all the conditions on the transition matrices that ensure an optimal solution in the average cost case. However, we have tried to concentrate on those that hold for general Markov decision processes, rather than for specific types. For inventory problems in particular Wijngaard [29, 30] has some useful conditions which guarantee an optimal policy.

(4) It would be useful to obtain counter-examples or proofs of the connections between C12 and C11, C16 and C12, C14 and C11 and C2 and C14, and one with compact action spaces to that involving C10 and C14.

8. REFERENCES

1. BATHER, J.A.: Optimal decision procedures for finite Markov chains: Part II - communicating systems, Adv. Appl. Prob. 5, 521-540 (1973).
2. CHUNG, K.L.: Markov chains with stationary transition probabilities, Springer, Berlin (1960).
3. DENARDO, E.V.: Contraction mappings in the theory underlying dynamic programming, S.I.A.M. Rev. 9, 165-177 (1967).
4. DERMAN, C.: Denumerable state Markovian decision processes - average cost criterion, Ann. Math. Stat. 37, 1545-1554 (1966).
5. DERMAN, C, VEINOTT, A.F.: A solution to a countable system of equations arising in Markovian decision processes, Ann. Math. Stat. 38, 582-584 (1967).
6. FEDERGRUEN, A, HORDIJK, A., TIJMS, H.C.: Recurrence conditions in denumerable state Markov Decision Problems: Report BW 81/77 Mathematisch Centrum, Amsterdam (to appear in the Proceedings of the International Conference on Dynamic Programming, Vancouver, 1977, Academic Press).
7. FEDERGRUEN, A., HORDIJK, A., TIJMS, H.C.: A note on

simultaneous recurrence conditions on a set of denumerable stochastic matrices. To appear in J. Appl. Prob.

8. FEDERGRUEN, A, TIJMS, H.C.: The optimality equation in average cost denumerable state semi-Markov decision problems, recurring conditions and algorithms, J. Appl. Prob. 15, 356-373 (1978).

9. FISHER, L.: On recurrent denumerable decision processes, Ann. Math. Stat. 39, 424-434 (1968).

10. FISHER, L., ROSS, S.M.: An example in denumerable decision processes, Ann. Math. Stat. 39, 674-675 (1968).

11. FLYNN, J.: Conditions for the equivalence of optimality criteria in dynamic programming, Ann. Stat. 4, 936-953 (1976).

12. HARRISON, J.: Discrete dynamic programming with unbounded rewards, Ann. Math. Stat. 43, 636-644 (1972).

13. HORDIJK, A.: Dynamic programming and Markov potential theory, Mathematical Centre Tract 51, Amsterdam (1974).

14. HORDIJK, A., SCHWEITZER, P.J., TIJMS, H.C.: The asymptotic behaviour of the minimal total expected cost for the denumerable state Markov decision model, J. Appl. Prob. 12, 298-305 (1975).

15. HORDIJK, A., SLADKY, K.: Sensitive optimality criteria in countable state dynamic programming, Maths. of O.R. 2, 1-14 (1977).

16. HOWARD, R.A.: Dynamic programming and Markov processes, Technology Press, Cambridge, Massachusetts (1960).

17. KUSHNER, H.: Introduction to stochastic control, Holt, Rinehart and Winston, New York (1971).

18. LIPPMAN, S.A.: On dynamic programming with unbounded rewards, Mgmt. Sci. 21, 1225-1233 (1975).

19. ROBINSON, D.R.: Markov decision chains with unbounded costs and applications to the control of queues, Adv. Appl. Prob. 8, 159-176 (1976).

20. ROSS, S.M.: Non-discounted denumerable Markovian decision models, Ann. Math. Stat. 39, 412-423 (1968).

21. ROSS, S.M.: Applied Probability Models with Optimisation Applications, Holden-Day, San Francisco (1970).

22. TAYLOR, H.M.: Markovian sequential replacement processes,

Ann. Math. Stat. 36, 1677-1694 (1965).

23. THOMAS, L.C.: Connectedness conditions used in finite state Markov decision processes, to appear in J. Math. Anal. Appl.

24. VAN NUNEN, J.A.E.E.: Contracting Markov Decision Processes, Mathematical Centre, Tract 71, Amsterdam (1976).

25. VAN NUNEN, J.A.E.E., WESSELS, J.: Markov decision processes with unbounded rewards, Markov Decision Theory, pp. 1-13, ed. by H.C. Tijms, J. Wessels, Mathematical Centre Tract 93, Amsterdam (1977).

26. VAN NUNEN, J.A.E.E., WESSELS, J.: A note on dynamic programming with unbounded rewards, Mgmt. Sci. 24, 576-580 (1978).

27. WESSELS, J.: Markov programming by successive approximations with respect to weighted supremum norms, J. Math. Anal. Appl. 58, 326-335 (1977).

28. WHITE, D.J.: Dynamic programming, Markov chains and the method of successive approximations, J. Math. Anal. Appl. 6, 373-376 (1963).

29. WIJNGAARD, J.: Recurrence conditions and the existence of average optimal strategies for inventory problems in a countable state space, Dynamische Optimierung, Bonner Mathematische Schrift Nr. 78, Bonn (1977).

30. WIJNGAARD, J.: Stationary Markovian Decision Problems and Perturbation Theory of Quasi-compact Linear Operators, Maths. of O.R. 2, 91-102 (1977).

VECTOR-VALUED MARKOVIAN DECISION PROCESSES WITH COUNTABLE STATE SPACE

N. Furukawa

(Department of Mathematics, Faculty of Science, Kyushu University, Japan)

1. INTRODUCTION

Mitten [2] studied dynamic programming problems in which the objective is not necessarily real-valued, and gave the notion, a preference order relation. Sobel [3] investigated infinite horizon stochastic decision problems with a preference order criterion similar to Mitten's, showing that some of the typical techniques in usual Markov decision processes are applicable. However, they require the preference order relation to be a complete-preorder relation. Viswanathan, Aggarwal and Nair [4] studied multiple criteria Markov decision processes in a multi-objective linear programming method.

In this paper we shall be concerned with the optimization of Markovian decision processes on infinite horizon in which the objective is measured by a vector-valued additive utility. The optimization is made according to a partial-order criterion determined by a convex cone. The main aims are to derive a characterization of an optimal policy and to give an algorithm for finding all of the optimal policies. In the formulation of our decision model, the state space is a countable set and the action space is a finite set. In a numerical example given in the final section, however, the state space is also finite.

In Section 2 we shall give the formulation of the decision model and the partial order criterion through a convex cone. In Section 3, a policy-improvement method will be given. Section 4 will be devoted to deriving the characterization of an optimal policy. In Section 5 we shall give an algorithm for finding all of the optimal policies and solve a numerical example.

2. PRELIMINARIES

Let R^p denote the p-dimensional Euclidian space. Let $K \subset R^p$ be a closed convex cone with vertex at the origin $\{O\}$ of R^p. We assume that $K \cap (-K) = \{O\}$, where $(-K) = \{-x | x \in K\}$. For any $x, y \in R^p$, we write $x \leq y$ if $y-x \in K$. Then the relation $\leq$ is defined as a partial-order relation.

Let Ω be a nonempty subset of R^p. A point $x \in \Omega$ is said to be a maximal point of Ω, if it holds that

$$({}^{\forall} y \in \Omega)(x \leq y \rightarrow x = y).$$

Let $e(\Omega)$ denote the set of all maximal points of Ω. It may happen that $e(\Omega) = \phi$ even if Ω is not empty, e.g. the case when Ω is an open set. It is easily verified that if Ω is a non-empty compact subset of R^p then $e(\Omega) \neq \phi$.

For a point x and a set U in R^p, their sum is defined by $x+U = \{z | z = x+u, u \in U\}$. For any countable set X, $M^p(X)$ denotes the set of all bounded p-vector functions defined on X.

A vector-valued Markovian decision model is specified by five elements S, A, (q^a_{ij}), (r^a_{ij}) and β : S is a countable set, the set of states of a system, A is a finite set, the set of feasible actions, (q^a_{ij}) is a one-step transition probability from i to j by taking an action a, the law of motion of the system, (r^a_{ij}) is an element of $M^p(S \times A \times S)$, the one-step reward vector function and $0 \leqq \beta < 1$, the discount factor. In this specification we should notice that A is independent of a state and (q^a_{ij}) independent of a time. Let F denote the set of all mappings from S into A. Then a stationary policy is defined as an infinite sequence $\pi = \{f, f, \ldots\}$ consisting of a common element f of the set F, and will be also denoted by f^∞. In this paper we shall confine ourselves to the stationary policies, and then a stationary policy is called simply a policy.

For each starting state i and each policy f^∞, an expected discounted total reward is given by

$$I(f^\infty)(i) = E_i^{f^\infty}\left[\sum_{n=1}^{\infty} \beta^{n-1} r_n\right],$$

where $E_i^{f^\infty}$ is the conditional expectation operator induced from the policy f^∞ given the starting state i, and r_n the one-step reward vector at the n-th stage (r_n is actually independent of the stage number n). Viewing $I(f^\infty)(i)$ as a function of $i \in S$, $I(f^\infty)$ is a mapping from S into R^p, precisely $I(f^\infty) \in M^p(S)$ by virtue of the discount factor and of the boundedness of r_n. Throughout this paper we shall use the notation I(f) in place of $I(f^\infty)$ as far as we have no confusion. A policy $f^{*\infty}$ is called optimal, if it holds that

$$(\forall f \in F)(\forall i \in S)(I(f^*)(i) \lesssim I(f)(i) \rightarrow I(f^*)(i) = I(f)(i)).$$

Considering every starting state, our optimality criterion is found to be a generalization of the Pareto optimal criterion in ordinary mathematical programming. Namely, let K be in particular the nonnegative cone in R^p, that is, $K = \{x = (x^1, x^2, \ldots, x^p) \in R^p \mid x^i \geqq 0,\ i=1,2,\ldots,p\}$, then an optimal policy in our sense turns out Pareto optimal at every state i. There is another consideration about the optimality criterion. In the case when p=1, let K be the nonnegative half line in R^1. Then the relation $\lesssim$ becomes the total order relation in R^1 in usual sense, and our criterion implies that an optimal policy is optimal in the conventional sense as defined by Blackwell in [1].

3. IMPROVEMENT OF POLICIES

The aim in this section is to give a policy-improvement algorithm in a vector-wise method. In Markovian decision processes associated with a real-valued utility, Howard's policy improvement routine is well known. Our algorithm is an extension of his method to the vector-valued case. We shall first need to study some properties of an operator peculiar to Markovian decision processes.

3.1 Lemma

For any point $x \in R^p$ and any set $U \subset R^p$, we have the following equality:

$$e((x+K) \cap U) = (x+K) \cap e(U). \tag{3.1}$$

(We admit both sides of the quality to be empty.)

Proof. Let z be any point of $e((x+K) \cap U)$. Then we have $x \leqslant z$, since $z \in (x+K)$. Let u be any point of U satisfying $z \leqslant u$. By the transitive law of $\leqslant$, we have $x \leqslant u$. Hence $u \in ((x+K) \cap U)$, which implies $u = z$ since z is a maximal point of the set $(x+K) \cap U$. The last result shows that $z \in e(U)$. We have thus proved

$$e((x+K) \cap U) \subset (x+K) \cap e(U).$$

In order to prove the converse implication, let z be any point of $(x+K) \cap e(U)$. Then trivially $z \in (x+K) \cap U$. Let p be any point of $(x+K) \cap U$ satisfying $z \leqslant p$. Then we have $z = p$, since $z \in e(U)$. Thus we have $z \in e((x+K) \cap U)$. This completes the proof.

3.1 Definition

With any $f \in F$, we associate the operator T_f, mapping $M^p(S)$ into $M^p(S)$, defined by

$$T_f u(i) = \sum_j (r_{ij}^{f(i)} + \beta u(j)) q_{ij}^{f(i)}, \quad i \in S.$$

In particular, with any $a \in A$ associate the operator T_a defined by

$$T_a u(i) = \sum_j (r_{ij}^a + \beta u(j)) q_{ij}^a, \quad i \in S.$$

In what follows, for any u, $v \in M^p(S)$ we shall use $u \leqslant v$ to mean $u(i) \leqslant v(i)$ for all $i \in S$. Then we have the following lemma which is a generalization of Theorem 3 (a) in [1].

3.2 Lemma

For any $f \in F$, T_f is monotone in the sense that if $u \leq v$ then $T_f u \leq T_f v$.

<u>Proof</u>. First we claim that if $w \in M^p(S)$ and $w(i) \in K$ for all $i \in S$ then $\sum_i w(i) p_i \in K$ for any probability $(p_1, p_2, \ldots)$ on S. This statement can be easily proved by using the assumption that K is a closed convex cone.

Now let u and v be elements of $M^p(S)$ satisfying $u \leq v$. Since $v(j) - u(j) \in K$ for all j, by the claim stated above it holds that

$$\sum_j (v(j) - u(j)) q_{ij}^{f(i)} \in K \text{ for all } i \in S.$$

This implies that

$$\beta \sum_j (v(j) - u(j)) q_{ij}^{f(i)} \in K \text{ for all } i \in S,$$

for K is a cone with vertex at the origin. From the last result we have $T_f v(i) - T_f u(i) \in K$ for all $i \in S$. This completes the proof.

For a family $\{u_\lambda\}$ of functions in $M^p(S)$, we define the function $\bigcup_\lambda u_\lambda$ by the relation

$$(\bigcup_\lambda u_\lambda)(i) = \bigcup_\lambda \{u_\lambda(i)\} \text{ for } i \in S.$$

Let S^p denote the set of all non-empty subsets of R^p, and let $F^p(S)$ the set of all mappings from S into S^p, i.e. any element of $F^p(S)$ is a set-valued mapping from S into R^p. For any $U \in F^p(S)$ we define the function e(U) on S by the relation

$$e(U)(i) = e(U(i)) \text{ for } i \in S.$$

So, if the right-hand side is not empty for every i, then e(U) is an element of $F^p(S)$. For any $U, V \in F^p(S)$ we define the function $U \cap V$ on S by the relation

$$(U \cap V)(i) = U(i) \cap V(i) \text{ for } i \in S.$$

As we have defined in the beginning of Section 2, K is a closed convex cone in R^p. But we shall also express the element of $\mathcal{F}^p(S)$, which makes every $i \in S$ correspond to the cone K, by the same notation K.

In the following lemma we shall find that we do not encounter the undesirable situation that the sets of maximal points we are concerned with are empty.

3.3 Lemma

(a) For any $u \in M^p(S)$, we have $e(\bigcup_{a \in A} T_a u) \in \mathcal{F}^p(S)$.

(b) For any $f \in F$, we have $(I(f) + K) \cap e\left[\bigcup_{a \in A} T_a I(f)\right] \in \mathcal{F}^p(S)$.

Proof. (a) is trivial from the finiteness of A. (b) follows directly from the finiteness of A and Lemma 3.1.

For any $u \in M^p(S)$ and $U \in \mathcal{F}^p(S)$, we shall use $u \in U$ to mean $u(i) \in U(i)$ for all $i \in S$. A policy improvement algorithm is given by the following theorem.

3.1 Theorem

Let $\{f_n\}$ be defined in the iterative manner:

(i) Take an arbitrary map $f_0 \in F$.

(ii) For each $n \geqq 0$ select a map $f_{n+1} \in F$ such that

$$T_{f_{n+1}} I(f_n) \in (I(f_n) + K) \cap e\left[\bigcup_{a \in A} T_a I(f_n)\right].$$

Then we have

(a) $I(f_0) \leqslant I(f_1) \leqslant \ldots I(f_n) \leqslant I(f_{n+1}) \leqslant \ldots,$

(b) if for some N it occurs that

$$(I(f_N) + K) \cap e\left[\bigcup_{a \in A} T_a I(f_N)\right] = \{I(f_N)\},$$

then we cannot improve f_N any more along the chain $\{I(f_n)\}$ starting from the initial map f_o in (i).

<u>Remark</u>. The existence of such f_{n+1} in (ii) is assured by Lemma 3.3, (b).

<u>Proof</u>. (a) for each n, f_{n+1} satisfies that $T_{f_{n+1}} I(f_n) \in I(f_n)+K$, which implies $I(f_n) \lesssim T_{f_{n+1}} I(f_n)$. By Lemma 3.2 we get inductively

$$I(f_n) \lesssim T_{f_{n+1}} I(f_n) \lesssim T^2_{f_{n+1}} I(f_n) \lesssim \ldots . \tag{3.3}$$

By the component-wise convergence of the sequence $\{T^m_{f_{n+1}} I(f_n)\}_{m=1,2,\ldots}$ of vector-valued functions, the limit as $m\to\infty$ of $\{T^m_{f_{n+1}} I(f_n)\}$ does exist and is equal to $I(f_{n+1})$. By (3.3) and the closedness of the cone K, we have $I(f_n) \lesssim I(f_{n+1})$.

(b) Easy.

In the case when $p = 1$ and K is the nonnegative half line, the above theorem is the very algorithm Howard has given. In that case, if we encounter the situation (b) then the policy f_N^∞ is sure to be optimal. However, this is not true in the case when $p \geqq 2$, namely, the situation (b) does not necessarily imply f_N^∞ to be optimal. Then it may happen that there exists some policy better than f_N^∞ at some state.

4. CHARACTERIZATION OF OPTIMAL POLICIES

For the processes with real-valued discounted utility, it has been shown by Blackwell [1] that a policy is optimal if and only if the expected return from the policy is a unique solution to the optimality equation. In this section we shall define an optimality equation fundamental in our decision problem and then make clear the relation between the equation and an optimal policy.

4.1 Definition

We define the operator Φ from $M^p(S)$ into $F^p(S)$ by

$$\Phi u = e(\bigcup_{a\in A} T_a u).$$

An element $u^*\in M^p(S)$ is said to be a fixed point of Φ, if it holds that $u^*\in\Phi u^*$ (i.e., $u^*(i)\in(\Phi u^*)(i)$ for all $i\in S$).

We call the equation $u\in\Phi u$ defined on $M^p(S)$ an optimality equation. If u^* is a fixed point of Φ, then it is said that u^* satisfies the optimality equation.

It is assured by Lemma 3.3 (a) that Φ is actually a mapping from $M^p(S)$ into $F^p(S)$. The operator Φ has a unique fixed point in the case $p = 1$, but in the case $p > 2$ the fixed point of Φ is not necessarily unique.

4.2 Definition

A fixed point u^* of Φ is said to be maximal, if it holds that for any fixed point u of Φ,

$$(\forall i\in S)(u^*(i)\lesssim u(i)\rightarrow u^*(i) = u(i)).$$

We shall now give a characterization of an optimal policy in the following theorem.

4.1 Theorem

A policy $f^{*\infty}$ is optimal if and only if its return $I(f^*)$ is a maximal fixed point of Φ (in other words, $I(f^*)$ is a maximal solution to the optimality equation).

<u>Proof</u>. Let $f^{*\infty}$ be optimal. Let i_0 be any point of S. Let V_{i_0} denote the set $\bigcup_{a\in A} T_a I(f^*)(i_0)$. Then $I(f^*)(i_0)$ belongs to V_{i_0}, since $T_{f^*(i_0)} I(f^*)(i_0) = I(f^*)(i_0)$. We shall show that $I(f^*)(i_0)$ is a maximal point of V_{i_0}. Suppose p to be a point of V_{i_0} such that $I(f^*)(i_0)\lesssim p$. Then p can be written as $p = T_{\hat{a}} I(f^*)(i_0)$ for some $\hat{a}\in A$. Define $\hat{f}$ by

$$\hat{f}(i) = \begin{cases} \hat{a} & \text{if } i = i_0, \\ f^*(i) & \text{if } i \neq i_0, \end{cases}$$

then we have $I(f^*) \lesssim T_{\hat{f}} I(f^*)$. By Lemma 3.2, inductively we get

$$I(f^*) \lesssim T_{\hat{f}} I(f^*) \lesssim T_{\hat{f}}^2 I(f^*) \lesssim \ldots \underset{(\lesssim)}{\nearrow} I(\hat{f}). \qquad (4.1)$$

Since $f^{*\infty}$ is optimal, (4.1) implies that

$$I(f^*) = I(\hat{f}) = T_{\hat{f}} I(f^*).$$

Particularly at i_0 we have

$$I(f^*)(i_0) = T_{\hat{f}} I(f^*)(i_0) = T_{\hat{a}} I(f^*)(i_0) = p.$$

Consequently $I(f^*)(i_0)$ is a maximal point of V_{i_0}. This result holds for every $i_0 \in S$, namely $I(f^*)$ is a fixed point of Φ.
Next, we shall show that $I(f^*)$ is maximal. Let u be any fixed point of Φ. Then for every i there exists $a_i \in A$ such that $u(i) = T_{a_i} u(i)$. Let $\tilde{f}$ denote the mapping which makes i correspond to a_i for each i, then it holds that $u = T_{\tilde{f}} u$. So we have $u = I(\tilde{f})$. Suppose that for some i, $I(f^*)(i) \lesssim u(i) = I(\tilde{f})(i)$. But, since f^* is optimal, we have $I(f^*)(i) = I(\tilde{f})(i) = u(i)$. This implies $I(f^*)$ is maximal.

Conversely, let $I(f^*)$ be a maximal fixed point of Φ. Suppose that there exist $g \in F$ and $i \in S$ such that

$$I(f^*)(i) \lesssim I(g)(i). \qquad (4.2)$$

Let Γ^g denote the subset $\{I(f);\ I(g) \lesssim I(f),\ f \in F\}$ of $M^p(S)$. Recall that in Section 3 we defined the order $u \lesssim v$ in $M^p(S)$ by $u(i) \lesssim v(i)$ for all $i \in S$. The order $\lesssim$ is easily shown to be a partial order in $M^p(S)$.

We now show that Γ^g is an inductively ordered set with respect to the partial order $\leqslant$ in $M^p(S)$. Take a linearly-ordered subset $\{I(f_\nu)\}_{\nu\in\Lambda}$ of Γ^g arbitrarily. Then, for each fixed $i\in S$, $\{I(f_\nu)(i)\}_{\nu\in\Lambda}$ is a monotone net in R^p. This monotone net is bounded in norm by the boundedness of the reward vector. Consequently, by virtue of our assumption: $K\cap(-K) = \{0\}$, the net $\{I(f_\nu)(i)\}_{\nu\in\Lambda}$ converges in norm to a point $z_i\in R^p$ and moreover we have

$$I(f_\nu)(i)\leqslant z_i \quad \text{for all } \nu\in\Lambda. \tag{4.3}$$

Let u denote the mapping which makes i correspond to z_i given as above for each $i\in S$, then u belongs to $M^p(S)$ and satisfies

$$\lim_\nu I(f_\nu)(i) = u(i) \quad \text{for each } i\in S. \tag{4.4}$$

It is easily shown that Γ^g is a closed subset of $M^p(S)$ endowed with the point-wise convergence topology because the action space is finite. Hence (4.4) implies there exists an element $f^*\in F$ such that $u = I(f^*)$. From (4.3), then, it follows that

$$I(f_\nu)\leqslant I(f^*) \quad \text{for all } \nu\in\Lambda.$$

Trivially $I(f^*)$ belongs to Γ^g. Thus we have proved that any linearly-ordered subset of Γ^g has an upper bound in Γ^g, namely, Γ^g is an inductively ordered set.

By virtue of Zorn's lemma, let $I(\bar{g})$ be a maximal element of Γ^g with respect to the partial order in $M^p(S)$. Since $I(\bar{g})\in\Gamma^g$, we have

$$I(f)\leqslant I(\bar{g}). \tag{4.5}$$

We shall prove that $I(\bar{g})$ is a fixed point of Φ. Suppose the contrary, then there is an element i_0 of S so that

$$I(\bar{g})(i_0)\not\in e\Big[\bigcup_{a\in A} T_a I(\bar{g})(i_0)\Big].$$

By Lemma 3.3 (b) we have

$$(I(\bar{g})(i_o) + K)\cap e\left[\bigcup_{a\in A} T_a I(\bar{g})(i_o)\right] \neq \phi ,$$

which means there is a point $p_{i_o} \in e\left[\bigcup_{a\in A} T_a I(\bar{g})(i_o)\right]$ such that $I(\bar{g})(i_o) \leqslant p_{i_o}$. Then $p_{i_o} \neq I(\bar{g})(i_o)$, because the one belongs to $e\left[\bigcup_{a\in A} T_a I(\bar{g})(i_o)\right]$ and the other does not belong to it. The point p_{i_o} can be expressed as $p_{i_o} = T_{a_{i_o}} I(\bar{g})(i_o)$ for some $a_{i_o} \in A$. Let

$$\tilde{S} := \{i\in S;\ I(\bar{g})(i) \notin e\left[\bigcup_{a\in A} T_a I(\bar{g})(i)\right]\}.$$

For each $i\in\tilde{S}$ then we can find a_i as above. Define $\hat{f}$ by

$$\hat{f}(i) = \begin{cases} a_i & \text{if } i\in\tilde{S} \\ \bar{g}(i) & \text{if } i\notin\tilde{S}. \end{cases}$$

Then we have

$$I(\bar{g}) \leqslant T_{\hat{f}} I(\bar{g}). \tag{4.6}$$

In (4.6) we should notice that

$$I(\bar{g}) \neq T_{\hat{f}} I(\bar{g}), \tag{4.7}$$

because at $i\in\tilde{S}$ we have

$$T_{\hat{f}} I(\bar{g})(i) = T_{a_i} I(\bar{g})(i) = p_i \neq I(\bar{g})(i).$$

By (4.6) we have $I(\bar{g}) \leqslant I(\hat{f})$, which implies $I(\bar{g}) = I(\hat{f})$ because $I(\bar{g})$ is a maximal element of Γ^g. Hence, again from (4.6), we have $I(\bar{g}) = T_{\hat{f}} I(\bar{g})$, which contradicts (4.7).

Now at i chosen in the beginning of the proof of the "if" part, by (4.2) and (4.5) we have

$$I(f^*)(i) \lesssim I(g)(i) \lesssim I(\bar{g})(i). \tag{4.8}$$

But, since $I(\bar{g})$ is a fixed point of Φ as we have proved and $I(f^*)$ is a maximal fixed point of Φ, (4.8) implies

$$I(f^*)(i) = I(\bar{g})(i) = I(g)(i).$$

Hence $f^{*\infty}$ is optimal. This completes the proof.

<u>Remark</u>. If in Theorem 3.1 the case (b) occurs at N, then it holds that $I(f_N) \in e\left[\bigcup_{a \in A} T_a I(f_N)\right] = \Phi I(f_N)$, which means $I(f_N)$ is a fixed point of Φ. On account of this, the algorithm given by Theorem 3.1 can be interpreted as one for finding fixed points of Φ as well as for improving policies.

5. ALGORITHM AND NUMERICAL EXAMPLE

After the consideration of results in Section 3 and 4 we can give an algorithm for finding all optimal policies as follows:

Step 1. Take an arbitrary map $f_0 \in F$.

Step 2. For each $n \geqq 0$ compute $I(f_n)$ and choose a map $f_{n+1} \in F$ such that

$$T_{f_{n+1}} I(f_n) \in (I(f_n) + K) \cap e\left[\bigcup_{a \in A} T_a I(f_n)\right].$$

Step 3. If for some N it occurs that

$$(I(f_N) + K) \cap e\left[\bigcup_{a \in A} T_a I(f_N)\right] = \{I(f_N)\},$$

then go to either one of the two cases:

(i) If $T_a I(f_N) \lesssim I(f_N)$ for all $a \in A$, then stop. (In this case f_N is a unique optimal policy.)

(ii) If there are i_0 and $a_{i_0} \in A$ such that

$$T_{a_{i_O}} I(f_N)(i_O) \nleq I(f_N)(i_O),$$

then delete all maps $f \in F$ satisfying $T_f I(f_N) \leqslant I(f_N)$ but not equal to f_N and define g, so that $g \neq f_N$, by

$$g(i_O) = a_{i_O} \text{ or } f_N(i_O) \text{ if } T_{a_{i_O}} I(f_N)(i_O) \nleq I(f_N)(i_O)$$

$$(5.1)$$

$$g(i) = f_N(i) \qquad \text{if otherwise.}$$

Step 4. In Step 1 put g in place f_O, and go to Step 2 and then to Step 3. (We shall call g given by (5.1) a doubtful map around the fixed point $I(f_N)$ or a doubtful map around f_N.)

Step 5. Continue the above steps until we finish the routine on all doubtful maps around the fixed points.

Step 6. (i) If there remains any uninvestigated map $h \in F$ in the course of Steps 1-5, then start again Step 1 putting h in place of f_O.

(ii) If there remains no map uninvestigated, then go to Step 7.

Step 7. Select all maximal ones among the fixed points obtained in the above steps. The policies corresponding to the selected maximal ones are all optimal, and there is no optimal policy other than these policies.

Now consider a three-state, three-objective Markov decision process. Let there be five alternatives of actions in each state and let the transition probabilities be

$$(q^1_{ij}) = \begin{pmatrix} 0.42 & 0.36 & 0.22 \\ 0.52 & 0.17 & 0.31 \\ 0.61 & 0.28 & 0.11 \end{pmatrix}$$

$$(q_{ij}^2) = \begin{pmatrix} 0.29 & 0.51 & 0.20 \\ 0.38 & 0.25 & 0.37 \\ 0.23 & 0.46 & 0.31 \end{pmatrix}$$

$$(q_{ij}^3) = \begin{pmatrix} 0.11 & 0.72 & 0.17 \\ 0.23 & 0.61 & 0.16 \\ 0.18 & 0.58 & 0.24 \end{pmatrix}$$

$$(q_{ij}^4) = \begin{pmatrix} 0.08 & 0.23 & 0.69 \\ 0.10 & 0.41 & 0.49 \\ 0.12 & 0.63 & 0.25 \end{pmatrix}$$

$$(q_{ij}^5) = \begin{pmatrix} 0.33 & 0.33 & 0.34 \\ 0.53 & 0.22 & 0.25 \\ 0.56 & 0.14 & 0.30 \end{pmatrix} .$$

The reward vectors are

$$r_{11}^1 = \begin{pmatrix} 12 \\ 7 \\ -5 \end{pmatrix} \qquad r_{12}^1 = \begin{pmatrix} 14 \\ 6 \\ 8 \end{pmatrix} \qquad r_{13}^1 = \begin{pmatrix} 28 \\ 35 \\ 4 \end{pmatrix}$$

$$r_{21}^1 = \begin{pmatrix} 35 \\ -8 \\ 20 \end{pmatrix} \qquad r_{22}^1 = \begin{pmatrix} -11 \\ 15 \\ 32 \end{pmatrix} \qquad r_{23}^1 = \begin{pmatrix} 17 \\ 22 \\ -2 \end{pmatrix}$$

$$r_{31}^1 = \begin{pmatrix} 42 \\ -3 \\ 0 \end{pmatrix} \qquad r_{32}^1 = \begin{pmatrix} 32 \\ 28 \\ 25 \end{pmatrix} \qquad r_{33}^1 = \begin{pmatrix} -6 \\ 25 \\ 15 \end{pmatrix}$$

$$r_{11}^2 = \begin{pmatrix} 25 \\ 40 \\ 0 \end{pmatrix} \quad r_{12}^2 = \begin{pmatrix} -5 \\ 15 \\ 25 \end{pmatrix} \quad r_{13}^2 = \begin{pmatrix} 18 \\ -12 \\ 15 \end{pmatrix}$$

$$r_{21}^2 = \begin{pmatrix} 45 \\ 29 \\ 28 \end{pmatrix} \quad r_{22}^2 = \begin{pmatrix} 28 \\ 33 \\ 32 \end{pmatrix} \quad r_{23}^2 = \begin{pmatrix} 14 \\ 26 \\ -4 \end{pmatrix}$$

$$r_{31}^2 = \begin{pmatrix} 20 \\ -18 \\ 30 \end{pmatrix} \quad r_{32}^2 = \begin{pmatrix} 36 \\ 26 \\ -12 \end{pmatrix} \quad r_{33}^2 = \begin{pmatrix} 24 \\ 42 \\ 0 \end{pmatrix}$$

$$r_{11}^3 = \begin{pmatrix} 53 \\ 58 \\ 16 \end{pmatrix} \quad r_{12}^3 = \begin{pmatrix} 40 \\ 67 \\ -3 \end{pmatrix} \quad r_{13}^3 = \begin{pmatrix} -8 \\ 83 \\ 38 \end{pmatrix}$$

$$r_{21}^3 = \begin{pmatrix} 31 \\ 14 \\ 24 \end{pmatrix} \quad r_{22}^3 = \begin{pmatrix} 23 \\ -3 \\ -2 \end{pmatrix} \quad r_{23}^3 = \begin{pmatrix} 2 \\ 0 \\ 1 \end{pmatrix}$$

$$r_{31}^3 = \begin{pmatrix} 18 \\ 27 \\ 0 \end{pmatrix} \quad r_{32}^3 = \begin{pmatrix} 25 \\ 41 \\ -2 \end{pmatrix} \quad r_{33}^3 = \begin{pmatrix} 10 \\ 19 \\ 17 \end{pmatrix}$$

$$r_{11}^4 = \begin{pmatrix} 26 \\ 49 \\ 15 \end{pmatrix} \quad r_{12}^4 = \begin{pmatrix} 17 \\ 28 \\ 23 \end{pmatrix} \quad r_{13}^4 = \begin{pmatrix} 0 \\ -3 \\ 0 \end{pmatrix}$$

$$r_{21}^4 = \begin{pmatrix} -5 \\ 16 \\ 42 \end{pmatrix} \quad r_{22}^4 = \begin{pmatrix} 28 \\ 18 \\ 38 \end{pmatrix} \quad r_{23}^4 = \begin{pmatrix} 42 \\ -14 \\ 15 \end{pmatrix}$$

$$r_{31}^4 = \begin{pmatrix} 16 \\ 28 \\ 2 \end{pmatrix} \quad r_{32}^4 = \begin{pmatrix} 32 \\ 52 \\ 30 \end{pmatrix} \quad r_{33}^4 = \begin{pmatrix} 40 \\ 16 \\ 17 \end{pmatrix}$$

$$r_{11}^5 = \begin{pmatrix} 50 \\ 27 \\ 24 \end{pmatrix} \quad r_{12}^5 = \begin{pmatrix} -10 \\ -4 \\ -12 \end{pmatrix} \quad r_{13}^5 = \begin{pmatrix} 26 \\ -3 \\ 38 \end{pmatrix}$$

$$r_{21}^5 = \begin{pmatrix} 35 \\ 42 \\ 52 \end{pmatrix} \quad r_{22}^5 = \begin{pmatrix} -2 \\ 31 \\ 48 \end{pmatrix} \quad r_{23}^5 = \begin{pmatrix} 52 \\ 28 \\ 37 \end{pmatrix}$$

$$r_{31}^5 = \begin{pmatrix} 18 \\ 16 \\ 0 \end{pmatrix} \quad r_{32}^5 = \begin{pmatrix} 30 \\ 28 \\ 41 \end{pmatrix} \quad r_{33}^5 = \begin{pmatrix} 22 \\ 15 \\ 25 \end{pmatrix}.$$

Let K be the nonnegative orthant in R^3, that is,

$$K = R_+^3 = \{(x_1,x_2,x_3) \in R^3;\ x_1 \geqq 0,\ x_2 \geqq 0,\ x_3 \geqq 0\}.$$

Let $\beta = 0.8$.

We shall write a stationary policy f^∞ by a simple notation f. Then all possible policies are the following 125 types numbered in lexicographic order:

$f_1(1)=1$ $f_2(1)=1$ $f_3(1)=1$ $f_4(1)=1$ $f_5(1)=1$

$f_1(2)=1$ $f_2(2)=1$ $f_3(2)=1$ $f_4(2)=1$ $f_5(2)=1$

$f_1(3)=1$, $f_2(3)=2$, $f_3(3)=3$, $f_4(3)=4$, $f_5(3)=5$,

$f_6(1)=1$ $f_7(1)=1$ $f_8(1)=1$ $f_9(1)=1$ $f_{10}(1)=1$

$f_6(2)=2$ $f_7(2)=2$ $f_8(2)=2$ $f_9(2)=2$ $f_{10}(2)=2$

$f_6(3)=1$, $f_7(3)=2$, $f_8(3)=3$, $f_9(3)=4$, $f_{10}(3)=5$.

. . . .

. . . .

$f_{121}(1)=5$ $f_{122}(1)=5$ $f_{123}(1)=5$ $f_{124}(1)=5$ $f_{125}(1)=5$

$f_{121}(2)=5$ $f_{122}(2)=5$ $f_{123}(2)=5$ $f_{124}(2)=5$ $f_{125}(2)=5$

$f_{121}(3)=1$ $f_{122}(3)=2$ $f_{123}(3)=3$ $f_{124}(3)=4$ $f_{125}(3)=5$

For the above example the numerical process is given as follows. In this process, for any $u,v \in M^p(S)$ $u<v$ means that $u \leq v$ and $u \neq v$, and for any $f,g \in F$ $f<g$ means $I(f)<I(g)$. If $I(f)$ is a fixed point of Φ, then we shall say f is a fixed point.

(i) Take f_1. Then $f_1<f_{74}$.

(ii) f_{74} is a fixed point. $f_2,f_3,f_4,f_5,f_7,f_8,f_9,f_{10},f_{12}$, $f_{13},f_{14},f_{15},f_{17},f_{18},f_{19},f_{20},f_{22},f_{23},f_{24},f_{25},f_{52},f_{53},f_{54},f_{55},f_{57}$, $f_{58},f_{59},f_{60},f_{62},f_{63},f_{64},f_{65},f_{67},f_{68},f_{69},f_{70},f_{72},f_{73},f_{75},f_{77}$, $f_{78},f_{79},f_{80},f_{82},f_{83},f_{84},f_{85},f_{87},f_{88},f_{89},f_{90},f_{92},f_{93},f_{94},f_{95}$, f_{97},f_{98},f_{99} and f_{100} are deleted. f_{46} and f_{121} are the doubtful maps around f_{74}.

(iii) $f_{46}<f_{49}$.

(iv) f_{49} is a fixed point. $f_{26},f_{27},f_{28},f_{29},f_{30},f_{36},f_{37},f_{38}$, $f_{39},f_{40},f_{47},f_{48}$ and f_{50} are deleted. $f_{34},f_{44},f_{109},f_{119}$ and f_{124} are the doubtful maps around f_{49}.

(v) $f_{121}<f_{124}$

(vi) f_{124} is a fixed point. $f_{101},f_{102},f_{103},f_{104},f_{105},f_{106}$, $f_{107},f_{108},f_{109},f_{110},f_{111},f_{112},f_{113},f_{114},f_{115},f_{122},f_{123}$ and f_{125} are deleted.

(vii) f_{34} is a fixed point. f_{31},f_{32},f_{33} and f_{35} are deleted.

(viii) f_{44} is a fixed point. f_{41},f_{42},f_{43} and f_{45} are deleted.

(ix) f_{119} is a fixed point. $f_{116}, f_{117}, f_{118}$ and f_{120} are deleted.

All doubtful maps have been investigated. But there are several maps which have not been investigated. Take f_6 among these maps.

(x) $f_6 < f_{71}$.

(xi) f_{71} is a fixed point. $f_{11}, f_{21}, f_{51}, f_{56}, f_{61}, f_{76}, f_{81}, f_{86}$ and f_{96} are deleted. f_{66} is the doubtful map around f_{71}.

(xii) f_{66} is a fixed point. f_{16} and f_{91} are deleted. Thus all maps have been investigated. The values of the fixed points we have obtained are as follows:

$$I(f_{34})(1)=\begin{pmatrix}104.54\\130.86\\89.54\end{pmatrix},\ I(f_{34})(2)=\begin{pmatrix}124.92\\143.70\\91.63\end{pmatrix},\ I(f_{34})(3)=\begin{pmatrix}131.34\\156.39\\97.71\end{pmatrix},$$

$$I(f_{44})(1)=\begin{pmatrix}119.51\\85.37\\111.65\end{pmatrix},\ I(f_{44})(2)=\begin{pmatrix}146.95\\76.76\\124.17\end{pmatrix},\ I(f_{44})(3)=\begin{pmatrix}147.02\\108.75\\120.86\end{pmatrix},$$

$$I(f_{49})(1)=\begin{pmatrix}102.41\\138.51\\140.26\end{pmatrix},\ I(f_{49})(2)=\begin{pmatrix}121.79\\154.96\\166.26\end{pmatrix},\ I(f_{49})(3)=\begin{pmatrix}129.12\\164.40\\150.81\end{pmatrix},$$

$$I(f_{66})(1)=\begin{pmatrix}163.69\\143.22\\75.12\end{pmatrix},\ I(f_{66})(2)=\begin{pmatrix}162.53\\82.30\\91.39\end{pmatrix},\ I(f_{66})(3)=\begin{pmatrix}164.70\\106.45\\72.13\end{pmatrix},$$

$$I(f_{71})(1)=\begin{pmatrix}162.82\\237.00\\109.33\end{pmatrix},\ I(f_{71})(2)=\begin{pmatrix}161.33\\211.46\\138.51\end{pmatrix},\ I(f_{71})(3)=\begin{pmatrix}163.94\\188.36\\102.01\end{pmatrix},$$

$$I(f_{74})(1)=\begin{pmatrix}160.97\\253.21\\128.42\end{pmatrix},\ I(f_{74})(2)=\begin{pmatrix}159.38\\228.58\\158.67\end{pmatrix},\ I(f_{74})(3)=\begin{pmatrix}159.83\\224.54\\144.61\end{pmatrix},$$

$$I(f_{119})(1)=\begin{pmatrix}141.96\\74.17\\112.28\end{pmatrix},\quad I(f_{119})(2)=\begin{pmatrix}153.66\\73.41\\124.35\end{pmatrix},\quad I(f_{119})(3)=\begin{pmatrix}153.94\\105.30\\121.05\end{pmatrix}$$

$$I(f_{124})(1)=\begin{pmatrix}135.29\\115.52\\136.88\end{pmatrix},\quad I(f_{124})(2)=\begin{pmatrix}142.89\\140.20\\164.09\end{pmatrix},\quad I(f_{124})(3)=\begin{pmatrix}146.36\\152.34\\149.04\end{pmatrix}.$$

From the above results it can be seen that $I(f_{34})$, $I(f_{44})$ and $I(f_{119})$ are not maximal and only five maps $f_{49}, f_{66}, f_{71}, f_{74}, f_{124}$ give maximal points. Thus we obtain the conclusion that optimal policies are no other than $f_{49}, f_{66}, f_{71}, f_{74}$ and f_{124}.

6. REFERENCES

1. Blackwell, D. "Discounted Dynamic Programming", *Ann. Math. Statist.*, **36**, 226-235, (1965).

2. Mitten, L. G. "Preference Order Dynamic Programming", *Management Science*, **21**, 43-46, (1974).

3. Sobel, M. J. "Ordinal Dynamic Programming", *Management Science*, **21**, 967-975, (1975).

4. Viswanathan, B., Aggarwal, V. V. and Nair, K. P. K. "Multiple Criteria Markov Decision Processes", Multiple Criteria Decision Making, ed. by Starr and Zeleny, North-Holland Publ. Comp., 263-272, (1977).

ADAPTIVE DUAL CONTROL APPROACH TO MARKOVIAN DECISION PROCESSES AND ITS APPLICATION

H. Myoken

(Nagoya City University)

ABSTRACT

This paper discusses the application of decision models based on learning control approach to Markov processes in which the transition probabilities corresponding to alternative decisions are not known with certainty. The processes are assumed to be finite, discrete-time and stationary. The rewards are time discounted. We consider the uncertain transition probability represented by the appropriate set with given level which is computed from the estimation of transition probability matrix every time. The asymptotic effects of the max-max, max-min and Bayes optimal policies are examined. It will be shown that the result of the present study has succeeded in reducing the computational requirement of the algorithm when compared with other approaches. The optimal decision rules based on the algorithm presented here have two purposes that might be conflicting: one is to increase the gain; the other is to estimate the uncertain transition probability matrix. Thus the decision processes can be treated from the viewpoint of adaptive dual control processes. Finally, an illustrative example is numerically presented based on the approach proposed in the paper.

MARKOV DECISION PROCESSES WITH UNKNOWN TRANSITION LAW: THE AVERAGE RETURN CASE

K. M. van Hee

(Eindhoven University of Technology)

ABSTRACT

In this paper we consider some problems and results in the area of Markov decision processes with an incompletely known transition law. We concentrate here on the average return under the Bayes criterion. We discuss easy-to-handle strategies which are optimal under some conditions. For detailed proofs we refer to a monograph published by the author.

1. INTRODUCTION

In this paper we review a part of van Hee (1978a), a monograph dealing with Markov decision processes in discrete time, with an incompletely known transition law. All proofs of statements given here can be found in this monograph. Moreover this monograph contains results for the discounted return case; some of these results are reviewed in van Hee (1978b).

We do not bother about measure theoretic problems and therefore we assume all sets to be countable or sometimes even finite, however we remark that in van Hee (1978a) the problems are treated in a general measure theoretic setting.

In general adaptive control of a Markov decision process with respect to the average return criterion is not a trivial problem as will be demonstrated in example 4. In fact a lot of the hard problems of statistical sequential analysis may be formulated in this way. On the other hand if some recurrence conditions are fulfilled the problem seems to be trivial. Indeed, the decision maker may learn about the system until he has enough information about the unknown parameter and afterwards he solves as a non-adaptive problem, because the return obtained in any finite number of stages is not of influence on the average return. However, the situation of having "enough information" occurs only in trivial models and therefore this idea is correct only in the limit situation.

Proving rigorously this expected result requires a complex mathematical apparatus as will be presented in this paper.

In this paper we also present "easy-to-handle" strategies that are optimal with respect to the average return criterion and sometimes even with respect to the discounted return criterion as demonstrated in van Hee (1978a).

We start with a description of the model and we discuss some of its properties. A Markov decision process (MDP) with unknown transition law is specified by a 5-tuple

$$(X,\ A,\ \Theta,\ P,\ r) \tag{1.1}$$

where X is the *state space*, A the *action space*, Θ the *parameter space*, P a *transition probability* from $X \times A \times \Theta$ to X and r the *reward function* (i.e. $r: X \times A \to \hat{R}$, where $\hat{R}$ is the set of real numbers). (We assume r to be bounded, if X or A is countable.) The parameter $\theta \in \Theta$ is unknown to the decision maker. At each stage 0, 1, 2,... the decision maker chooses an action $a \in A$ where he may base his choice on the sequence of past states and actions.

A *strategy* π is a sequence $\pi = (\pi_0, \pi_1, \pi_2, \ldots)$ where π_0 is a transition probability from X to A and π_n a transition probability from $(X \times A)^n \times X$ to A ($n \geqq 1$).

The set of all strategies is denoted by Π.

A strategy is called *stationary* if there is a function $f : X \to A$ such that $\pi_n(\{f(x_n)\}|x_0, a_0, x_1, a_1, \ldots, x_n) = 1$.

According to the well-known Ionescu Tulcea theorem (cf. Neveu (1965)) we have for each starting state $x \in X$, each strategy $\pi \in \Pi$ and each parameter $\theta \in \Theta$ probability $P^{\pi}_{x,\theta}$ on

$$\Omega := (X \times A)^{\hat{N}} \qquad (\hat{N} := \{0,\ 1,\ 2, \ldots\}) \tag{1.2}$$

and a random process $\{(X_n, A_n),\ n \in \hat{N}\}$ where

$$X_n(\omega) := x_n,\quad A_n(\omega) := a_n \text{ if } \omega = (x_0, a_0, x_1, a_1, \ldots) \in \Omega \tag{1.3}$$

(The expectation with respect to $P^{\pi}_{x,\theta}$ is denoted by $E^{\pi}_{x,\theta}$)

The *average return* is defined by

$$g(x, \theta, \pi) := \liminf_{N\to\infty} \frac{1}{N} \sum_{n=0}^{N-1} E^{\pi}_{x,\theta} \left(r(X_n, A_n) \right) \qquad (1.4)$$

It only happens in non-interesting cases that there is a strategy $\pi' \in \Pi$ such that $g(x, \theta, \pi') \geq g(x, \theta, \pi)$ for all $x \in X$, $\theta \in \Theta$ and $\pi \in \Pi$.

So we cannot use this as an optimality criterion. We have chosen the Bayes criterion [for a motivation cf. van Hee (1978a)].

Fix some probability q on Θ. (Such a probability is called a *prior distribution*.) The *Bayesian average return* with respect to q is defined by

$$g(x, q, \pi) := \liminf_{N\to\infty} \frac{1}{N} \sum_{n=0}^{N-1} \left\{ \sum_{\theta} q(\theta)\, E^{\pi}_{x,\theta} \left(r(X_n, A_n) \right) \right\} \qquad (1.5)$$

(Note that the definitions (1.4) and (1.5) are consistent if we identify $\theta \in \Theta$ with the distribution that is degenerate at θ).

The set of all probabilities on Θ will be denoted by W. A strategy π' is called *ε-optimal* in $(x, q) \in X \times W$ if

$$g(x, q, \pi') \geq g(x, q, \pi) - \varepsilon \text{ for all } \pi \in \Pi \qquad (1.6)$$

(a 0-optimal strategy is simply called optimal).

The Bayes criterion allows us to consider the parameter as a random variable Z with range Θ and distribution q on Θ. On $\Theta \times \Omega$ we have the probability $P^{\pi}_{x,q}$ determined by

$$P^{\pi}_{x,q}\, (Z \in B, (X_0, A_0, X_1, A_1, \ldots) \in C) = \sum_{\theta \in B} P^{\pi}_{x,\theta}\, (C)\, q(\theta) \qquad (1.7)$$

for events C in Ω.

We compute the so-called *posterior distributions* Q_n of Z in the following way:

$$Q_n\ (B):=P^{\pi}_{x,q}\left(Z \varepsilon B \mid X_0, A_0, X_1, A_1, \ldots, X_n, A_n\right) \qquad (1.8)$$

(Note that Q_n is determined $P^{\pi}_{x,q}$ - a.s.)

A simple example may clarify these concepts. Consider a sequence of independent Bernoulli trials $X_1, X_2, X_3, \ldots$ with $\mathbb{P}_\theta\left[X_i = 1\right] = 1 - \mathbb{P}_\theta\left[X_i = 0\right] = \theta$. Let $\theta \varepsilon \Theta = [0,1]$ be the unknown parameter and let q be the prior distribution on Θ. In the Bayesian set up we may define a probability space $\hat{\Omega} := \Theta \times \{0,1\}^{\infty}$ and random variables $Z, X_1, X_2, X_3, \ldots$ on $\hat{\Omega}$ such that $Z(\omega) = \omega_0$, $X_i(\omega) = \omega_i$ $i = 1,2,3,\ldots$ and $\omega = (\omega_0, \omega_1\ \omega_2, \ldots) \in \hat{\Omega}$.

The conditional distribution Q_n of Z given $X_1, X_2, \ldots, X_n$ is computed by

$$Q_n(d\theta)(\omega) =$$

$$\prod_{i=1}^{n} \theta^{X_i(\omega)} \cdot (1-\theta)^{1-X_i(\omega)}\ q(d\theta) \Big/ \int \prod_{i=1}^{n} t^{X_i(\omega)} \cdot (1-t)^{1-X_i(\omega)}\ q(dt)$$

This conditional distribution is usually called the posterior distribution.

Define the probability $T_{x,a,x'}(q)$ on Θ by

$$T_{x,a,x'}\ (q)\ (\theta):=\frac{P(x' \mid x,a,\theta)\ q(\theta)}{\sum_{\theta'} p(x' \mid x,a,\theta')\ q(\theta')},\ (x,x' \varepsilon X,\ \theta \varepsilon \Theta,\ a \varepsilon A)$$

(1.9)

It is possible to choose versions of the posterior distributions such that $Q_0 = q$ and $Q_{n+1} = T_{X_n,A_n,X_{n+1}}(Q_n)$.

As indicated by Bellman (cf. Bellman (1961)) and proved in a very general setting in Rieder (1975) this decision model is equivalent to a MDP with a known transition law, specified by a 4-tuple

$$(X \times W,\ A,\ \overline{P},\ \overline{r}) \tag{1.10}$$

where $X \times W$ is state space, A the action space, $\overline{P}$ the transition law defined by

$$\overline{P}(x',\ T_{x,a,x'}(q)\,|\,x,q,a) \ := \sum_{\theta} q(\theta)\ P(x'\,|\,x,a,\theta) \tag{1.11}$$

and $\overline{r} : X \times W \times A \to \hat{R}$, the reward function, is defined by

$$\overline{r}(x,q,a) \ := r(x,a) \tag{1.12}$$

Note that the state (x,q) of the new model (1.10) consists of the original state $x \in X$ and the "information state" $q \in W$. It turns out that each state (x,q) and each strategy $\overline{\pi}$ for the new model, define a probability $\overline{P}^{\overline{\pi}}_{x,q}$ and a random process $\left\{(X_n,\ Q_n,\ A_n),\ n \in N\right\}$ on $\overline{\Omega} := (X \times W \times A)^{\hat{N}}$. Where $X_n(\overline{\omega}) := x_n$, $Q_n(\overline{\omega}) := q_n$ and $A_n(\overline{\omega}) := a_n$ where $\overline{\omega} = (x_0,q_0,a_0,x_1,q_1,a_1,\ldots) \in \overline{\Omega}$.

The original model (1.1) and the new model (1.10) have the following relationship:

$$E^{\pi}_{x,q}\left[r(X_n,A_n)\right] = \overline{E}^{\overline{\pi}}_{x,q}\left[\overline{r}(X_n,\ Q_n,A_n)\right] \tag{1.13}$$

where $\overline{\pi}$ is the strategy for model (1.10), which is defined by

$$\overline{\pi}_n(a_n\ |\ x_0,q_0,a_0,\ldots,x_n,q_n) \ := \pi_n(a_n|x_0,a_0,\ldots x_n) \tag{1.14}$$

Hence models (1.1) and (1.10) are equivalent and therefore we use the notations of model (1.1).

So we are dealing with a Markov decision process with known transition law again. However this new MDP has some odd properties. At first the state space is infinite even if the state space of the original model is finite. Further the new MDP is *transient* in general, i.e. in most cases $Q_n \neq Q_m$, $P^{\pi}_{x,q}$ - a.s. for all $n \neq m$. In section 2 we show by an example that even if X and A are finite sets there need not be an optimal strategy.

In the next section we introduce strategies that are easy-to-handle, at least if X, A and Θ are finite sets, and we consider conditions guaranteeing these strategies to be optimal. These conditions imply that the posterior distributions Q_n converge to degenerate distributions, which property is used explicitly to prove the optimality.

We conclude this section by introducing a parameter structure that is quite general and that facilitates formulating some results.

From now on we assume that we are dealing with the following structure:

(i) $X = \hat{X} \times Y$ (1.15)

(ii) R is a transition probability from $\hat{X} \times A \times Y$ to $\hat{X}$

(iii) $K_1, K_2, K_3, \ldots, K_n$ is a partition of $\hat{X} \times A$,

(iv) $\Theta = \prod_{i=1}^{n} \Theta_i$, $\theta = (\theta_1, \theta_2, \theta_3, \ldots, \theta_n)$

(v) $P(x', y' \mid x, y, a, \theta) = R(x' \mid x, a, y') \cdot p_i(y' \mid \theta_i)$ if and only if $(x,a) \in K_i$ where $p_i(. \mid \theta_i)$ is a probability on Y.

As an illustration of this structure let $\hat{X} = \{x_1, x_2, \ldots, x_n\}$, $A = \{a_1, a_2, \ldots, a_m\}$ and let $K_{i,j} = \{(x_i, a_j)\}$, $i=1,2,\ldots,n$ and $j = 1, 2, \ldots, m$. Then for all pairs of state and action there is an unknown parameter in the transition law. This situation occurs when all transition probabilities are unknown.

Hence we have factorized the original transition law. If $(x,a) \in K_i$ then the transition to the next state x', y' depends on θ only through its i-th component θ_i. We present below some examples having this structure. It is straightforward to verify that

$$T_{(x,y),a,(x',y')}(q)(\theta) = \sum_{i=1}^{\infty} 1_{K_i}(x,a) \frac{p_i(y'|\theta_i)\, q(\theta)}{\sum_{\theta} p_i(y'|\theta_i)\, q(\theta)} \tag{1.16}$$

(provided that the denominator does not vanish). Here 1_B represents the indicator function of the set B.

Although this parameter structure seems to be rather complicated there are practical situations where this structure occurs in a natural way.

The state of the system at stage n is $X_n = (\hat{X}_n, Y_n)$. The state component Y_n is called the *supplementary state variable*. It can be proved that if $\theta \in \Theta$ is known then it is sufficient to consider $\hat{X}_n$ instead of X_n (see van Hee (1978a) page 52).

So $\hat{X}_n$ has to be considered as the original state variable if the parameter is known, while Y_n only occurs since it contains information concerning the unknown parameter.

From now on we are dealing with this parameter structure and we shall consider only $\hat{X}_n$ and $\hat{X}$. To facilitate notations we omit the circumflex from now on.

Example 1

Consider an inventory control model without backlogging. If the demand distribution is known, the inventory level may be chosen as the state variable. However if the demand distribution is unknown the sequence of successive inventory levels does not reflect the sequence of successive demands and therefore we have to consider the demand in each period as a *supplementary state variable*. Here X is the set of inventory levels, i.e. $X = (0,\infty)$ and Y is the set of possible demands. The transition function is

$$R(x'|x,a,y') = 1 \text{ if } x' = \max\{a-y',0\} \text{ and } a \geq x$$

$$= 0 \text{ otherwise}$$

(Hence the action a is the inventory level after ordering.)

The sets $K_2, K_3, \ldots$ are empty and $p_i(.|\theta_1)$ is the demand distribution with unknown parameter $\theta_1 \in \Theta_1$.

Example 2

Consider a waiting line model with bulk arrivals. At each time point 0,1,2,... a group of customers arrives and the distribution of the size of the group is unknown. The service distribution is exponential with parameter a and controllable by the parameter a. Let y' be the number of customers arriving in some period. Let x be the queue length at the beginning of that period and x' at the end. Then, if $c := x + y' - x' \geq 0$ we have

$$R(x' | x,a,y') = \frac{a^c}{c!} e^{-a}$$

and if $c < 0$ then $R(x' | x,a,y') = 0$. Further $K_2, K_3, \ldots$ are empty and $p_i(y' | \theta_1)$ is the probability of a group of size y'.

Example 3

Consider a linear system with random disturbances. The state at stage n is X_n and the disturbance at stage n is Y_n.

Then

$$X_{n+1} = C_1 X_n + C_2 A_n + Y_{n+1}$$

where X,Y and A are Euclidean spaces and C_1 and C_2 suitable matrices. Assume that $\{Y_n, n \in N\}$ forms a sequence of i.i.d. random variables with an incompletely known distribution.

If only the sequence $(X_0, X_1, X_2, \ldots)$ is observable to the controller then he may reconstruct the sequence of supplementary state variables.

2. OPTIMAL STRATEGIES

We start this section with an example showing that there need not be an optimal strategy even if X and A are finite sets.

Example 4

Let $X = \{1,2,3,4,5,6\}$, $A = \{1,2,3\}$, $Y = \{0,1\}$, $\Theta_1 = (0,1)$. (Note that Θ_1 is not countable here but if we replace Θ_1 by a countable subset of $(0,1)$ the same arguments are valid, however notations become more difficult.)

Further let $R(x'|x,a,y')$ be defined by:

$R(3|3,a,1) = R(4|3,a,0) = R(4|4,a,1) = R(3|4,a,0) =$

$R(5|5,a,1) = R(6|5,a,0) = R(6|6,a,0) = R(5|6,a,1) = 1$

for all $a \in A$, and let

$R(1|1,1,1) = R(2|1,1,0)=1$; $R(3|1,2,y) = R(5|1,3,y) =$

$= R(3|2,2,y) = R(5|2,3,y)=1$ for all $y \in Y$.

Finally $K_2, K_3, \ldots$ are empty and $p_1(1|\theta) = \theta$, $p_1(0|\theta) = 1-\theta$.

The example can be represented in the diagram:

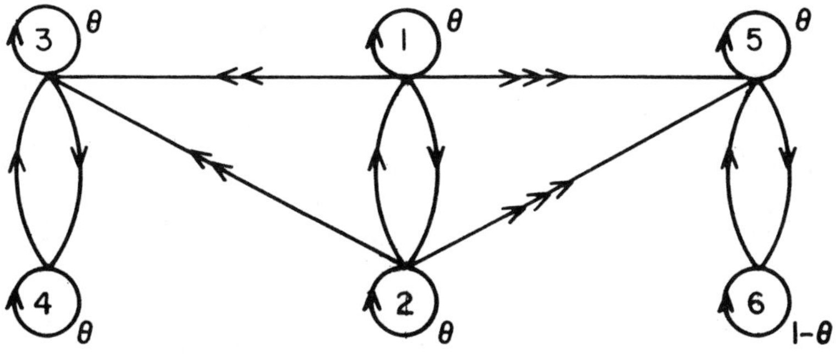

Only in states 1 and 2 the chosen action has effect. The rewards obtained are $r(3) = r(5) = 7$ and $r(4) = r(6) = 3$ for all actions and in the other states no rewards are obtained.

The average return in the sub-chain $\{3,4\}$ is $\frac{1}{2}(7+3) = 5$ and in the sub-chain $\{5,6\}$: $7\theta + 3(1-\theta) = 4\theta+3$. Consider a starting state $x \in \{1,2\}$.

It is easy to verify that for known $\theta \in \Theta_1$ the optimal action is a maximizer of the function $5.\delta(2,a) + \{4\theta+3\}\,\delta(3,a)$ (where δ is the Kronecker function), $a \in \{2,3\}$. It is also straight-forward to verify that if we have to choose one of the actions 2 or 3 and if q is the prior distribution then the maximizer of $5.\delta(2,a) + \{4.\int\theta q(d\theta) + 3\} \,.\, \delta(3,a)$, $a \in \{2,3\}$ is the best choice.

Let π^n be the strategy that chooses action 1 the first n times and the maximizer of the function

$$5\delta(2,a) + \{4 \int \theta . Q_n(d\theta) + 3\}\ \delta(3,a),\ a \in \{2,3\}$$

thereafter, where Q_n is the posterior distribution at time n if the system starts in state 1 with prior distribution q.

Then the *Bayesian average return* in states 1 and 2 is

$$E_q\left[\max \{5,4 \int \theta\ Q_n(d\theta) + 3\}\right]$$

(this expectation does not depend on the starting state and the strategy).

Note that

$$E_q\left[\max \{5,4 \int \theta\ Q_{n+1}(d\theta) + 3 \mid Q_1,\ldots,Q_n\right] \geqq$$

$$\geqq \max \{5,4\ E_q\left[\int \theta\ Q_{n+1}(d\theta) \,|\, Q_1,\ldots Q_n\right] + 3\} = \max \{5,4 \int \theta\ Q_n(d\theta)+3\}$$

with equality if and only if $5 \geqq 4 \int \theta\ Q_{n+1}(d\theta) + 3\}$, P_q - a.s. However if q gives positive mass to the set $\{\theta \in \Theta \mid \theta > \frac{1}{2}\}$ the equality never holds. Hence in this case the strategy π^n is worse than π^{n+1} and consequently there is no optimal strategy.

We first introduce two assumptions:

(i) r is bounded on $X \times A$ (2.1)

(ii) there are bounded functions g and h on Θ, and $X \times \Theta$ respectively such that $h(x,\theta) + g(\theta) = \sup_{a \in A} L(x,a,\theta)$

where $L(x,a,\theta) = \sum_{i=1}^{\infty} 1_{K_i}(x,a) \{ r(x,a) + \sum_{y'} \sum_{x'} R(x' | x,a,y') \cdot p_i(y' | \theta_i) h(x',\theta) \}$

For models with known parameter value these conditions are the well-known conditions considered in Derman (1966) and Ross (1968). In that case there exist stationary optimal strategies, at least if A is finite. Moreover the optimal average return is $g(\theta)$.

In Ross (1968) several situations are given where (2.1) (ii) holds for fixed parameter value θ.

For instance, if X and A are finite sets, and for each stationary strategy the resulting process is an irreducible Markov chain, then (2.1) (ii) is valid. The strategies which we consider, choose at each stage n a maximizer of

$$\sum_{\theta} L(X_n,a,\theta)\, Q_n(\theta), \quad a \in A \qquad (2.2)$$

We call these rules *Bayesian equivalent rules* because we are maximizing the "Bayesian equivalent" of the function we have to maximize in case the parameter is known.

(If there are no maximizers one may choose an ε_n-maximizer where $\varepsilon_1, \varepsilon_2, \varepsilon_3, \ldots$ is a decreasing sequence with $\lim_{n\to\infty} \varepsilon_n = 0$). In case Θ is finite $L(x_1,a_1,.)$ can be computed, however if Θ is infinite one needs approximations using perturbation theory, cf Schweitzer (1968). As far as the author knows there are no good algorithms for this problem in the average return case. For the discounted case see van Hee (1978a) section 6.2.

In example 4 we already encountered these strategies. So it is clear by the example that these strategies are not optimal in general. However we give at the end of this section conditions guaranteeing these strategies to be optimal.

In van Hee (1978a) we consider these strategies also for the discounted total return case and we prove there that these strategies are optimal for the linear system with quadratic costs (in discrete time) and also for some inventory control models. There we also consider bounds on the discounted total return of these strategies.

In Fox and Rolph (1973), Mandl (1974) and Georgin (1978) another heuristic strategy is considered, which turns out to be optimal in a lot of situations. This strategy can be formulated in the following way:

> "At each stage estimate the unknown parameter θ using the available data, by $\hat{\theta}$. Then compute an optimal (stationary) strategy for the model where the parameter is known and equal to $\hat{\theta}$. Then use the corresponding action in the actual state. Repeat this procedure at each stage."

Hence, if we consider Bayes estimates, then the method proposed by the other authors may be formulated in the following way:

choose at each stage n a maximizer of

$$L\left(X_n, a, \sum_{\theta} \theta\, Q_n(\theta)\right), \ a \in A \tag{2.3}$$

(here we assumed the parameter set as a sub-set of $\hat{R}$).

If $\sum_{\theta} \theta\, Q_n(\theta)$ does not belong to Θ then one may choose a parameter $\tilde{\theta} \in \Theta$ that maximizes $Q_n(.)$.

To prove the optimality of Bayesian equivalent rules we need the following limit theorem: (for a proof see van Hee (1978a) th. 2.4).

<u>Theorem 1</u>

If $$\sum_{n=0}^{\infty} 1_{K_i}(X_n, A_n) = \infty \quad (P_{x,q}^{\pi} - \text{a.s.})$$

then $$\lim_{n\to\infty} \sum_{\theta} f(\theta_i)\, Q_n(\theta) = f(Z_i) \quad (P_{x,q}^{\pi} - \text{a.s.}) \quad i = 1,2,3,\ldots$$

for all bounded functions f on Θ,

(where $Z = (Z_1, Z_2, Z_3, \ldots)$ cf (1.15) (i.v.)

Note that theorem 1 gives a sufficient condition for the consistency of the Bayes estimation procedure.

We introduce the function ϕ on $X \times A \times \Theta$.

$$\phi(x,\theta,a) := L(x,a,\theta) - h(x,\theta) - g(\theta)$$

Note that $\phi(x,\theta,a) \leqq 0$ for all $x \in X, \theta \in \Theta$, $a \in A$.

Further we extend ϕ to a function on $X \times A \times W$ and likewise the functions h and g:

$$\text{(i)} \quad \phi(x,q,a) := \sum_{\theta} q(\theta)\, \phi(x,\theta,a) \qquad (2.4)$$

$$\text{(ii)} \quad h(x,q) := \sum_{\theta} q(\theta)\, h(x,\theta)$$

$$\text{(iii)} \quad g(q) := \sum_{\theta} q(\theta)\, g(\theta)$$

Note that these definitions are consistent with (2.1) (ii) if we identify θ with the distribution that is degenerate at θ.

The next theorem provides a sufficient condition for a strategy π to be optimal. (For a proof see van Hee (1978a) th. 4.1).

<u>Theorem 2</u>

If $$\liminf_{N\to\infty} \frac{1}{N} \sum_{n=0}^{N-1} E_{x,q}^{\pi} \left[\phi(X_n, Q_n, A_n)\right] = 0$$

then

$$\liminf_{N\to\infty} \frac{1}{N} \sum_{n=0}^{N-1} E_{x,q}^{\pi} \left[r(X_n, A_n)\right] = g(q)$$

and

π is optimal.

Theorems 1 and 2 are the key tools for proving optimality of the Bayesian equivalent rules.

As an illustration we shall consider two practical examples from more general theorems. For the first example we provide a proof to demonstrate the technique.

Theorem 3

Let X and A be finite sets, let $M_1, M_2, \dots M_m$ be a partition of X and let $K_i = M_i \times A$, $i=1,\dots,m$ ($K_i=\phi$ for $i > m$).

Let the Markov chain $\{X_n, n \in \hat{N}\}$ be irreducible for each stationary strategy and each parameter value.

Then a strategy π^* that chooses at stage n a maximizer of the function

$$\phi(X_n, Q_n, a),\ a \in A$$

is optimal.

(Note that a maximizer of $\phi(X_n, Q_n, a)$ is a maximizer of $\sum_\theta L(X_n, a, \theta) . Q_n(\theta)$)

Proof.

Let A_n be the action at stage n under strategy π^*. Hence

$$0 \geqq \phi(X_n, Q_n, A_n) = \max_{a\varepsilon A} \sum_\theta Q_n(\theta)\ \phi(X_n, \theta, a) =$$

$$= \max_{f\varepsilon F} \sum_\theta Q_n(\theta) . \phi\left(X_n, \theta, f(X_n)\right)$$

$$\geqq \max_{f\varepsilon F} \sum_\theta Q_n(\theta) \min_{x\varepsilon X} \phi(x, \theta, f(x))\Big),$$

where F is the set of all functions from X to A.

Note that F represents the set of all stationary strategies for the model with known parameter value.

Since the Markov chain $\{X_n, n \in \hat{N}\}$ is irreducible for each stationary strategy it can be proved (cf. lemma 4.7 in van Hee (1978a)) that the number of visits to each set M_i is almost surely infinite for all strategies. Therefore the condition of theorem 1 is fulfilled for all i and so we have

$$\lim_{n\to\infty} \sum_{\theta} Q_n(\theta) \min_{x\varepsilon X} \phi\big(x,\theta,f(x)\big) = \min_{x\varepsilon X} \phi\big(x,Z,f(x)\big)$$

Since there is a stationary optimal strategy for the model with known parameter value (cf. Ross (1968)) we have

$$\max_{f\varepsilon F} \min_{x\varepsilon X} \phi\big(x,\theta,f(x)\big) = 0 \text{ for all } \theta \varepsilon \Theta.$$

Therefore we find

$$0 \geqq \liminf_{n\to\infty} \phi(X_n,Q_n,A_n) \geqq \lim_{n\to\infty} \max_{f\varepsilon F} \sum_{\theta} Q_n(\theta) \min_{x\varepsilon X} \phi\big(x,\theta,f(x)\big) = 0$$

and so

$$\lim_{N\to\infty} \frac{1}{N} \sum_{n=0}^{N-1} E_{x,q}^{\pi^*} \big[\phi(X_n,Q_n,A_n)\big] = 0$$

Application of theorem 2 gives the desired result.

In this example we assumed that the information we obtain after the transition about the unknown parameter does not depend on the action chosen. In the next example we relax this assumption.

Here we assume:

(i) $M_1,\ldots,M_m$ is a partition of X, $N_1,\ldots,N_n$ is a partition of A. (2.5)

(ii) For each stationary strategy the Markov chain $\{X_t, t \varepsilon \hat{N}\}$ is irreducible

(iii) The partition $K_1, K_2,\ldots$ of $X \times A$ consists of the sets $\{M_i \times N_j, i=1,\ldots,m, j=1,\ldots,n\}$.

Before we consider the strategy that turns out to be optimal, we first introduce the concept of a *sequence of density zero*. A sequence $S = (s_1, s_2,\ldots)$ is said to be of density zero if

$$\limsup_{k\to\infty} \frac{1}{k} \max \{i \in N \mid s_i \leqq k\} = 0$$

Examples of such sequences are: $(s_i = 2^i, i \in \hat{N})$ and $(s_i = i^2, i \in \hat{N})$

In theorem 4 we consider a strategy that is inspired by an idea in Mallows and Robbins (1964). In Fox and Rolph (1973) this idea is used in a similar way for Markov renewal programs.

The idea is that we make use of *forced choice actions* to guarantee that we return to each set K_i infinitely often, which is necessary to apply theorem 1. However we do this with a frequency that is so low as not to influence the Bayesian average return. (In fact the concept of a sequence of density zero is used here.)

Now we are ready to formulate in an informal way the strategy $\hat{\pi}$ that will be optimal. In (2.6) this strategy is sketched:

Fix (forced choice) actions $a_1, a_2, \ldots, a_n$ such that $a_i \in N_i$ (2.6)

Let $t_i(n)$ be the number of visits to the set M_i by stage n.

If $X_n \in M_i$ and $t_i(n) \in S$ for some $i \in \{1,\ldots,m\}$ then the next action in the sequence $(a_1,\ldots,a_n)$ is chosen

Otherwise, if $t_{X_n}(n) \notin S$, then a maximizer of the function

$$\phi(X_n, Q_n, a),\ a \in A$$

is chosen.

Hence $\hat{\pi}$ uses the same actions as the strategy π^* in theorem 3 except for stages where $t_{X_n}(n) \in S$. Then the forced choice actions are chosen in order.

The proof of theorem 4 is rather technical, although the idea is simple (see van Hee (1978a) th. 4.10).

Theorem 4

Let (2.5) hold. The strategy $\hat{\pi}$ defined in (2.6) is optimal.

The result is also true in a more abstract model. In fact it completes results of Mandl (1974) and it generalizes work of Rose (1975).

3. REFERENCES

Bellman, R. "Adaptive control processes: a guided tour", Princeton (N.Y.), Princeton University Press (1961).

Derman, C. "Denumerable state Markovian decision processes - average cost criterion", *Ann. Math. Statist.*, **37**, 1545-1554 (1966).

Fox, B.L. and Rolph, J.E. "Adaptive policies for Markov renewal programs", *Ann. Math. Statist.*, **35**, 846-856 (1973).

Georgin, J.P. "Estimation et controle des chaînes de Markov sur des espaces arbitraire", In: Lecture Notes in Mathematics 636, Springer-Verlag, Berlin (1978).

van Hee, K.M. "Bayesian control of Markov chains", Amsterdam, *Mathematical Centre Tracts*, **95** (1978a).

van Hee, K.M. "Markov decision processes with unknown transition law: the discounted case", Third Formator Symposium on Math. Methods for the Analysis of Large Scale Systems, Edited by J. Benes, Academia Prague (1978).

Mallows, C.L. and Robbins, H. "Some problems of optimal sampling strategy", *J. Math. Anal. Appl.*, **8**, 90-103 (1964).

Mandl, P. "Estimation and control in Markov chains", *Advances in Appl. Probability*, **6**, 40-60 (1974).

Neveu, J. "Mathematical foundations of the calculus of probability", San Francisco, Holden-Day (1965).

Rieder, U. "Bayesian dynamic programming", *Advances in Appl. Probability*, **7**, 330-348 (1975).

Rose, J.S. "Markov decision processes under uncertainty - average return criterion", unpublished report (1975).

Ross, S.M. "Arbitrary state Markovian decision processes", *Ann. Math. statist.*, **39**, 2118-2122 (1968).

Schweitzer, P.J. "Perturbation theory and finite Markov chains", *J. Appl. Probability*, **5**, 401-413 (1960).

PIECEWISE LINEAR MARKOV DECISION PROCESSES WITH AN APPLICATION TO PARTIALLY OBSERVABLE MODELS

K. Sawaki

(Nanzan University)

ABSTRACT

This paper applies policy improvement and successive approximation or value iteration to a general class of Markov decision processes with discounted costs. In particular, a class of Markov decision processes, called piecewise-linear, is studied. Piecewise-linear processes are characterized by the property that the value function of a process observed for one period and then terminated is piecewise-linear if the terminal reward function is piecewise-linear. Partially observable Markov decision processes have this property.

An algorithm based on policy improvement is developed to compute simple approximations to an optimal policy and the optimal value function. This algorithm has the property that only simple functions and policies are generated, so that they are easily represented in a computer.

1. INTRODUCTION

The combined theories of dynamic programming and Markov decision processes have been applied to many areas as inventory, queueing and machine maintenance problems.

This paper develops a theory for a simple class of dynamic programming models with the property that only piecewise-linear cost functions and piecewise constant policies are involved, as well as an algorithm which yields policies that are both "simple and ε-optimal". This algorithm includes policy improvement [1], [2], [3], [7] and successive approximation [1], [4] as special cases. The formulation of our general model is motivated by consideration of the special structure which partially observable models [6], [9], [11] possess.

In Section 2 piecewise linear Markov decision processes with an abstract state and finite action set over an infinite horizon will be discussed. Section 3 will discuss several examples having piecewise linear property. Section 4 explicitly develops the algorithm basedon modified policy improvement.

2. PIECEWISE LINEAR MARKOV DECISION PROCESSES

We shall formulate an optimal control problem with discounted costs and with complete observation over an infinite horizon under the setting of [1]. The state space Ω is an arbitrary set, say a non-empty Borel subset of a separable Banach space X. Let A be the finite set of actions and a is an element of A. For each pair $(x,a) \varepsilon \Omega \times A, q(\cdot|x,a)$ is the one step transition probability of the system on the Borel subsets of Ω. The immediate cost $c(x,a)$ is a bounded Borel measurable function on $\Omega \times A$.

When the system is in state x and action a is chosen, then we incur a cost $c(x,a)$. We define a policy to be a sequence $\{\delta_n, n = 1,2,3,...\}$, where δ_n tells us what action to choose at the n-th period as a Borel measurable function of the history $H = (x_1,a_1,...x_n)$ of the system up to period n. Let Δ be a family of policies. A policy $\delta = (\delta,\delta,...)$ which is independent of time n is called stationary. Our discounted total cost $V^\delta(x)$ at an initial state x under a policy δ is written as

$$V^\delta(x) = E\{\sum_{n=1}^{\infty} \beta^{n-1} c(X_n,\delta_n(X_n))|X_1 = x\} \tag{1}$$

where $\{X_n : n = 1,2,...\}$ is a Markov chain with probability transition function $q(\cdot|x, \delta_n(x))$. The discount factor is denoted by β and $0 \leq \beta < 1$. The function V^δ defined by (1) is called the cost of policy δ.

Define the optimal cost function V^* by

$$V^*(x) = \inf_{\delta \varepsilon \Delta} V^\delta(x) \text{ for all } x \varepsilon \Omega. \tag{2}$$

It is well known (see [1]) that there exists an optimal stationary policy δ^* with $V^{\delta^*} = V^*$ and also that V^* satisfies

$$V^*(x) = \min_{a\varepsilon A} \{c(x,a) + \beta \int_\Omega V^*(x')q(dx'|x,a)\} \text{ for all } x \varepsilon \Omega.$$

Therefore Δ can be restricted, hereafter, into the set of stationary policies. Let $B(\Omega)$ be the set of all bounded Borel measurable functions on Ω. The norm defined by $\|v\| = \sup\{|v(x)| : x\varepsilon\Omega\}$ for $v \varepsilon B(\Omega)$ makes $B(\Omega)$ a Banach space. For u and v in $B(\Omega)$ we write $u \leq v$ if $u(x) \leq v(x)$ for all $x\varepsilon\Omega$.

For each $\delta\varepsilon\Delta$ define $U_\delta: B(\Omega) \to B(\Omega)$ by

$$(U_\delta v)(x) = c(x, \delta(x)) + \beta\int_\Omega v(x')q(dx'|x, \delta(x)) \text{ for } v \varepsilon B(\Omega),$$

$x \varepsilon \Omega$ and also define $U_* : B(\Omega) \to B(\Omega)$ by $U_* v = \inf\{U_\delta v : \delta\varepsilon\Delta\}$. If $\delta(x) = a$ for each $x \varepsilon \Omega$, then we write $U_a = U_\delta$. An operator $U : B(\Omega) \to B(\Omega)$ is <u>monotone</u> if $u \leq v$ implies $Uu \leq Uv$ and is a contraction if for some $\beta \varepsilon [0,1)$ $\|Uu - Uv\| \leq \beta\|u-v\|$ for each $u,v \varepsilon B(\Omega)$. Denardo [4] verifies that U_* and U_δ are monotone contraction operators. By Banach's fixed point theorem, for each $\delta \varepsilon \Delta$ there is a unique $V^\delta \varepsilon B(\Omega)$ such that $U_\delta V^\delta = V^\delta$. Similarly Denardo shows that $V^* = U_* V^*$. If $\|V^\delta - V^*\| \leq \varepsilon$ then δ is an <u>ε-optimal policy</u>, and if $\|V - V^*\| \leq \varepsilon$ then V is an <u>ε-optimal cost function</u>.

Any set of the form $\{x \varepsilon \Omega : \ell_{ij}(x) < (\text{or} \leq) d_j, j=1,\dots,n_i\}$, $i=1,2,\dots,m$ where ℓ_{ij} defined on $X \supset \Omega$ is a linear functional and d_j is a real number is called a <u>convex polyhedron</u>. A collection $P = \{E_1, E_2, \dots, E_m\}$ of subsets of $\Omega \subset X$ is a <u>partition</u> of Ω if $E_i \cap E_j = \phi$ for $i \neq j$ and if $\bigcup_{i=1}^m E_i = \Omega$. Each member of a partition P is a <u>cell</u>. If each cell of a partition is a convex polyhedron, then the partition is called simple. The <u>product</u> of two partitions P^1 and P^2 is $P^1 \cdot P^2 = \{E \cap D : E \varepsilon P^1, D \varepsilon P^2\}$. The product of $P^1 \cdot P^2 \dots P^m$ is defined inductively by $\prod_{i=1}^m P^i = P^m \cdot \prod_{i=1}^{m-1} P^i$. Plainly, the finite product of simple partitions is again simple. A function $v \in B(\Omega)$ is piecewise linear (abbreviated, hereafter, by p.w.) if there exists a simple partition $\{E_1, E_2, \ldots, E_m\}$ of Ω such that $v(x) = v_i(x)$ for all $x \in E_i, i=1,2,\ldots,m$ and each v_i is the restriction to E_i of a linear function on X. A policy δ is piecewise (p.w.) constant if there is a simple partition $\{E_1, E_2, \ldots, E_m\}$ of Ω such that $\delta(x) = a_i$ for all $x \in E_i$, $i=1,2,\ldots,m$. The paper Denardo and Rothblum [5] discusses affine (but not piecewise) dynamic programs.

Although V^* is not necessarily p.w. linear and δ^* is not necessarily p.w. constant, we will show for a class of Markov decision processes having the structure described in the following assumption that there are ε-optimal p.w. linear cost functions and p.w. constant policies.

Asssumption I (A.I.) $(U_a V)(x)$ is p.w. linear on Ω for each a, provided that V is p.w. linear on Ω.

The following theorem shows how the structure in Assumption I implies that U_* and U_δ preserve the p.w. linearity of value functions and the p.w. constant of policies.

Theorem 1. Suppose that (A.I.) holds and that V is p.w. linear. Then

(i) $U_\delta V$ is p.w. linear whenever δ is p.w. constant;

(ii) $U_* V$ is p.w. linear; and

(iii) there exists a p.w. constant policy δ such that $U_\delta V = U_* V$.

Proof.

(i) Suppose that δ is p.w. constant with respect to a simple partition $\{E_i\}$. Let E_i be an arbitrary but fixed cell from the partition and suppose that $\delta(x) = a$ for $x \in E_i$. Then

$$(U_\delta V)(X) = (U_a V)(x) \text{ for } x \in E_i.$$

From (A.I.), $U_a V$ is p.w. linear for each $a \in A$. Hence $U_\delta V$ is p.w. linear on each cell E_i, and is consequently p.w. linear on Ω.

(ii) (iii) The functions $U_a V$ are p.w. linear by (A.I.). Suppose that $U_a V$ is p.w. linear with respect to the simple partition P^a. Let $P = \prod_{a \in A} P^a$. Then P is finer than each P^a and so each $U_a V$ is p.w. linear with respect to P. For each $F \in P$ and $a \in A$, there is some linear functional α_F^a *such that*

$$(U_a V)(x) = \alpha_F^a(x) \text{ for } x \in F.$$

For each $F \in P$, define the sets G_F^b, $b \in A = \{1,=,\ldots,p\}$, by $G_F^b = \{x\colon \alpha_F^b x < \alpha_F^a x,\ a=1,2,\ldots,b-1 \text{ and } \alpha_F^a x \leq \alpha_F^a x,\ a=b+1,\ldots,p\}$. Then $\{G_F^a : a \in A\} = P^F$ is a partition of F and $\hat{P} = \prod_{F \in P} P^F$ is a partition of Ω with the property that

$$(U_* V)(x) = \alpha_F^a(x) \text{ if } x \in G_F^a \in \hat{P}.$$

The policy δ defined by $\delta(x) = a$ for $x \in G_F^a \in \hat{P}$ satisfies $U_\delta V = U_* V$.

<u>Corollary</u>. Suppose that (A.I.) holds and that $V^0 \in B(\Omega)$ is p.w. linear.

(i) Define $V^n(x) = (U_\delta V^{n-1})(x)$, $n=1,2,\ldots$, for p.w. constant δ.

(ii) Define $V^n(x) = (U_* V^{n-1})(x)$, $n=1,2,\ldots$.

Then V^n is p.w. linear and there exists a p.w. constant stationary policy, δ_n, satisfying $U_{\delta_n} V^{n-1} = U_* V^{n-1}$.

We next consider the effects of iterating monotone contraction mappings such as U_* and U_δ, citing some results of Denardo [4].

<u>Lemma 1.</u> Suppose that U is a contraction mapping on $B(\Omega)$ with contraction coefficient $\beta < 1$. Let $V^0 \in B(\Omega)$ be given and

define the functions $V^n, n=1,2,\ldots$ by

$$V^n(x) = (UV^{n-1})(x).$$

Then

(i) $\{V^n\}$ converges in norm to the fixed point $\hat{V}$ of U; i.e., $U\hat{V} = \hat{V}$. Now assume that U is also monotone.

(ii) If $V^1 \leq V^0$, then $\{V^n\}$ is monotonically decreasing to $\hat{V}$.

(iii) If $V^1 \geq V^0$, then $\{V^n\}$ is monotonically increasing to $\hat{V}$.

Remark. The fixed point $\hat{V}$ need not to be p.w. linear since the cells in the limiting partition are not necessarily finite in number nor polyhedral.

3. EXAMPLES

Model 1. Let $X = R^N$ and Ω be a convex polyhedron in R^N such that $q(\cdot|x,a)$ is a probability measure on Ω for each $(x,a) \in \Omega \times A$. Assume that $c(x,a) = c^a \cdot x$ (the inner product of two vectors c^a and x). Also assume that for each convex polyhedron $B \subset \Omega$

$$q^a(B,x) = \int_B x' \, q(dx' \,|x,a)$$

is p.w. linear in x with respect to a simple partition $P^a(B)=\{E_j(a,B), j=1,2,\ldots m_{a,B}\}$ for each a. These two assumptions imply (A.I.).

We explicitly check that (A.I.) is satisfied. Let $a \in A$ be arbitrary but fixed and suppose that V is p.w. linear with respect to a simple partition $\{E_i, i=1,2,\ldots,m\}$. Let $P^a = \prod_{i=1}^{m} P^a(E_i) = \{\tilde{E}_j^a; j=1,2,\ldots,r\}$, the product partition, which is again simple.

$$(U_a V)(x) = c^a \cdot x + \beta \int_\Omega V(x')q(dx' | x,a)$$

$$= c^a \cdot x + \beta \sum_{i=1}^{m} \int_{E_i} V_i x' \, q(dx' | x,a)$$

$$= c^a \cdot x + \beta \sum_{i=1}^{m} V_i q^a(E_i,x)$$

$$= \left[c^a + \beta \sum_{i=1}^{m} V_i \lambda_{il}^a\right] x \text{ for } x \in \tilde{E}_j^a$$

where $\lambda_{il}^a . x = q^a(E_i,x)$ for $x \in E_\ell(a,E_i)$ and the index ℓ depends on i for each $a \in A$. $U_a V$ is linear on each $\tilde{E}_j^a$. Hence $U_a V$ is p.w. linear with respect to the simple partition $P^a = \{\tilde{E}_j^a, j=1,2,\ldots,r\}$, which satisfies (A.I.).

Model 2. A partially observable Markov Decision Process (Sawaki and Ichikawa [9], Dynkin [6]).

Consider a Markov decision process (called the core process) with state space $\{1,2,\ldots,N\}$, with action set A, with probability transition matrices p^a and with immediate cost vectors h^a. Let z_n be the state at the n-th transition. Assume that the process $\{z_n, n = 1,2,\ldots\}$ cannot be observed, but at each transition a signal θ is transmitted to the decision maker. The set of possible signals Θ is assumed to be finite. For each n, given that $Z_n = j$ and that action a is to be implemented, the signal θ_n is independent of the history of the signals and actions $\{\theta_0,a_0,\theta_1,a_1,\ldots,\theta_{n-1},a_{n-1}\}$ prior to the n-th transition and has conditional probability denoted by $\gamma_{j\theta}^a = P[\theta_n = \theta | Z_n = j, a]$.

Let $X = R^N$ and $\Omega = \{x_1,x_2,\ldots,x_N): \sum_{i=1}^{N} x_i = 1, x_i \geq 0, \forall_i\}$.

Define the i-th component of X_n, the random variable of x, to be

$$P[Z_n = i|\theta_0, a_0, \theta_1, \ldots, \theta_{n-1}, a_{n-1}, \theta_n], i=1,=,\ldots,N.$$

It can be shown (see Dynkin [6]) that

$$P[Z_{n+1} = j|\theta_0, a_0, \theta_1, \ldots, \theta_n, a_n, \theta_{n+1}] = P[Z_{n+1} = j|\theta_{n+1}, a_n, X_n].$$

Thus X_n represents a sufficient statistic for the complete past history $\{\theta_0, a_0, \ldots, a_{n-1}, \theta_n\}$. It follows that $\{X_n : n=0,1,2,\ldots\}$ is a Markov process (see Dynkin [6]), called the observed process. Its immediate cost is $c(x,a) = h^a \cdot x$. Its action set is A. Its probability transition function is determined by the following calculation. For each measurable subset $B \subseteq \Omega$, $x \in \Omega$, and $a \in A$,

$$\begin{aligned}
q(B|x,a) &= P[X_{n+1} \in B|X_n = x,\ a_n = a] \\
&= \sum_\theta P[X_{n+1} \in B|\theta_{n+1} = \theta,\ X_n = x,\ a_n = a] \cdot P[\theta_{n+1} = \theta|X_n = x,\ a_n = a] \\
&= \sum_\theta P[X_{n+1} \in B|\theta_{n+1} = \theta,\ X_n = x,\ a_n = a] \cdot \sum_j P[\theta_{n+1} = \theta|Z_{n+1} = j,\ X_n = x, a] \cdot P[Z_{n+1} = j|X_n = x,\ a_n = a] \\
&= \sum_\theta P[X_{n+1} \in B|\theta_{n+1} = \theta,\ X_n = x,\ a_n = a] \cdot \sum_j \gamma^a_{j\theta} \sum_i P[Z_{n+1} = j|Z_n = i,\ X_n = x,\ a_n = a] P[Z_n = i|X_n = x,\ a_n = a] \\
&= \sum_\theta P[X_{n+1} \in B|\theta_{n+1} = \theta,\ X_n = x,\ a_n = a] \sum_j \gamma^a_{j\theta} \sum_i P^a_{ij} x_i \\
&= \sum_\theta P[X_{n+1} \in B|\theta_{n+1} = \theta,\ X_n = x,\ a_n = a]\ \underline{1} P^a(\theta) x
\end{aligned}$$

where $\underline{1} = (1,1,\ldots,1)$ and $P^a(\theta) = \left[P^a_{ij}(\theta)\right] = \left[P^a_{ji}\gamma^a_{i\theta}\right]$. Define the vector $T(x|\theta,a)$ by

$$T(x|\theta,a) = \frac{P^a(\theta)x}{\underline{1}P^a(\theta)x}.$$

Note that $T(X_n|\theta,a) = X_{n+1}$, and that

$$P[X_{n+1} \varepsilon B | \theta_{n+1} = \theta, X_n = x, a_n = a] = \begin{cases} 1 \text{ if } T(x|\theta,a) \varepsilon B \\ 0 \text{ if } T(x|\theta,a) \notin B. \end{cases}$$

So,

$$q(B|x,a) = \sum_{\theta \varepsilon \Phi^a(B,x)} \underline{1}\, P^a(\theta)x$$

where $\Phi^a(B,x) = \{\theta: T(x|\theta,a) \varepsilon B\}$.

Finally, we show that the observed process $\{X_n\}$ is a special case of Model 1; i.e., $q^a(B,x) = \int_B x' q(dx'|x,a)$ is p.w. linear in x for each convex polyhedral set $B \subseteq \Omega$ and action $a \varepsilon A$. Using the previously computed $q(B|x,a)$ we have

$$\begin{aligned} q^a(B,x) &= \int_B x' q(dx'|x,a) \\ &= \sum_{\theta \varepsilon \Phi^a(B,x)} T[x|\theta,a]\, \underline{1}\, P^a(\theta)x \\ &= \sum_{\theta \varepsilon \Phi^a(B,x)} \frac{P^a(\theta)x}{\underline{1}P^a(\theta)x}\, \underline{1}\, P^a(\theta)x \\ &= \sum_{\theta \varepsilon \Phi^a(B,x)} P^a(\theta)x \end{aligned}$$

which can be shown to be p.w. linear (see Brumelle and Sawaki [2]).

Model 3. A Machine Replacement Model with Partially Observable States (Sawaki and Ichikawa [9])

This model is an application of model 2 to reliability. A machine consists of n internal components. The state of the machine is the number of working components which is governed by probability transition matrices P. The machine produces constant finished products (say M units) and the machine cannot be inspected, that is, the state of the machine is unobservable. Let θ be the number of defective products out of M finished products. Assume that the conditional probabilities of finding θ given the machine in state i are given by $P\{\theta|X_t = i\} = \gamma_{i\theta}$ $i=0,1,\ldots,n,$ $\theta=0,1,\ldots,M,$

when the machine is not replaced. Assume that $\gamma_{0\theta} > 0$ for each θ. Thus, the only available information about the state is the posterior probability x_i and θ. If the probability at time t is $x = (x_0,\ldots,x_n)$ and θ has been observed, then it will be $T(x|\theta)$ given in Model 2. Let c_i be the operating cost if the machine is in state i and let $c = (c_0,\ldots,c_n)$. The replacement cost is $R > 0$ and the daily expected operating cost is

$$cx = \sum_{i=0}^{n} c_i x_i$$

which is linear in x. It is known from Blackwell [1] that the minimal expected total discounted cost $V^*(x)$ is the unique solution to the optimal equation

$$V^*(X) = \min \{cx + \beta\sum_{\theta} P(\theta|x)V^*(T(x|\theta)),\ R+cn+\beta \sum_{\theta} P(\theta|e)V^*(T(e|\theta))$$

where $P(\theta|x) = x\,P\Gamma_\theta\,\underline{1}$ with $\Gamma_\theta = \text{diag}\,[\gamma_{j\theta}]$, $\underline{1} = (1,1,1\ldots1)\Gamma$ and $e = (0,0,0,\ldots1)$.

Model 4. A classical linear economic model (Sawaki [8])

Let x be a price vector of N commodities (or N securities) in the market and assume that a new price vector x' can be written as

$$x' = P_\theta^a x$$

where P_θ^a is an N × N matrix depending on the present economic situation θ and on an economic alternative a. Let $P[\theta|x,a]$

be the conditional probability of θ forecasted, given x and a. Assume that there exists a simple partition $\{E_i\}$ of the set of price vectors x such that

$$P[\theta \mid x,a] = P^a_{\theta i} \text{ for } x \in E_i,$$

which is p.w. constant with respect to $\{E_i\}$. Therefore, the model belongs to the class of model 1, provided the immediate cost is well defined.

4. ALGORITHM

If the state space X is uncountable, or even countably infinite, then this procedure is difficult to implement on a computer. However, if the Markov decision process has the structure of (A.I.) and V is p.w. linear, then $U^n_\delta V$ is p.w. linear and each δ^n constructed as in theorem 1(iii) is p.w. constant. In this case, the cost functions and policies can be specified by a finite number of items - the inequalities describing each cell of a simple partition and the corresponding action or linear function.

A commonly proposed method for solving Markov decision problems is policy improvement (Howard [2]) which converges "super-linearly" in many situations. Since the successive approximation method converges only linearly, it is desirable to adapt the policy improvement method to our model. Our version of policy improvement includes the successive approximation and policy improvement as special cases.

In the remainder of this section, we discuss the algorithm in general terms, choosing the parameters $\{k_n\}$ which specify the degree of approximation of V^δ in the n-th iteration, terminating the algorithm, and a proof that the algorithm converges.

Algorithm for finding an ε-optimal policy under (A.I.).

Start with a p.w. constant policy δ^0 and a p.w. linear function $y^0 \in B(\Omega)$ satisfying $y^0 \geq U_{\delta^0} y^0$.

An iteration of the algorithm is described as follows: $n = 0,1,2,\ldots$. At the start of the n-th iteration, we have a p.w. constant policy δ^n and a p.w. linear function $y^n \in B(\Omega)$ satisfying $y^n \geq U_{\delta^n} y^n$.

(i) Compute $U_{\delta^n}^{k_n} y^n$ where the integer k_n is the number of iterations of U_{δ^n} which are to be performed.

(ii) Set $y^{n+1} = U_{\delta^n}^{k_n} y^n$ and find a policy δ^{n+1} such that

$$U_{\delta^{n+1}} y^{n+1} = U_* y^{n+1}.$$

(iii) If $\|y^n - y^{n+1}\| \leq (1 - \beta)\varepsilon$, then stop with y^n ε-optimal and δ_n ε-optimal. Moreover, $V^* \leq V^{\delta_n} \leq y^{n+1}$.

(iv) If $\|y^n - y^{n+1}\| > (1 - \beta)\varepsilon$, then increases n by 1 and perform another iteration.

To start, the algorithm needs a p.w. constant policy δ and a p.w. linear function y satisfying $y \geq U_\delta y$. There is no difficulty in finding a p.w. constant policy; for example, $\delta(x)=a$ for all $x \in \Omega$ is satisfactory and one can choose $y^0(x)=M/(1-\beta)$ for each $x \in \Omega$ which satisfies $y^0 \geq U_\delta y^0$, where $M=max\{|c(x,a)| : x\in\Omega, a\in A\}$. Note that if each $k_n = 1$ in Step (i), the algorithm is successive approximation and that if each $k_n = \infty$, the algorithm is reduced into policy improvement.

Theorem 2. For each iteration, $n = 0,1,2,\ldots$, in the algorithm,

$$y^n \geq U_{\delta^n} y^n \geq U_{\delta^n}^2 y^n \geq \ldots \geq U_{\delta^n}^{k_n} y^n = y^{n+1}.$$

In other words, $\{y^n\}$ is a decreasing sequence.

Proof. First, it is true for $n = 0$. Since $y^0 \geq U_{\delta^0} y^0$ and since U_{δ^0} is monotone, it follows that

$y^0 \geq U_{\delta^0} y^0 \geq U^2_{\delta^0} y^0 \geq \ldots \geq U^{k_0}_{\delta^0} y^0 = y^1 \geq U_{\delta^0} y^1$. By definition δ_1 satisfies $U_{\delta_1} y^1 = U_* y^1$. However, $U_* y^1 \leq U_{\delta^0} y^1 \leq y^1$, and so not only is the Theorem established for $n = 0$, but we have also shown that $U_{\delta^1} y^1 \leq y^1$.

Now suppose $U_{\delta^n} y^n \leq y^n$. The same argument as in the first paragraph establishes the Theorem for n and also that $U_{\delta^{n+1}} y^{n+1} \leq y^{n+1}$. Hence the proof is completed by induction.

Corollary. $y^n \geq V^*$ for $n = 1,2,\ldots$.

Proof. For an arbitrary n, $y^n \geq U_{\delta^n} y^n \geq U_* y^n$. Since U_* is monotone, $y^n \geq U^j_* y^n$ for each j. By Lemma 1, $U^j_* y$ decreases monotonically and converges to V^* as $j \to \infty$. Consequently, $y^n \geq V^*$ and the proof is complete.

We next show that if the algorithm terminates then it will provide an ε-optimal cost function and an ε-optimal policy.

Theorem 3. If $\|y^n - y^{n+1}\| \leq (1 - \beta)\varepsilon$, then $\|y^n - V^*\| \leq \varepsilon$, i.e. y^n is ε-optimal. Moreover, δ_n is also ε-optimal and $V^* \leq V^{\delta_n} \leq y^n$.

Proof. Note that $U_{\delta^n} y^n = U_* y^n$ and that by the previous corollary $y^n \geq V^*$.

$$\|y^n - V^*\| \leqq \|y^n - U_* y^n\| + \|U_* y^n - U_* V^*\|$$

$$\leqq \|y^n - U_{\delta^n} y^n\| + \beta\|y^n - V^*\|$$

$$\leqq \|y^n - U^m_{\delta^n} y^n\| + \beta\|y^n - V^*\| \text{ for } m=1,2,\ldots,$$

because $y^n \geq U_{\delta^n} y^n \geq U^m_{\delta^n} y^n$ for $m = 1,2,\ldots$. (Theorem 2).

Thus $\quad (1-\beta)\|y^n - V^*\| \leq \|y^n - U^m_{\delta^n} y^n\| = \|y^n - y^{n+1}\| \leq (1-\beta)\varepsilon,$

and so $\|y^n - V^*\| \leq \varepsilon$.

The last statement in the Theorem follows by Theorem 2 and Corollary.

Theorem 4. Suppose that $\{y_n\}$ is a sequence of costs generated by the algorithm.

(i) y^n converges pointwise to $y \in \beta(\Omega)$.

(ii) $y = U_\star y$, i.e. y is optimal.

In other words, the algorithm converges.

Proof.

(i) First of all we shall show that $\{y^n\}$ is bounded below. By Theorem 2 we have $y^n \geq U^m_{\delta^n} y^n$ for each $m = 1,2,\ldots$. It is well known (see [1] and [4]) that $U^m_{\delta^n} y^n \to V^{\delta^n}$ as $m \to \infty$. Therefore $y^n \geq V^{\delta^n}$. Since the cost $c(x,a)$ is bounded below, i.e. $|c(x,a)| \leq M$ for all x,a, $|V^{\delta^n}| \leq \frac{M}{1-\beta}$. Hence $y^n(x) \geq \frac{-M}{1-\beta}$ for all x. From Theorem 2 y^n is a decreasing sequence. Hence y^n converges pointwise.

(ii) By a choice of y^0 and Theorem 2 we know that

$$y^n \geq U_{\delta^n} y^n \geq U_\star y^n. \tag{3}$$

To show the other way we have

$$\begin{aligned} y^n &= U^m_{\delta^{n-1}} y^{n-1} && \text{(By definition of } y^n\text{)} \\ &\le U_{\delta^{n-1}} y^{n-1} && (U^m_\delta y \le Uy,\ \forall y \in B(\Omega)) \qquad (4) \\ &= U_* y^{n-1} && \text{(By definition of } \delta^{n-1}\text{)}. \end{aligned}$$

Then, from (3) and (4), we obtain

$$U_* y^n \le y^n \le U_* y^{n-1}.$$

From the statement (i) $y^n \searrow y$. Since a contraction mapping U_* is continuous, $U_* y^n \to U_* y$. Therefore, we must have

$$U_* y = y$$

which completes the proof.

Remark. Since the algorithm involves only p.w. linear and constant functions, the cost functions and policies can be specified by a finite number of items - the inequalities describing each cell of a simple partition and the corresponding action or linear function. Therefore, they are easily able to be represented in a computer.

4. ACKNOWLEDGEMENT

The author would like to express his sincere gratitude to Professor Shelby Brumelle, The University of British Columbia, for his many helpful comments and suggestions. Also, special thanks go to Professor Yoshio Iihara for his suggestions and encouragement, and to Professor D.J. White for his many comments and careful reading.

5. REFERENCES

1. Blackwell, D. "Discounted Dynamic Programming", *Annals of Mathematical Statistics*, **36**, 226-235, (1965).

2. Brumelle, S. L. and Sawaki, K. "Generalized Policy Improvement for Simple Dynamic Programs", Working Paper 546, Faculty of Commerce, University of British Columbia, Vancouver, (1978).

3. Brumelle, S. L. and Puterman, M. L. "On the Convergence of Newton's Method for Operators with Supports", Operations Research Center Report ORC 76-12, University of California, Berkeley, (1976).

4. Denardo, E. V. "Contraction Mappings in the Theory Underlying Dynamic Programming", *SIAM Review*, **9**, 165-177, (1967).

5. Denardo, E. V. and Rothblum, U. G. "Affine Dynamic Programming", presented at the 1977 International Conference on Dynamic Programming, University of British Columbia, (1977).

6. Dynkin, E. B. "Controlled Random Sequences", Theory of Probability and Its Applications X, 1-14, (1965).

7. Howard, R. A. "Dynamic Programming and Markov Processes", Wiley, New York, (1960).

8. Sawaki, K. "Piecewise Linear Markov Decision Processes with An Application into Partially Observable Models", presented at the 1978 International Conference on Markov Decision Processes, University of Manchester, (1978).

9. Sawaki, K. and Ichikawa, A. "Optimal Control for Partially Observable Markov Decision Processes Over an Infinite Horizon", *Journal of Operations Research Society of Japan*, **21**, 1-16, (1978).

10. Smallwood, R. D. and Sondik, E. J. "Optimal Control of Partially Observable Markov Processes Over a Finite Horizon", *Operations Research*, **21**, 1071-1088, (1973).

11. Sondik, E. J. "The Optimal Control of Partially Observable Markov Processes Over the Infinite Horizon: Discounted cost", *Operations Research*, **26**, 282-304, (1978).

THE OPTIMALITY OF ISOTONE STRATEGIES FOR MARKOV DECISION PROBLEMS WITH UTILITY CRITERION

Chelsea C. White, III*

(Department of Engineering Science and Systems, University of Virginia)

ABSTRACT

This paper presents sufficient conditions for the existence of specially structured ε- optimal and optimal strategies for Markov decision processes with utility criterion. Three different utility criteria are examined. It is shown that under weak conditions, separably-summarized utility criteria induce memoryless ε-optimal strategies. Conditions are then presented which imply that the set of memoryless, isotone strategies are complete for the summarized utility case. Finally, a stationary Markov decision process possessing a utility criterion slightly more general than the separable form and several additional assumptions are shown to imply the existence of memoryless, stationary, isotone optimal strategies.

1. INTRODUCTION

The research that is presented in this paper has been motivated by the following question: how does criterion definition affect optimal strategy structure for Markov decision processes? In answering a very specific aspect of this question, we utilize the structured strategy results due to Porteus [9] and Kreps and Porteus [8] in order to determine sufficient conditions for the existence of specially structured ε-optimal and optimal strategies for Markov decision processes with utility criteria. The issue of temporal resolution of uncertainty in the context of an expected utility criterion is ignored here; see [7] for further discussion. As a preliminary result, the most general utility criterion considered, the separably-summarized utility, is shown under weak conditions to imply the existence of ε-optimal strategies which are memoryless in the sense that the present state of the controlled

*This research was supported by NSF Grant ENG 76-15774

Markov chain acts as a sufficient statistic with respect to the set of all history remembering strategies. This result is similar to a result presented in [6] and justifies restriction to memoryless strategies for the two other utility criteria considered, the summarizable utility and a utility function presented in [5], which is a slight generalization of the stage invariant, separable utility function. Conditions due to Topkis [10, 11] are then introduced, and results in [11] are slightly generalized which imply the existence of (i) optimal isotone strategies for the summarizable utility case and (ii) optimal isotone, stationary strategies for the generalized, separable utility case.

2. PROBLEM DEFINITION AND NOTATION

Consider the following Markov decision problem with expected utility criterion. Let the state process $\{s_t, t=0,1,\ldots\}$ have as its state space the countable poset (partially ordered set) $(S_t, \leqslant_S)$ at stage t and evolve according to the transition probability $p_t(s_t, d_t, s_{t+1})$ where d_t represents the decision selected at stage t, $p_t(s_t, d_t, s_{t+1}) \geqslant 0$, and $\sum_{s_{t+1}} p_t(s_t, d_t, s_{t+1}) = 1$. (The dependence of $\leqslant_S$ on t will remain implicit for notational brevity.)

The decision selected at stage t is restricted to be a member of the poset $(D_t(s_t), \leqslant_D)$ and is allowed to be chosen on the basis of all former decisions and all past and present realizations of the state process. Thus, the admissible policy space at stage t, Δ_t, is the set of all functions $\delta: H_t \to D_t$, $\delta(h_t) \varepsilon D_t(s_t)$, where $D_t = \underset{s \varepsilon S_t}{X} D_t(s)$, $H_t = Q_0 x \ldots x Q_{t-1} x S_t$, and $Q_t = \{(s,d): s\varepsilon S_t, d \varepsilon D_t(s)\}$. Let $\Pi = \Delta_0 x \Delta_1 x \ldots$ denote the associated admissible strategy space.

Let the extended real-valued function u, defined on $H = H_\infty$, be called the utility function. The problem is to select a strategy $\pi^* \varepsilon \Pi$ such that $E^{\pi^*}(u|h_0) \geqslant E^{\pi}(u|h_0)$ for all $\pi \varepsilon \Pi$ and $h_0 \varepsilon H_0$, where $E^{\pi}(\cdot|h_t)$ is the expectation operator conditioned on the history $h_t \varepsilon H_t$ and strategy π.

In order to conform to the notation found in [8], the following definitions are made. Let $V_t = H_t^{[-\infty,+\infty]}$ be the space of admissible value functions (where in general A^B represents the set of all functions mapping set A into set B). Since a finite horizon Markov decision process with general utility structure is equivalent (c.f. [4]) to a Markov decision process incurring only a terminal charge and having a state space $S_t = H_t$ for all t and since a characteristic of the approach to be taken is that infinite stage problems are indirectly defined as limits of finite stage problems (c.f. [8]), we define the one transition value function b_t as

$$b_t(h_t, d_t, v) = \sum_{s_{t+1}} p_t(s_t, d_t, s_{t+1}) v(h_{t+1}),$$

where $h_{t+1} = (h_t, d_t, s_{t+1})$. The concomitant single stage transition operators $B_{t\delta}([B_{t\delta}v](h_t) = b_t[h_t, \delta(h_t), v])$, $B_t(\pi)$, and A_t, the multistage operators $B_t^n(\pi)$ and A_t^n, the finite horizon value and optimal value functions $v_t^n(\pi)$ and f_t^n, and the infinite horizon value and optimal value functions $v_t^*(\pi)$, $v_t(\pi)$, and f_t are all defined in the Appendix.

Throughout the entire development, we will assume that $\{\bar{u}_t\}$ is the terminal value function sequence where

$$\bar{u}_t(h_t) = \sup\{u(h) : h \varepsilon H \text{ and } h_t(h) = h_t\},$$

and where in general $h_t(h_\tau)$ is the projection of $h_\tau \varepsilon H_\tau$ into H_t for $\tau > t$. Assume also that u is such that $\lim_t \bar{u}_t(h_t(h)) = u(h)$ for all $h \varepsilon H$. Observe that $\bar{u}_t(h_t(h_{t+1})) \geq \bar{u}_{t+1}(h_{t+1})$ for all $h_{t+1} \varepsilon H_{t+1}$ for all t, a fact that will be of substantial importance in the development of our results.

The intent of this paper is to determine sufficient conditions which imply the existence of ε- optimal strategies having certain desirable characteristics. These characteristics include memorylessness (dependence of h_t only through s_t), isotonicity (nondecreasing monotonicity in the state), and stationarity. Our first general class of problem formulations are now examined.

3. SEPARABLY-SUMMARIZED UTILITY

Definition: A utility function u is said to be separably summarized if for every t=0,1,... there exist real-valued functions α_t and β_t defined on $Q_0 x \ldots x Q_{t-1}$, where $\beta_t \geq 0$, and a real-valued function u_t defined on $Q_t x Q_{t+1} x \ldots$ such that $u=\alpha_t+\beta_t u_t$.

Several examples of separably-summarized utility functions, including the usual additive form, are given in [6].

We describe the set of all memoryless policies at stage t as $\Delta_t^*=\{\delta\varepsilon\Delta_t:\delta(h_t)=\delta(h_t') \text{ if } s_t=s_t'\}$, where the set of all memoryless strategies is $\Pi^*=\Delta_0^* x \Delta_1^* x \ldots$. The related structured reward function set which will be examined is

$$V_t^* = \{v\varepsilon V_t: v=\alpha_k+\beta_k v_k \text{ for all } k<t,$$

$$\text{where } v_k:Q_k x \ldots x Q_{t-1} x S_t \rightarrow R\}.$$

Our main results can be stated following the next weak technical assumption.

Assumption:

A1: For all t, if $\{v^n\}$ is a nonincreasing sequence of functions in V_{t+1}, then $B_{t\delta}\left[\lim_n v^n\right] = \lim_n B_{t\delta} v^n$, for all $\delta\varepsilon\Delta_t$.

Observe, for example, the nonincreasing restriction in A1 can be deleted if the set of all $s_{t+1}\varepsilon S_{t+1}$ such that $p_t(s_t,d_t,s_{t+1})\neq 0$ is finite for each pair $(s_t,d_t)\varepsilon Q_t$.

Theorem 1. For the separably-summarized case, the following results hold:

(i) $f_t^n = \sup_{\pi\varepsilon\Pi^*} v_t^n(\pi)\varepsilon V_t^*$ and $f_t^n=A_t f_{t+1}^n$ for all t and n>t.

(ii) $f_t = \lim_n f_t^n = \sup_{\pi\varepsilon\Pi^*} v_t(\pi)\varepsilon V_t^*$ and $f_t = A_t f_{t+1}$ for all t,

(iii) if π is such that $B_t(\pi)f_{t+1} = A_t f_{t+1}$, for all t, then π is optimal (i.e. every conserving strategy is optimal), and

(iv) there exists an ε-optimal memoryless strategy for each $\varepsilon>0$.

Proof: It follows from Theorems 1, 5 and 6 in [8] that the above results hold if five hypotheses are true: regularity, preservation and ε-attainment, decreasing closure, decreasing monotone convergence, and negativity (see the Appendix for their definitions). Regularity follows directly from the definitions and assumptions and from the easily proven fact that $\bar{u}_t \varepsilon V_t^*$ for all t. A straightforward algebraic argument shows that preservation holds.

ε-attainability is proved by noting there is a $\delta\varepsilon\Delta_t^*$ such that for given $\varepsilon>0$ and for each real-valued function v_t on $Q_t x S_{t+1}$,

$$\sum_{s_{t+1}} p_t[s_t,\delta(s_t),s_{t+1}]v_t[s_t,\delta(s_t),s_{t+1}]+\varepsilon$$

$$\geq \sup_{d_t\varepsilon D_t(s_t)} \{\sum_{s_{t+1}} p_t(s_t,d_t,s_{t+1})v_t(s_t,d_t,s_{t+1})\},$$

uniformly on S_t.

Decreasing closure is trivially satisfied; decreasing monotone convergence follows from A1. For negativity, it is sufficient to show that $B_{t\delta}\bar{u}_{t+1} \leq \bar{u}_t$ for all $\delta\varepsilon\Delta_t$, which is straightforward.

Q.E.D.

Observe that the notion of partial order on the sets S_t and $D_t(s)$ have not been utilized. Conditions involving these partial orders could be given which imply the existence of optimal isotone strategies. These conditions, however, are cumbersome and difficult to verify in practice. The next two sections consider two special cases of the separably-summarized utility function which admit relatively simply described sufficient conditions for the existence of optimal isotone strategies.

4. SUMMARIZED UTILITY

In this section, we present conditions which imply the existence of an optimal isotone strategy for a problem having a summarized utility function. Such a utility function is defined as follows:

Definition. The utility function u is said to be summarized if for each t, there exists a real valued function u_t defined on $Q_t x Q_{t+1} x \ldots$ such that $u=u_t$; i.e. $u(s_0,d_0,\ldots) = u_t(s_t,d_t,\ldots)$.

A summarized utility function is a special case of a separably-summarized utility function with the restriction that $\alpha_t=0$ and $\beta_t=1$ for all t. Observe that if $u = \sum_{t=0}^{\infty} r_t(s_t,d_t)$, the usual additive utility function, then $u_t = s_t' + \sum_{k=t}^{\infty} r_k(s_k,d_k)$ is summarized if $s_t' = \sum_{k=0}^{t-1} r_k(s_k,d_k)$ and if s_t' is augmented onto state s_t. Other examples of summarized utility functions are given in [6].

The previous section, and the fact that summarizable utility functions are separably-summarized, guarantee that it is sufficient to assume the following throughout the remainder of this section:

$$\Delta_t = S_t^{D_t}$$

$$V_t = S_t^{[-\infty,+\infty]},$$

where $b_t(s_t,d_t,v) = \sum_{s_{t+1}} p_t(s_t,d_t,s_{t+1})v(s_{t+1})$. Our interest in this section is to present conditions which imply the existence of optimal isotone strategies; hence,

$$\Delta_t^* = \{\delta\varepsilon\Delta_t : \delta \text{ is isotone on } S_t\},$$

where we define isotonicity as follows:

Definition: A function $\delta\varepsilon\Delta_t$ is isotone on S_t if and only if $s\leq_S s'$ implies $\delta(s)\leq_D\delta(s')$.

This definition of a structured policy will naturally be associated with

$$V_t^* = \{v\varepsilon V_t: v \text{ is isotone on } S_t\}.$$

The six assumptions given below will be shown to be sufficient for the existence of optimal strategies in Π^*. We present these assumptions after several important definitions.

Definitions:

(a) A poset that contains a least upper bound and a greatest lower bound for each pair of its elements is said to be a lattice. A subset of a lattice is a sublattice if it is also a lattice with respect to the relation on the original lattice.

(b) A real-valued function f defined on a lattice S is said to be supermodular on S if

$$f(x\wedge y) + f(x\vee y) \geq f(x) + f(y)$$

for all x and y in S, where $x\vee y$ and $x\wedge y$ are a least upper bound and a greatest lower bound, respectively, of the points x and y.

(c) A subset K of a poset $(S,\leq)$ is said to be increasing if $s\varepsilon K$ and $s\leq s'$ imply $s'\varepsilon K$.

Supermodularity, through the more easily recognized and conceptualized concept of isotone differences, is closely related to the economic concept of complementary products. Supermodular functions on lattices have several interesting qualitative correspondences to convex functions on convex sets. These correspondences are more fully discussed in [10] and its references.

Assumptions:

B1: For each t and $v\varepsilon V_{t+1}$, there exists a $\delta\varepsilon\Delta_t$ such that $B_{t\delta}v=A_t v$.

B2: $\bar{u}_t\varepsilon V_t^*$ for all t.

B3: For each t, $\sum_{s_{t+1}\varepsilon K} p_t(s_t,d_t,s_{t+1})$ is isotone in s_t and supermodular in (s_t,d_t) on Q_t for each increasing subset K of S_{t+1}.

B4: For each t, if for s, $s'\varepsilon S_t$ and $s\leq_S s'$, then $D_t(s)\subseteq D_t(s')$.

B5: For each t, Q_t is a sublattice.

B6: For each t, if $\{v^n\}$ is a nonincreasing sequence of functions in V_{t+1}, then $B_{t\delta}[\lim_n v^n]=\lim_n B_{t\delta}v^n$ for all $\delta\varepsilon\Delta_t$.

B1 will be used to prove the existence of an optimal strategy. A number of well-known conditions suffice for B1 to hold, e.g. finiteness of the decision sets or compactness of the decision sets with appropriate smoothness conditions imposed on the function b_t. The isotonicity assumption in B3 is a generalization of Derman's increasing failure rate assumption [3]. The supermodularity assumption in B3 appears to be quite reasonable in a diverse variety of contexts such as advertising and pricing decisions [1] and queueing control [12]. B4 implies that as the state of the system increases (with respect to $\leq_S$), the number of options available to the decision-maker does not decrease. B5 and B6 are weak technical assumptions required for the use of results in [10,11] and [8].

We can now state the main result of this section.

Theorem 2. For the summarized utility case, assume B1 through B6 hold. Then,

(i) $f_t^n = \sup_{\pi\varepsilon\Pi_*} v_t^n(\pi)\ \varepsilon V_t^*$ and $f_t^n=A_t f_{t+1}^n$ for all t and n>t,

(ii) $f_t = \lim_n f_t^n=\sup_{\pi\varepsilon\Pi_*} v_t(\pi)\varepsilon V_t^*$ and $f_t=A_t f_{t+1}$ for all t,

(iii) if π is such that $B_t(\pi)f_{t+1}=A_t f_{t+1}$, for all t, then π is optimal, and

(iv) there exists an optimal (structured) strategy in Π^*

Proof: This proof will follow the same outline as the proof of Theorem 1. Regularity follows from the definitions and B2.

For preservation, assume $v \in V^*_{t+1}$. The isotonicity assumption in B3 then implies that $b_t(s_t,d_t,v)$ is isotone in s_t on Q_t (c.f. [2]). Thus, for

$$s_t \lesssim_S s_t', \quad \sup_{d_t \in D_t(s_t)} b_t(s_t,d_t,v) \leq \sup_{d_t \in D_t(s_t)} b_t(s_t',d_t,v)$$

$$\leq \sup_{d_t \in D_t(s_t')} b_t(s_t',d_t,v),$$

where the last inequality follows from B4.

For attainability, Theorem 6.3 in [11], the isotonicity of v, the supermodularity assumption in B3, and B5 imply that $b_t(s_t,d_t,v)$ is supermodular on Q_t. Theorem 1.5 in [11] then implies that $\sup\{b_t(s_t,d_t,v): d_t \in D_t(s_t)\}$ is supermodular on S_t. This result, the assumption that the supremum is achieved for each $s_t \in S_t$, and Lemma 1.1 in [11] imply the existence of a $\delta \in \Delta^*_t$ such that $B_{t\delta}v = A_t v$.

Decreasing closure is again trivially satisfied; decreasing monotone convergence follows from B6. Proof of negativity follows as before.

Q.E.D.

5. STATIONARITY

The last section presented conditions which imply the existence of optimal strategies which are isotone, i.e. composed of isotone policies. We now wish to examine a subclass of Markov decision problems with separably-summarized utility functions which admit optimal strategies which are both isotone and stationary. When the existence of an optimal strategy is assured, Kreps has given conditions which guarantee that an optimal stationary strategy exists (c.f. [5] Corollary 2). We now make the following assumptions to (essentially) conform to the problem statment presented in [5].

Assume for all t that:

(i) $S = S_t$

(ii) $D(s) = D_t(s)$ for all $s \in S$

(Observe, therefore, that $Q=Q_t$)

(iii) $p(s,d,s') = p_t(s,d,s')$ for all $s,s' \in S$, $d \in D(s)$

(iv) there exist real-valued functions g and γ defined on Q such that

$$u(s_0,d_0,s_1,d_1,\ldots)=g(s_0,d_0)+\gamma(s_0,d_0)u(s_1,d_1,\ldots)$$

for all $h \in H$.

Observe that our construction differs slightly from that presented in [5] in that (a) the notion of a summary descriptor is (implicitly) augmented onto the state (to redefine the notion of state) and (b) decisions and the utility function are (explicitly) allowed to be based on past actions selected (the development in [5] would consider this latter issue by augmenting the decision onto the state).

Observe also that (iv) implies the existence of real-valued functions G_t and Γ_t, both defined on $Qx\ldots xQ$(t times), such that

$$u(s_0,d_0,\ldots,s_t,d_t,\ldots) = G_t(s_0,d_0,\ldots,s_{t-1},d_{t-1})$$

$$+\Gamma_t(s_0,d_0,\ldots,s_{t-1},d_{t-1})u(s_t,d_t,\ldots)$$

where $G_0=0$, $\Gamma_0=1$, and

$$G_{t+1}(s_0,d_0,\ldots,s_t,d_t) = G_t(s_0,d_0,\ldots,s_{t-1},d_{t-1})$$

$$+g(s_t,d_t)\Gamma_t(s_0,d_0,\ldots,s_{t-1},d_{t-1})$$

and

$$\Gamma_{t+1}(s_0,d_0,\ldots,s_t,d_t)=\gamma(s_t,d_t)\Gamma_t(s_0,d_0,\ldots,s_{t-1},d_{t-1}).$$

Hence, u is separably-summarized with $\alpha_t=G_t$, $\beta_t=\Gamma_t$, and $u_t=u$. It is therefore sufficient to restrict consideration to $V=V_t=S^{[-\infty,+\infty]}$ and $\Delta=\Delta_t=S^D$.

Define the real-valued function $\bar{u}_t^{\ell}$ on $Q^{t-\ell}\times S$ as

$$\bar{u}_t^{\ell}(s_\ell,d_\ell,\ldots,d_{t-1},s_t) =$$

$$\sum_{k=\ell}^{t-1} g(s_k,d_k)\prod_{m=\ell}^{k-1}\gamma(s_k,d_k)+\bar{u}_0(s_t)\prod_{m=\ell}^{t-1}\gamma(s_m,d_m).$$

We observe that $\bar{u}_t^{0}=\bar{u}_t$ and that $\bar{u}_t^{\ell}=g(s_\ell,d_\ell)+\gamma(s_\ell,d_\ell)\bar{u}_t^{\ell+1}$ for $\ell<t$, where $\bar{u}_t^{t}=\bar{u}_0$, since (it is easily shown) $\bar{u}_t$ satisfies the recursion

$$\bar{u}_t(s_0,d_0,\ldots,s_{t-1},d_{t-1},s_t)=g(s_0,d_0)$$

$$+\gamma(s_0,d_0)u_{t-1}(s_1,d_1,\ldots,s_{t-1},d_{t-1},s_t).$$

It is thus sufficient to examine the (usual) "expected utility to be accrued" problem. For this problem,

$$b(s,d,v)=b_t(s,d,v)=g(s,d)+\gamma(s,d)\sum_{s'} p(s,d,s')v(s'),$$

where the terminal value function for all t is $\bar{u}_0$. Observe also that, for example, if $\bar{u}_0$ is uniformly bounded on S and if $|\gamma(s,d)|<1$ for all $(s,d)\varepsilon Q$, then $\bar{u}_t$ converges to a separable utility function.

Results in [5] will be referenced which guarantee an optimal stationary strategy exists. We therefore concentrate on isotonicity and define $\Delta^*=\{\delta\varepsilon\Delta:\delta$ is isotone on $S\}$ and $V^*=\{v\varepsilon V$: v is nonnegative and isotone on $S\}$. Consider the following assumptions:

C1: For all $v\varepsilon V$ there exists a $\delta\varepsilon\Delta$ such that $B_\delta v=Av$.

C2: $\bar{u}_0\varepsilon V^*$.

C3: $\sum_{s'\varepsilon K} p(s,d,s')$ is isotone in s and supermodular in (s,d) on Q for each increasing subset K of S.

C4: g(s,d) and $\gamma(s,d)$ are nonnegative and supermodular in (s,d) and isotone in s on Q.

C5: If $s, s' \varepsilon S$ and $s \leq_S s'$, then $D(s) \subseteq D(s')$.

C6: Q is a sublattice.

C7: If $\{v^n\}$ is a nonincreasing sequence of functions in V, then $B_\delta\left[\lim_n v^n\right] = \lim_n B_\delta v^n$ for all $\delta\varepsilon\Delta$.

Observe that since expected utility to-be-accrued is now being examined, $\bar{u}_O$ represents the terminal value function for each finite horizon problem.

We can now state the main result for the stationary Markov decision problem.

Theorem 3: Assume C1 through C7 hold. Then,

(i) $f=f_t=\lim_n f_t^n= \sup_{\pi\varepsilon\Pi}{}_* v_t(\pi)\varepsilon V^*$,

(ii) $f = Af$,

(iii) If π is such that $B_t(\pi)f=Af$ for all t, then π is optimal, and

(iv) there exists an optimal stationary, (structured) strategy in Π^*.

Proof: Again, this proof will follow the same outline as the proof of Theorem 1. Regularity follows from the definitions and C2; showing that preservation holds is straightforward.

For attainability, the supermodularity of the sum $\Sigma_{s'} p(s,d,s')v(s')$ on Q follows as in the proof of Theorem 2; clearly, this sum is nonnegative. The supermodularity and nonnegativity of the product $\gamma(s,d)\Sigma_{s'} p(s,d,s')v(s')$ follows from C4 and a straightforward adaptation of the proof of Theorem 3.1 in [10]. Supermodularity and nonnegativity are preserved under summation; hence, b(s,d,v) is supermodular on Q for $v\varepsilon V^*$. Attainability then follows as in the proof of Theorem 2.

The decreasing closure assumption is trivially satisfied; C7 gives decreasing monotone convergence. Negativity hold if $B_\delta \bar{u}_O \leq \bar{u}_O$ for all $\delta\varepsilon\Delta$, which is proved as follows. Since $\bar{u}_1 \leq \bar{u}_O$,

$$\bar{u}_1(s_O,d_O,s_1) = g(s_O,d_O)+\gamma(s_O,d_O)\bar{u}_O(s_1) \leq \bar{u}(s_O)$$

for all d_0 and s_1; hence, $b(s_0,d_0,\bar{u}_0) \leq \bar{u}_0(s_0)$ for all $(s_0,d_0) \varepsilon Q$, implying the result.

The existence of an optimal strategy which is stationary follows from Corollary 2 [5] and the existence of an optimal strategy (c.f.(iii)).

Q.E.D.

6. CONCLUSIONS

Sufficient conditions have been presented for the existence of memoryless ε-optimal strategies for a Markov decision problem with separably-summarized utility criterion. Two special cases of this utility criterion were examined further. For the summarized utility case, the set of memoryless, isotone strategies were shown to possess an optimal strategy under certain isotonicity and supermodularity assumptions on the cost and dynamic structure of the problem. Additional assumptions were then shown to imply the existence of optimal memoryless, isotone, stationary strategies for a utility criterion introduced by Kreps [5]. The computational implications of these results is a topic of future research.

7. REFERENCES

1. Albright, S. C. and Winston, W. "Markov decision models of advertising and pricing decisions", *Operations Research*, **27**, 668-681, (1979).

2. Barlow, R. E. and Proschan, F. "Theory of maintained systems: distribution of time to first system failure", *Math. of O.R.*, **1**, 1-27, (1976).

3. Derman, C. "On optimal replacement rules when changes in state are Markovian", in Mathematical Optimization Techniques, R. Bellman (ed.), U. of Calif. Press, Berkeley, (1963).

4. Kreps, D. M. "Decision problems with expected utility criteria, I: upper and lower convergent utility", *Math. of O.R.*, **2**, 45-53, (1977).

5. Kreps, D. M. "Decision problems with expected utility criteria, II: stationarity", *Math. of O.R.*, **2**, 266-274, (1977).

6. Kreps, D. M. "Markov decision problems with expected utility criterion", Tech. Rep. No. 29, Dept. of Opns. Res., Stanford U., Stanford, Calif., (1975).

7. Kreps, D. M. and Porteus, E. L. "Temporal resolution of uncertainty and dynamic choice theory", *Econometrica*, **46**, 185-200, (1978).

8. Kreps, D. M. and Porteus, E. L. "On the optimality of structured policies in countable stage decision processes II: positive and negative problems", *SIAM J. Appl. Math.*, **32**, 457-466, (1977).

9. Porteus, E. L. "On the optimality of structured policies in countable stage decision processes", *Management Science*, **22**, 148-157, (1975).

10. Topkis, D. M. "Minimizing a subadditive function on a lattice", *Opers. Res.*, **26**, 305-321, (1978).

11. Topkis, D. M. "Applications of minimizing a subadditive function on a lattice", working paper, (1977).

12. Winston, W. "Optimality of monotonic policies for a queuing system with variable arrival and service rates", Tech. Rep. Grad. School of Business, Indiana U., Bloomington, Ind., (1976).

8. APPENDIX

Several definitions presented elsewhere [8,9] which are used in the main body of the paper are now listed for the convenience of the reader.

For a given $b_t(h_t, d_t, s_{t+1})$,

(i) $B_{t\delta}$ is such that $[B_{t\delta}v](h_t) = b_t[h_t, \delta(h_t), v]$.

(ii) $B_t(\pi)$ is such that $B_t(\pi)v = B_{t\delta_t}v$, where $\pi = \{\delta_0, \ldots, \delta_t, \ldots\} \varepsilon \Pi$.

(iii) A_t is such that $A_t v = \sup_{\delta \varepsilon \Delta_t} B_{t\delta} v = \sup_{\pi \varepsilon \Pi} B_t(\pi) v$.

(iv) $B_t^n(\pi)$ is such that $B_n^n(\pi)v = v$ and $B_t^n(\pi)v = B_t(\pi)B_{t+1}^n(\pi)v$ if $t = 0, 1, \ldots, n-1$.

(v) A_t^n is such that $A_n^n v = v$ and $A_t^n v = A_t A_{t+1}^n v$ $t = 0, 1, \ldots, n-1$.

(vi) $v_t^n(\pi)$ is such that $v_n^n(\pi) = \bar{u}_n$ and $v_t^n(\pi) = B_t(\pi) v_{t+1}^n(\pi)$, $t = 0, 1, \ldots, n-1$, where $\bar{u}_n$ is defined in the Problem

Definition and Notation section.

(vii) $f_t^n = \sup_{\pi\varepsilon\Pi} v_t^n(\pi)$.

(viii) $v_t^*(\pi)$ is such that $v_t^*(\pi)=\lim_n \sup v_t^n(\pi)$ taken pointwise.

(ix) $v_t(\pi)=\lim_n v_t^n(\pi)$ when the limit exists.

(x) $f_t = \sup_{\pi\varepsilon\Pi} v_t^*(\pi)$.

The following five definitions are used in the proofs:

(i) Regularity. For every t, the following hold: $\Phi\neq\Delta_t^*\subseteq\Delta_t$; for each $h_t\varepsilon H_t$ and $d_t\varepsilon D_t(s_t)$, there exists a $\delta\varepsilon\Delta_t$ such that $\delta(h_t)=d_t$; $V_t^*\subseteq V_t\subseteq H_t^{[-\infty,+\infty]}$, $B_{t\delta}:V_{t+1}\to V_t$ for all $\delta\varepsilon\Delta_t$; $B_{t\delta}$ is isotone for all $\delta\varepsilon\Delta_t$; there exists a positive real number α_t such that $B_{t\delta}$ is α_t-Lipschitzian on V_{t+1} for all $\delta\varepsilon\Delta_t$; and $\bar{u}_t\varepsilon V_t^*$.

(ii) Preservation and ε-attainment. For all t, $A_t:V_{t+1}^*\to V_t^*$. For all t, $v\varepsilon V_{t+1}^*$, and $\varepsilon>0$, there exists $\delta\varepsilon\Delta_t^*$ such that $B_{t\delta}v+\varepsilon\geq A_t v$.

(iii) Decreasing closure. For every t, if $\{v^n\}$ is a nonincreasing sequence of functions in $V_t(V_t^*)$, then $\lim_n v^n\varepsilon V_t(V_t^*)$.

(iv) Decreasing monotone convergence. For every t, if $\{v^n\}$ is a nonincreasing sequence of functions in V_{t+1}, then

$B_{t\delta}[\lim_n v^n] = \lim_n B_{t\delta}v^n$ for all $\delta\varepsilon\Delta_t$.

(v) Negativity. For every t, $A_t\bar{u}_{t+1}\leq\bar{u}_t$.

DYNAMIC PROGRAMMING AND AN UNDISCOUNTED, INFINITE HORIZON, CONVEX STOCHASTIC CONTROL PROBLEM

Roger Hartley

(Department of Decision Theory, University of Manchester)

1. INTRODUCTION

In this paper we consider a stochastic control problem from the point of view of the theory of Markov decision processes. The model investigated has linear dynamics in discrete time with convex costs and we will be interested in the long term expected average cost criterion. Such a model could be regarded as a Markov decision problem with rather general state and action spaces and we may ask such typical questions as: is there an optimal policy; is there an optimal stationary policy; are the functional equations of dynamic programming satisfied? Conditions which are sufficient (but not necessary) to guarantee affirmatory answers to these questions are available for problems with uncountably infinite state and action spaces (see Tijms [10]) but unfortunately these conditions are not satisfied by the model at hand. The main obstacles arise from the unboundedness of the cost function and the difficulty in establishing ergodicity of policies (or a similar condition). However, we shall show that under reasonable and easily verified conditions on the cost function, the dynamics and the random noise, we can establish the desired results. Our proof follows similar lines to those often pursued in studying dynamic programming with the average cost criterion *viz.* we approach this case as the limit of the discounted case with the discount factor tending to one. Although theorems are available to justify such a procedure directly (e.g. see Ross [9]) they do not apply here. It is necessary to pursue somewhat unconventional routes which exploit strongly our assumptions and the structure of the model. We hope that the development that follows will partially justify our contention that the viewpoint and techniques of Markov decision theory have some value in the analysis of stochastic control problems even when the straightforward "plugging in" of standard theorems is not possible.

In the next section we describe our model in detail and state the main theorem. The next three sections are concerned with proving the main theorem. In the first of these sections we quote the main theorem for discounted optimality and prove some basic results. The following section is devoted to the topic of switching policies and their properties. The proof of the main theorem itself is given in section five whilst in the final section we discuss extensions of theorem 1 and related results.

2. THE MODEL AND THE MAIN THEOREM

We will suppose that the state space is $\mathbb{R}^m$ and the action space is $\mathbb{R}^n$. The time parameter t is a non-negative integer and we will write $x_t \in \mathbb{R}^m$ for the state at time t and $u_t \in \mathbb{R}^n$ for the action taken at time t. The transition law can then be written

$$x_{t+1} = Ax_t + Bu_t + v_t \qquad \text{for } t \geq 0 \tag{1}$$

where A and B are suitably dimensioned matrices and v_t is the realisation of the random variable V_t where $V_o, V_1, \ldots$ is a sequence of independent, identically distributed random variables, each being a copy of the random variable V.

The <u>history</u> of the process up to time t is the sequence of states and actions occurring prior to t, together with x_t. The set of all histories up to time t (a finite dimensional Euclidean space) is denoted by H_t. A <u>non-anticipative policy</u> π is a sequence of mappings $u_o, u_1, \ldots$ where u_t: $H_t \rightarrow \mathbb{R}^n$, is a Baire function. Such a policy, π together with the transition law (1) and the known distribution of the sequence $V_o, V_1, \ldots$ induces a stochastic process in the state which we shall denote $\{X_t | t \geq 0\}$. An important subset of the set of non-anticipative policies is the class of Markov policies in which the mapping u_t depends only on x_t and not on the remainder of the history at time t. In this case $\pi = (u_o(.), u_1(.), \ldots)$ a sequence of Baire functions from $\mathbb{R}^m$ to $\mathbb{R}^n$ and $\{X_t | t \geq 0\}$ becomes a non-homogeneous Markov chain with transitions given by

$$X_{t+1} = AX_t + Bu_t(X_t) + V_t \qquad \text{for } t \geq 0$$

An important subset of the set of Markov policies are those in which there exists some Baire function $u: \mathbb{R}^m \to \mathbb{R}^n$ such that $u_t = u$ for $t = 0,1,\ldots$ In this case we say that π is stationary and write $\pi = u^\infty$. We note that the stochastic process generated by a stationary policy is a time-homogeneous Markov chain.

We shall write E_x^π for expectations with respect to the process $\{X_t \mid t \geq 0\}$ when conditioned on $X_o = x$.

The return function $r: \mathbb{R}^m \times \mathbb{R}^n \to \mathbb{R}$ is convex and therefore continuous. Consequently, for any $x \in \mathbb{R}^m$ and any Markov policy $\pi = (u_o, u_1, \ldots)$ we can define the long run average expected cost, starting in state x and using policy π, by

$$\rho(x,\pi) = \limsup_{t\to\infty} \{E_x^\pi \sum_{i=0}^{t} r(X_i, u_i(X_i))/(t+1)\}$$

Our objective is to minimise ρ over Markov policies and we will say that such a policy π^* is optimal if

$$\rho(x,\pi^*) = \inf_{\pi} \rho(x,\pi)$$

where the infinium is over Markov policies.

One case of this model which has been intensively studied is that in which $r(x,u)$ is a positive, semi-definite, quadratic form. However, this form for r drastically simplifies the analysis and permits efficient computational schemes to be devised. None of this simplicity carries over into the more general problem. A good treatment of the quadratic case may be found in Kushner [7].

Before stating our main theorem we will present three conditions on the return function and the dynamics.

(A1) The return function r(x,u) is <u>inf-compact</u>, i.e. for any $\gamma\in\mathbb{R}$ the level set

$$L_\gamma = \{(x,u) | r(x,u) \leq \gamma\} \tag{2}$$

is compact. In our case, continuity of r means that this can be weakened to the requirement that L_γ be bounded, if it is non-empty. A standard result in convex analysis allows us to weaken this still further to the stipulation that r should have some bounded non-empty level set [7]. In the case when r is quadratic, inf-compactness is equivalent to being positive definite.

(A2) The pair (A,B) is <u>controllable</u>, i.e. there exists an integer $p(\leq m)$ such that

$$\text{rank } (A^{p-1}B, A^{p-2}B, \ldots, AB, B) = m \tag{3}$$

This condition is equivalent to the assumption that if the random noise V_t is omitted from the transition law (1) then for any x_o we can choose $u_o, \ldots, u_{p-1}$ so that $x_p = 0$ (see, e.g. Barnett [1]). A further equivalent form is given in lemma 2 below.

(A3) The random variable V has <u>bounded support</u>, i.e. there is a bounded set S such that $V\in S$ w.p.1. Although this assumption excludes such interesting and important distributions as gaussian, in practice we can truncate such a distribution to a bounded region with very little effect on the probability density function and such a procedure would be also necessary in realistic computational schemes.

It is worth remarking that all three assumptions are easy to verify which is far from the rule in Markov decision theory (cf. the paper by L. C. Thomas in this volume).

Theorem 1

Suppose (A1), (A2) and (A3) are satisfied. Then

(i) There exists a stationary policy, say $\hat{u}^\infty$, which is optimal.

(ii) $\rho(x,\hat{u}^\infty)$ is independent of x and equal to $\hat{\gamma}$, say.

(iii) there exists an inf-compact, convex function $g:\mathbb{R}^m \to \mathbb{R}$ which satisfies

$$\hat{\gamma} + g(x) = \inf_u E_V\{r(x,u) + g(Ax+Bu+V)\} \tag{4}$$

where E_V indicates expectation with respect to the distribution of V. Furthermore the infimum in (4) is achieved at $\hat{u}(x)$ for all $x \in \mathbb{R}^m$ and $\hat{\gamma}$ is a minimal solution of (4) in the sense that, if (γ',g') satisfies (4) where g' is any real-valued convex function which is bounded below, then $\gamma' \geq \hat{\gamma}$.

3. SOME BASIC RESULTS

Our proof of theorem 1 is based on a consideration of the limiting form of the dynamic programming functional equations for the corresponding discounted problem and so we commence with a statement of the known results for such problems.

For any real α satisfying $0 < \alpha < 1$, $x \in \mathbb{R}^m$ and Markov policy π we can define the expected discounted return starting in state x, using policy π when the discount factor is α, by

$$\rho_\alpha(x,\pi) = E_x^\pi \sum_{t=0}^{\infty} \alpha^t r(X_t,u_t(X_t))$$

where $\pi = (u_0,u_1,\ldots)$. We define $g_\alpha: \mathbb{R}^m \to \mathbb{R}$, by

$$g_\alpha(x) = \inf_\pi \rho_\alpha(x,\pi)$$

where the infimum is over all Markov policies. We will say that a Markov policy π^* is <u>α-optimal</u> if

$$\rho_\alpha(x,\pi^*) = g_\alpha(x).$$

We then have the following result which is proved in [4].

<u>Theorem 2</u>

Assume (A1). Then

(a) g_α is inf-compact and convex.

(b) $$g_\alpha(x) = \inf_u E_v\{r(x,u) + \alpha g_\alpha((Ax+Bu+v)\} \qquad (5)$$

(c) there is a Baire function $\hat{u}_\alpha : \mathbb{R}^m \to \mathbb{R}^n$ such that the infimum in (5) is achieved at $\hat{u}_\alpha(x)$ for all $x \in \mathbb{R}^m$ and $\hat{u}_\alpha^\infty$ is an α-optimal policy.

The proof of theorem 1 proceeds by a series of lemmas the first of which proves useful in considering limits of inf-compact functions.

<u>Lemma 1</u>

Let $\{f_i | i = 1,2,\ldots\}$ be a sequence of inf-compact, convex functions on $\mathbb{R}^m$ and suppose that $\lim_{i\to\infty} f_i$ exists and is inf-compact. Then

$$\lim_{i\to\infty} \inf_x f_i(x) = \inf_x \lim_{i\to\infty} f_i(x)$$

(We remark that our result implies the existence of the limit on the left-hand side.)

<u>Proof</u>

By a standard argument (independent of inf-compactness

and convexity) we see that

$$\limsup_{i\to\infty} \inf_x f_i(x) \leq \inf_x \lim_{i\to\infty} f_i(x) \tag{6}$$

To establish the opposite inequality we first note that the inf-compactness of a function implies that its infimum is achieved. Let us define a sequence $\{x_i^* | i = 1,2,...\} \subseteq \mathbb{R}^m$ by choosing x_i^* to be any point at which f_i achieves its infimum. We claim that the sequence $\{x_i^* \; i = 1,2,...\}$ is bounded. To establish this let us define $f = \lim_{i\to\infty} f_i$ and suppose, without losing generality, that $f(x) \geq f(0)$ for all $x \in \mathbb{R}^m$. (Note that, since f is inf-compact, its infimum is achieved.) Now choose any $\beta > 0$. Since f is inf-compact there exists a real σ such that

$$f(x) \geq \beta + f(0) \quad \text{if} \quad ||x|| = \sigma \tag{7}$$

Now, for any $\beta' \in (0,\beta)$ we can choose i_o such that, for all $i \geq i_o$

$$f_i(0) \leq \beta' + f(0)$$

Suppose that there were a subsequence $\{x_{i_j}^* | j = 1,2,...\}$ of $\{x_i^* | i = 1,2,...\}$ such that $i_j \geq i_o$ and $||x_{i_j}^*|| > \sigma$ for $j = 1,2,...$ Then

$$f_{i_j}(\sigma x_{i_j}^* / ||x_{i_j}^*||) \leq (\sigma/||x_{i_j}^*||) f_{i_j}(x_i^*)$$
$$+ (1-\sigma/||x_{i_j}^*||)\; f_{i_j}(0) \leq f_{i_j}(0) \quad [\text{since } f_{i_j}(x_{i_j}^*) \leq f_{i_j}(0)]$$
$$\leq \beta' + f(0)$$

Now the fact that the surface of a sphere is a compact set would allow us to select a sub-subsequence $\{x_{k_j}^* | j = 1,2,...\}$ such that

$$\sigma x_{k_j}^* / ||x_{k_j}^*|| \to \hat{x}$$

But the controllability condition (3) guarantees that we can find matrices $N_o^*, \ldots N_{p-1}^*$ such that

$$A^{p-1}BN_o^* + \ldots + BN_{p-1}^* = -A^p$$

which means that $x_p = 0$ for any x_o. □

4. SWITCHING POLICIES

Define the non-anticipative policy π^* by $\pi^* = (u_o^*, u_1^*, \ldots)$ where, for any integer $t \geq 0$, if $t = kp+s$ where $0 \leq s \leq p-1$, we put $u_t^* = N_s^* x_{kp}$. The next lemma specifies some useful properties of process $\{X_t\}$ generally by π^*.

Lemma 3

Assume (A2) and (A3). There is a random vector W^*, a convex function $f^*: \mathbb{R}^m \to \mathbb{R}$ and a real number κ^*, which is a lower bound on f^*, satisfying the following properties

(i) W^* has bounded support

(ii) X_{kp} is distributed as W^* for $k = 1,2,\ldots$ and any X_o.

(iii) $$E_x^{\pi^*} r(X_t, u_t^*(X_t)) \leq f^*(x) \quad \text{for } t = 0,1,\ldots,p-1$$
$$\leq \kappa^* \quad \text{for } t = p, p+1, \ldots$$

Proof

A simple calculation shows that

$$X_t = M_t^* X_o + W_t^* \qquad \text{for } t = 0,1,\ldots, p-1$$

$$X_{kp+s} = M_s^* W^* + W_s^* \quad \text{for } k = 1,2,\ldots;\ s = 0,1,\ldots,p-1$$

where the matrices $M_o^*, M_1^*, \ldots, M_{p-1}^*$ are defined recursively by for some $\hat{x}\in\mathbb{R}^m$ satisfying $||\hat{x}|| = \sigma$. But convergence of finite convex functions is uniform on compact sets [7] and we could deduce that $f(\hat{x}) \leq \beta' + f(0)$ contradicting (7) and justifying our claim.

We have seen that there is a closed bounded set S such that $x_i^* \in S$ for all i. By the uniformity result alluded to above this means that for any $\epsilon > 0$ we can find i_1 such that, for all $i \geq i_1$,

$$f_i(x_i^*) \geq f(x_i^*) - \epsilon$$

$$\geq f(0) - \epsilon$$

which gives

$$\liminf_{i\to\infty} f_i(x_i^*) \geq f(0) - \epsilon$$

Since this is true for any $\epsilon > 0$ it is also valid if $\epsilon = 0$ which, together with (6) proves the lemma. □

The next result explores the concept of controllability and shows that the controls required to drive the state vector to the origin can be chosen as linear functions of x_o.

<u>Lemma 2</u>

Assume (A2). Then there exist (n x m)-matrices $N_o^*,\ldots,N_{p-1}^*$ such that if we define $x_{i+1} = Ax_i + Bu_i$ and put $u_i = N_i^* x_o$ for $i = 0,1,\ldots,p-1$ then $x_p = 0$ whatever the value of x_o.

<u>Proof</u>

For any (n x m)-matrices $N_o,\ldots,N_{p-1}$ if we put $u_i = N_i x_o$ for $i = 0,1,\ldots,p-1$ we must have

$$x_p = A^p x_o + A^{p-1}BN_o x_o + \ldots. + BN_{p-1}x_o$$

$$M_o^* = I_m \qquad \text{(the (m x m) unit matrix)}$$

$$M_t^* = AM_{t-1}^* + BN_{t-1}^* \qquad \text{for } t = 1,2,\ldots,p-1$$

and the random vectors $W_o^*, W_1^*, \ldots, W_p^*$ are defined recursively by

$$W_o^* = 0$$

$$W_t^* = AW_{t+1}^* + V_{t-1} \qquad \text{for } t = 1,2,\ldots, p-1$$

and W* is a copy of W_p^* which is independent of W_s^* in the expression for X_{kp+s}. These expressions show that, since $V_o, \ldots, V_{p-1}$ all have bounded support and W* is the image of $V_o, V_1, \ldots, V_{p-1}$ under some linear map, W* must have bounded support. We note that W_t^* also has bounded support for t = 0,1,...,p-1.

Let us define, for t = 0,1,...,p-1 and $x \in \mathbb{R}^m$

$$f_t(x) = E_{W_t^*}\, r(M_t^* x + W_t^*, N_t^* x)$$

$$\kappa^* = \max_{t=0,1,\ldots,p-1} E_{W^*} f_t(W^*)$$

We observe that the bounded support of W_t^* and W* ensures the $f_t(x)$ and κ^* are finite. We can then define f* by

$$f^*(x) = \max \{f_o^*(x), f_1^*(x), \ldots, f_{p-1}^*(x), \kappa^*\}$$

and properties (ii) and (iii) are readily verified from our assumptions and the convexity of r. □

The next stage of the argument uses π^* to construct another policy. Let us fix x, $y \in \mathbb{R}^m$ and let $\{Y_t^\alpha | t = 0,1,\ldots\}$ be the Markov chain generated by the α-optimal policy $\hat{u}_\alpha^\infty$, conditioned on $Y_o^\alpha = y$. Let $D_o \subset \mathbb{R}^m$ be the set

$$D_o = \{z | \underline{r}(z) \leq 2p\ \kappa^*\} \tag{8}$$

where

$$\underline{r}(z) = \inf_w r(z,w) \tag{9}$$

and κ^* is the number quoted in the previous lemma. Our policy is specified in terms of a given sequence $(E_o, E_1, \ldots, E_{p-1})$ of subsets of $\mathbb{R}^m$ and we will show in the next lemma how the sets $E_o, \ldots, E_{p-1}$ may be chosen in order that the policy should have certain desirable properties. We can define an extended integer-valued random variable I^α by

$$I^\alpha = \inf \{i | Y^\alpha_{2pi} \in D_o,\ \hat{u}_\alpha(Y^\alpha_{2pi+s}) \in E_s,\ s = 0,1,\ldots,p-1\}$$

where $I^\alpha = \infty$ if $Y^\alpha_{2pi} \in D_o$, $\hat{u}(Y^\alpha_{2pi+s}) \in E_s$ for no i. We can then define a random variable

$$\tau^\alpha = 2pI^\alpha + p$$

and note that τ^α is a stopping time.

The non-anticipative policy $\pi_\alpha(y) = (u_o, u_1, \ldots)$ is defined for any sample path as follows, using the notation of lemma 2. Put

$$u_t = u^*_t \qquad \text{for } t = 0,1,\ldots,2p-1$$

$$= u^\alpha_t(x_t) \qquad \text{for } t \geq \tau_\alpha$$

If $2p \leq t < \tau^\alpha$ and $t = 2pk+s$, for $s = 0,1,\ldots,p-1$ put

$$u_t = \hat{u}_\alpha(y^\alpha_t) + N^*_s(x_{2pk} - y^\alpha_{2pk})$$

$$\text{if } y^\alpha_{2pk} \in D_o \text{ and } \hat{u}_\alpha(y^\alpha_{2pk+\ell}) \in E_\ell \text{ for } \ell = 0,1,\ldots,s$$

$= 0 \qquad$ if $y^{\alpha}_{2pk} \in D_o$ and some $\hat{u}\ (y^{\alpha}_{2pk+\ell}) \notin E_\ell$

for $\ell = 0,1,\ldots,s$

$= u^*_t \qquad$ if $y^{\alpha}_{2pk} \notin D_o$

and for $s = p,p+1,\ldots,2p-1$ put

$$u_t = u^*_t$$

where $x_t(y^{\alpha}_t)$ denotes the value of X^{α}_t (and Y^{α}_t) at time t. We see that $\pi_\alpha(y)$ is non-anticipative since u_t is a function only of x_t, and $y^{\alpha}_{t'}$ for $t' \leqslant t$. The former is known and the observed sequence of $\{x_{t'} | t' \leqslant t\}$ means that the values of the noise vectors can be calculated for $t' < t$ and hence, knowing $\hat{u}_\alpha$, we can find $y^{\alpha}_{t'}$.

Since the construction of policy $\pi_\alpha(y)$ is rather involved we will try to explain its structure in informal terms. If $E_o, E_1, \ldots, E_{p-1} = \mathbb{R}^m$ then the policy uses π^* until Y^{α}_{2pk} first enters D_o. It then "switches" to $\hat{u}^{\infty}_{\alpha}$ in the sense that the controls are chosen to drive the "difference" process $Y^{\alpha}_t - X_t$, which is deterministic, to 0 in p steps. Thereafter the processes coincide and the same controls are used for each process, i.e. $\hat{u}_\alpha$. Up until the stage at which switching takes place the total discounted loss so far, excluding only the first 2p stages, incurred by policy $\pi_\alpha(y)$ is no greater than the loss in-by $\hat{u}_\alpha$. This follows from our choice of D_o. Our objective is to arrange for the difference between the total discounted loss up to the t'th stage using $\pi_\alpha(y)$ and that using $\hat{u}^{\infty}_{\alpha}$ to be uniformly bounded in t and α and we can achieve this by making the expected loss incurred using $\pi_\alpha(y)$ during the switching steps bounded. It is for this reason that we introduce the sets $E_o, E_1, \ldots, E_{p-1}$. If, during the switching steps, we find $\hat{u}_\alpha(y^{\alpha}_{2pk+s}) \notin E_s$ then we "abort" the switch by choosing $u_t = 0$ for p-s-1 steps and then reverting to π^*. In the next lemma we shall demonstrate that it is possible to find sets $E_o, E_1, \ldots, E_{p-1}$ which are all bounded and also have the property that if we "abort", as described above, the total discounted loss up to stage $2p(k+1)$ using $\pi_\alpha(y)$ is no greater than that using $\hat{u}_\alpha^\infty$. The boundedness of $E_o, E_1, \ldots, E_{p-1}$ allows us to bound the loss incurred in switching.

Lemma 4

Assume (A1) - (A3). Fix $x, y \in \mathbb{R}^m$ and let $\pi_\alpha(y)$ be the policy described above. Let $\alpha_o \in (0,1)$. Then the sets $E_o, E_1, \ldots, E_{p-1}$ can be chosen so that for all $\alpha \in (\alpha_o, 1)$ we have, for some constant K (which may depend on α_o)

$$E_x^{\pi_\alpha(y)} \sum_{t=0}^{\infty} \alpha^t r(X_t, U_t) \leq g_\alpha(y) + 2pf^*(x) + K - \underline{r}(y) \qquad (10)$$

where $\{U_t | t = 0,1,\ldots\}$ is the stochastic process of controls generated using $\pi_\alpha(y)$ when $X_o = x$ and f^* is defined in lemma 3 and $\underline{r}$ is defined in (9).

Proof

We will first describe a recursive scheme for obtaining a sequence of sets $E_o, E_1, \ldots, E_{p-1}$ and then show that (10) is satisfied for the resulting $\pi_\alpha(y)$.

Suppose that for some $s = 0,1,\ldots,p-1$ we are given D_o and $E_o, E_1, \ldots, E_{s-1}$ a sequence of non-empty, bounded sets. We shall show how to determine E_s and prove that it is non-empty and bounded. We first note that the compactness of D_o is a equivalent to $\underline{r}$ being inf-compact and this can be shown directly or by appealing to results in [3]. The non-emptiness of D_o is trivial.

For $z, d_o, e_o, e_1, \ldots, e_{s-1} \in \mathbb{R}^m$ define a non-anticipative policy $\bar{\pi} = (\bar{u}_o, \bar{u}_1, \ldots)$ as follows (if $s=0$, $\bar{\pi}$ depends on no other variables).

$$
\begin{aligned}
\bar{u}_t(x) &= e_t + N_t^*(z - d_o) && \text{if } 0 \leqslant t \leqslant s-1 \\
&= 0 && \text{if } s \leqslant t \leqslant p-1 \\
&= u_t^* && \text{if } \quad t \geqslant p.
\end{aligned}
$$

This allows us to define, for $s = 0, 1, \ldots, p$,

$$\phi_s(z, d_o, e_o, e_1, \ldots, e_{s-1}) = E_z^{\bar{\pi}} \sum_{t=0}^{2p-1} r(X_t, \bar{u}_t(X_t))$$

and we note that, since every control is a linear function of x, ϕ_s is a convex function. Further it follows easily from the linearity of the controls, the finite support of the random noise V and the continuity of r that ϕ_s is finite-valued. Hence it must be continuous [7]. This allows us to define a real number μ_s by

$$\mu_s = \sup \phi_s(z, d_o, e_o, e_1, \ldots, e_{s-1})$$

where the supremum is over $z \in S'$, $d_o \in D_o$, $e_\ell \in E_\ell$ for $\ell = 0, 1, \ldots, s-1$, with S' any bounded set such that $W^* \in S'$ w.p.1. We may now put

$$E_s = \{w \in \mathbb{R}^n \mid \inf_z r(z, w) \leqslant \mu_s \alpha_o^{-s}\}$$

The compactness of E_s follows from a similar argument to that used to establish compactness of D_o. We remark that $\bar{\pi}$ mimics the switching controls of $\pi_\alpha(y)$ when a "switch" is aborted at the s'th step.

We will prove (10) by partitioning the sum. Firstly we have

$$E_x^{\pi_\alpha(y)} \sum_{t=0}^{2p-1} \alpha^t r(X_t,U_t) = E_x^{\pi^*} \sum_{t=0}^{2p-1} \alpha^t r(X_t,u_t^*(X_t))$$

$$\leq 2pf^*(x)$$

$$\leq 2pf^*(x) + E_y^\alpha \sum_{t=0}^{p-1} \alpha^t r(Y_t^\alpha,u_t(Y_t^\alpha)) - r(y,u_\alpha(y))$$

$$\leq E_y^\alpha \sum_{t=0}^{p-1} \alpha^t r(Y_t^\alpha,u_t(Y_t^\alpha)) + 2pf^*(x) - \underline{r}(y) \quad (11)$$

where the second line follows from lemma 3 and $\alpha < 1$, $r \geq 0$.

Now for $k = 1,2,...$ we have

$$E_x^{\pi_\alpha(y)} [\sum_{t=2kp}^{2kp+2p-1} \alpha^t r(X_t,U_t) | \tau < 2kp] =$$

$$E_y^\alpha [\sum_{t=2kp}^{2kp+2p-1} \alpha^t r(X_t,\hat{u}_\alpha(X_t)) | \tau < 2kp] \quad (12)$$

and

$$E_x^{\pi_\alpha(y)} [\sum_{t=2kp}^{2kp+2p-1} \alpha^t r(X_t,U_t) | \tau > 2kp, Y^\alpha_{2kp} \notin D_o]$$

$$= E_{W^*} E_{W^*}^{\pi^*}[\sum_{t=2kp}^{2kp+2p-1} \alpha^t r(X_t,u_t^*)]$$

$$\leq 2p\kappa^*\alpha^{2pk}$$

$$\leq E_y[\alpha^{2pk}\ r(Y^\alpha_{2pk},\hat{u}_\alpha(Y^\alpha_{2pk}))\,|\,\tau>2kp,\ Y^\alpha_{2kp}\notin D_o]$$

$$\leq E^\alpha_y[\sum_{t=2kp}^{2kp+2p-1} \alpha^t r(Y^\alpha_t,\hat{u}_\alpha(Y^\alpha_t))\,|\,\tau>2kp,\ Y^\alpha_{2kp}\notin D_o] \qquad (13)$$

where the equality follows from the fact that $\tau > 2kp$ means, by lemma 3 (ii), that X_{2kp} is distributed as W* for $k\geq 1$. The first inequality is a consequence of lemma 3 (iii) and the second inequality is immediate from the definition of D_o. We have used our assumption that $r\geq 0$ for the third inequality.

Next, if s = 0,1,...p-1 we will define A_s to be the event that $\tau > 2kp$, $\hat{u}_\alpha(Y^\alpha_{2kp+\ell}) \in E_\ell$, $\ell = 0,1,\ldots,s-1$ and $\hat{u}_\alpha(Y^\alpha_{2kp+s}) \notin E_s$. Then we have

$$E^{\pi_\alpha(y)}_x [\sum_{t=2kp}^{2kp+2p-1} \alpha^t r(X_t,U_t)\,|\,A_s]$$

$$\leq \alpha^{2kp}\ E_{W*}\ E^\alpha_y[\phi_s(W*,Y^\alpha_{2kp},\ \hat{u}_\alpha(Y^\alpha_{2kp}),\ldots$$

$$\ldots,\ \hat{u}_\alpha(Y^\alpha_{2kp+s-1}))\,|\,A_s]$$

$$\leq \alpha^{2kp+s}\ \alpha_o^{-\varepsilon}\ \mu_s$$

$$\leq E^\alpha_y[\alpha^{2kp+s}\ r(Y^\alpha_{2kp+s},\hat{u}_\alpha(Y^\alpha_{2kp+s}))\,|\,A_s]$$

$$\leq E_y[\sum_{t=2kp}^{2kp+2p-1} \alpha^t r(Y^\alpha_t,\hat{u}_\alpha(Y^\alpha_t))\,|\,A_s] \qquad (14)$$

where the first inequality follows from the fact that X_{2kp} is distributed as W* and by comparing $\overline{\pi}$ and $\pi_\alpha(y)$. The second inequality is a consequence of the way we have defined μ_s and the third inequality follows from the definition of E_s.

Finally, we have the case that switching takes place without aborting.

$$E_x^{\pi_\alpha(y)}\left[\sum_{t=2kp}^{2kp+2p-1} \alpha^t r(X_t,U_t) \,\middle|\, \tau = 2kp+p\right]$$

$$= \alpha^{2kp} E_{W*} E_y^\alpha [\phi_p \quad (W^*, Y_{2kp}^\alpha, \hat{u}(Y_{2kp}^\alpha), \ldots$$

$$\ldots, \hat{u}_\alpha(Y_{2kp+p-1}^\alpha)) \mid \tau = 2kp+p]$$

$$\leq \alpha^{2kp}$$

$$\leq \mu_{p-1} + E_y^\alpha\left[\sum_{t=2kp}^{2kp+2p-1} \alpha^t r(Y_t^\alpha, \hat{u}_\alpha(Y_t^\alpha)) \,\middle|\, \tau = 2kp+p\right] \tag{15}$$

Observing that the conditioning events in (12), (13), (14) and (15) for $s = 0,1,\ldots,p-1$ partition the sample space, we may combine the inequalities to obtain, for $k = 1,2,\ldots$

$$E_x^{\pi_\alpha(y)} \sum_{t=2kp}^{2kp+2p-1} \alpha^t r(X_t,U_t) \leq p_k \mu_p \quad +$$

$$+ E_y^\alpha \sum_{t=2kp}^{2kp+2p-1} \alpha^t r(Y_t^\alpha, \hat{u}_\alpha(Y_t^\alpha))$$

where p_k is the probability that $\tau = 2pk+p$. Summing this over k, adding (11) and writing K for $\mu_p \Pr(\tau<\infty)$ we achieve the inequality (10), thus completing the proof. □

We would like to appeal to theorem 2 to conclude that the left hand side of (10) is bounded below by $g_\alpha(x)$. Unfortunately this cannot be done since $\pi_\alpha(y)$ is not a Markov policy. However, the next lemma states that the result we require is still true.

<u>Lemma 5</u>

Assume (A1), (A2), (A3). Then, for any $x,y \in \mathbb{R}^m$,

$$g_\alpha(x) \leq E_x^{\pi_\alpha(y)} \sum_{t=0}^{\infty} \alpha^t r(X_t,U_t)$$

where we use the notation of lemma 4.

Proof

Theorem 2(b) asserts (inter alia) that for any $x \in \mathbb{R}^m$, $u \in \mathbb{R}^n$,

$$g_\alpha(x) \leq E_V[r(x,u) + \alpha g_\alpha(Ax+Bu+V)]$$

By iterating this inequality it follows that for any k = 1, 2,...

$$g_\alpha(x) \leq E_x^{\pi_\alpha(y)} \sum_{t=0}^{2kp-1} \alpha^t r(X_t,U_t) + \alpha^{2kp} E_x^{\pi_\alpha(y)} g_\alpha(X_{2kp}) \tag{16}$$

But

$$\begin{aligned} E_x^{\pi_\alpha(y)} g_\alpha(X_{2kp}) &= E_x^{\pi_\alpha(y)}[g_\alpha(X_{2kp}) \mid \tau < 2kp]\Pr(\tau < 2kp) \\ &\quad + E_x^{\pi_\alpha(y)}[g_\alpha(X_{2kp}) \mid \tau > 2kp]\Pr(\tau > 2kp) \\ &= E_y^\alpha[g_\alpha(Y_{2kp}^\alpha) \mid \tau < 2kp]\Pr(\tau < 2kp) \\ &\quad + E_{W*}[g_\alpha(W*) \mid \tau > 2kp]\Pr(\tau > 2kp) \\ &\leq E_y^\alpha g_\alpha(Y_{2kp}^\alpha) + (1-\alpha)^{-1} E_{W*} f*(W*) \end{aligned} \tag{17}$$

where we have used our assumption that $r \geq 0$, which implies $g_\alpha \geq 0$, and the inequality $g_\alpha(x) \leq \rho_\alpha(x,\pi^*) \leq (1-\alpha)^{-1} f^*(x)$, from lemma 3(iii) (using a simple argument based on Theorem 2)

Now theorem 2 also tells us that for any $z \in \mathbb{R}^m$

$$g_\alpha(z) = E_V[r(z,\hat{u}_\alpha(z)) + \alpha g_\alpha(Az+B\hat{u}_\alpha(z)+V)]$$

and iterating this gives, for any k = 1, 2,...

$$\begin{aligned} g_\alpha(y) &= E_y^\alpha \sum_{t=0}^{2kp-1} \alpha^t r(Y_t^\alpha,\hat{u}_\alpha(Y_t^\alpha)) + \alpha^{2kp} E_y^\alpha g_\alpha(Y_{2kp}^\alpha) \\ &= \rho_\alpha(y,\hat{u}_\alpha^\infty). \end{aligned}$$

Taking limits as $k \to \infty$ shows that

$$\alpha^{2kp} E_y^\alpha g_\alpha(Y_{2kp}^\alpha) \to 0 \tag{18}$$

Hence, if we let $k \to \infty$ in (16) and use (17) and (18) we obtain the inequality of the lemma. □

For any $x \in (0,1)$ and $x \in \mathbb{R}^m$ we will define

$$\Delta g_\alpha(x) = g_\alpha(x) - g_\alpha(0).$$

Lemma 6

There exists a sequence $\{\alpha_i | i \geq 1\} \subseteq (0,1)$ and an inf-compact, convex function $g : \mathbb{R}^m \to \mathbb{R}$ such that $\alpha_i \to 1$ and $\Delta g_{\alpha_i} \to g$ (pointwise) as $i \to \infty$. Furthermore, there is a continuous function $h ; \mathbb{R}^m \to \mathbb{R}$ such that $|\Delta g_{\alpha_i}(x)| \leq h(x)$ for $i = 1,2,...$ and any $x \in \mathbb{R}^m$.

Proof

From lemmas 4 and 5 we conclude that for any $x,y \in \mathbb{R}^m$, $\alpha \in (\alpha_0,1)$

$$g_\alpha(x) \leq g_\alpha(y) + 2p\, f^*(x) + K - \underline{r}(y) \tag{19}$$

for some $\alpha_0 \in (0,1)$. Since r is bounded below, this means that there is some constant K' such that

$$-2f^*(0) - K' \leq \Delta g_\alpha(x) \leq 2pf^*(x) + K' \tag{20}$$

Consequently, for any sequence of α_j's in $(\alpha_0,1)$ with limit 1 as $j \to \infty$, the real number sequence $\{\Delta g_{\alpha_j}(x)\}$ is bounded for each x and by a result of convex analysis [7, p.91] this guarantees the existence of a finite convex function g and a sub sequence $\{\alpha_i | i \geq 0\}$ such that $\alpha_i \to 1$ and $\Delta g_{\alpha_i} \to g$ pointwise as $i \to \infty$.

To show that g is inf-compact we observe that (19) implies that for all $x \in \mathbb{R}^m$

$$\Delta g_\alpha(x) \geq \underline{r}(x) - 2pf^*(0) - K$$

and thus

$$g(x) \geq \underline{r}(x) - 2pf^*(0) - K$$

and hence, for any $\gamma \in \mathbb{R}$

$$\{x|g(x)\leq\gamma\} \subseteq \{x|\underline{r}(x) \leq \gamma+2pf^*(0)+K\}$$

which means that g has bounded non-empty level sets since $\underline{r}$ is inf-compact.

The last assertion of the lemma is established by taking

$$h(x) = K + 2p \max\{f^*(x), f^*(0)\}$$

for any $x \in \mathbb{R}^m$ and inequality $|\Delta g_\alpha| \leq h$ is immediate from (20). □

5. PROOF OF THEOREM 1

We have now assembled all the necessary machinery to prove the main theorem.

We start by re-writing the discounted optimality equation (5) as

$$(1-\alpha_i)g_{\alpha_i}(0) + \Delta g_{\alpha_i}(x) =$$

$$\inf_u E_V\{r(x,u) + \alpha_i \Delta g_{\alpha_i}(Ax+Bu+V)\} \tag{21}$$

where $\{\alpha_i|i\geq 1\}$ is the sequence defined in lemma 6. Using the results of that lemma together with lemma 1 and the dominated convergence theorem (with h of lemma 6 as dominating function and recalling that V has bounded support) we may deduce that the right hand side of (21) has as its limit as $i \to \infty$ the right hand side of the average cost functional equation (4). Hence $(1-\alpha_i)\Delta g_{\alpha_i}(0)$ has a limit, say γ, as $i \to \infty$ and the limit of (21) is (4) with γ replacing $\hat{\gamma}$.

To show that $\gamma = \hat{\gamma}$ we first claim that the function defined on $\mathbb{R}^m \times \mathbb{R}^n$ by

$$(x,u) \rightarrow E_V\{r(x,u) + g(Ax+Bu+V)\} \quad (22)$$

is convex and inf-compact. This may be deduced directly from the definition or by appealing to results in [3]. It follows from a result in [4] that there is a Baire function $\hat{u} : \mathbb{R}^m \rightarrow \mathbb{R}^n$ with the property that, for any $x \in \mathbb{R}^m$,

$$E_V\{r(x,\hat{u}(x)) + g(Ax+B\hat{u}(x)+V)\} = \inf_u E_V\, r(x,u) + \\ +g(Ax+Bu+V)\}.$$

This means that, if we write $\hat{\pi}$ for the policy $\hat{u}^{\infty}$, then

$$\begin{aligned} g(x) &= E_V\{r(x,\hat{u}(x)) + g(Ax+B\hat{u}(x)+V)\} - \gamma \\ &= E_x^{\hat{\pi}}\, r(X_o,\hat{u}(X_o)) + E_x^{\hat{\pi}}\, g(X_1) - \gamma \\ &= E_x^{\hat{\pi}}\, r(X_o,\hat{u}(X_o)) + E_x^{\hat{\pi}}\, r(X_1,\hat{u}(X_1)) + E_x^{\hat{\pi}}\, g(X_2) - 2\gamma \\ &= \ldots\ldots \\ &= E_x^{\hat{\pi}} \sum_{i=0}^{t} r(X_i,\hat{u}(X_i)) + E_x^{\hat{\pi}}\, g(X_{t+1}) - (t+1)\gamma \quad (23) \end{aligned}$$

Now g is inf-compact and bounded below (by ν, say) and so we have

$$\gamma \geq (t+1)^{-1} E_x^{\hat{\pi}} \sum_{i=0}^{t} r(X_i,\hat{u}(X_i)) + (\nu - g(x))/(t+1)$$

Letting $t \rightarrow \infty$ gives $\gamma \geq \rho(x,\hat{\pi})$.

To obtain the reverse inequality we observe that for any Markov policy $\pi = (u_o, u_1,\ldots)$ we have, by definition of g_α,

$$\begin{aligned} \gamma = \lim_{i\to\infty} (1-\alpha_i) g_{\alpha_i}(0) &\leq \lim_{i\to\infty} (1-\alpha_i) E_o^{\pi} \sum_{t=0}^{\infty} \alpha_i^t r(X_t,u_t(X_t)) \\ &= \lim_{i\to\infty} (1-\alpha_i) \sum_{t=0}^{\infty} \alpha_i^t E_o^{\pi} r(X_t,u_t(X_t)) \\ &\leq \limsup_{t\to\infty} (t+1)^{-1} \sum_{i=0}^{t} E_o^{\pi} r(X_t,u_t(X_t)) \end{aligned}$$

$$\leq \rho(x,\pi)$$

where the second line follows from the monotone convergence theorem and the third line from an Abelian theorem [11]. Putting $\pi = \hat{\pi}$ in this inequality shows that $\gamma = \rho(x,\hat{\pi}) = \hat{\gamma}$ and hence that $\hat{\pi}$ is an optimal policy. This proves (i), (ii) and the first part of (iii).

To complete the proof we will suppose that (γ',g') satisfies (4) where g' is real-valued, convex and bounded below. This implies that the function defined on $\mathbb{R}^m \times \mathbb{R}^n$ by (22) above, with g' in place of g, is convex and real-valued. We may now appeal to theorem 2 of [6] to conclude that for any $\varepsilon > 0$ there exists a Baire function $u_\varepsilon : \mathbb{R}^m \to \mathbb{R}^n$ such that for any $x \in \mathbb{R}^m$

$$\gamma' + g'(x) \geq E_V\{r(x,u_\varepsilon(x)) + g'(Ax+Bu_\varepsilon(x)+V)\} - \varepsilon$$

Iterating this inequality gives, for any integer $t \geq 0$,

$$\gamma'+g'(x)/(t+1) \geq (t+1)^{-1}E_x^{\pi_\varepsilon} \sum_{i=0}^{t} r(X_i,u_\varepsilon(X_i)) +$$

$$E_x^{\pi_\varepsilon} g'(X_t)/(t+1) - \varepsilon$$

where we have written π_ε for u_ε^∞. Letting $t \to \infty$, this gives

$$\gamma' \geq \rho(x,\pi_\varepsilon) - \varepsilon \geq \gamma - \varepsilon$$

and since ε is arbitrary we may conclude that $\gamma' \geq \gamma$. □

6. EXTENSIONS AND RELATED RESULTS

In this section we will describe several ways in which our results may be generalised.

Firstly, we can relax the requirement (A1) that r be inf-compact to the requirement that it be duotonic. This concept is formally defined in [3]. An informal definition may be given by saying that, restricted to any one-dimensional subspace of $\mathbb{R}^{m+n}$, the function r be inf-compact or constant. In the case of a quadratic function this relaxes positive definiteness to positive semi-definiteness. If r is duotonic, theorem 1 still holds.

A proof may be found in [2].

Secondly, we can allow that prior to each stage we have a forecast of the noise vector at that stage. Knowing the distribution of noise conditional on any forecast and the distribution of forecasts enables us to write down a generalisation of the functional equation (4) and prove the corresponding version of theorem 1. More details may be found in [2]. By applying Jensen's inequality to these results it can be shown that the long run average expected cost for the corresponding deterministic problem (i.e. when the random variable V is replaced by its expectation) is a lower bound on $\hat{\gamma}$ in theorem 1.

Finally, the assumption of bounded support (A3) can be dropped, but only at the cost of introducing extra assumptions on r and V. However, these extra assumptions are weak enough to allow several interesting cases of r and V including quadratic r and any V with finite variance. For further details we refer the reader to [5].

7. REFERENCES

[1] S. BARNETT, Introduction to Mathematical Control Theory, Oxford University Press, London, (1975).

[2] G.D. BASTERFIELD, University of Manchester Ph.D. Thesis, forthcoming.

[3] R. HARTLEY, Inequalities for a class of sequential stochastic decision processes, Proceedings of Oxford International Conference on Stochastic Programming, Ed. M.J. Dempster, Academic Press, New York, forthcoming.

[4] R. HARTLEY, Inequalities for a class of stochastic control problems, Decision Theory Note No. 62. University of Manchester.

[5] R. HARTLEY, Linear, convex, undiscounted, discrete-time stochastic control and dynamic programming, Decision Theory Note No. 63, University of Manchester.

[6] R. HARTLEY, Inequalities in completely convex stochastic programming, Decision Theory Note No.50, University of Manchester.

[7] H.J. KUSHNER, Introduction to Stochastic Control, Holt, Rinehart and Winston, New York (1971).

[8] R.T. ROCKAFELLAR, Convex Analysis, Princeton University, N.J., (1972).

[9] S.M. ROSS, Arbitrary state Markov decision processes, Annals of Mathematical Statistics, 39, pp.2118-2122, (1968).

[10] H. TIJMS, On Dynamic Programming with Arbitrary State Space, Compact Action Space and the Average Return as Criterion, Research Report, Mathematical Centre, Amsterdam, (1975).

[11] D.V. WIDDER, The Laplace Transform, Princeton University Press, Princeton N.J., (1946).

OPTIMAL CONTROL OF SYSTEMS SUBJECT TO RANDOM JUMP DISTURBANCES AND THEIR REPRESENTATION BY MARKOV DECISION PROCESSES

R. C. H. Cheng

(University of Wales Institute of Science and Technology)

ABSTRACT

The optimal control of systems subject to random Markov disturbances is considered. To date solution procedures have been based on optimal control theory and dynamic programming principles which yield differential conditions satisfied by the optimal solution. These conditions are implicit in form and hence are hard to use in practice. In this paper it is shown that under suitable conditions such a problem can be discretised and be approximated by a Markov Decision Process. This opens the way to practical solution procedures based on algorithms such as the policy iteration method which appears to be particularly efficient in such situations.

1. INTRODUCTION

A number of authors have considered the optimal control of stochastic systems of the form

$$\frac{dx}{dt} = f(x(t),u(t),J(t)) \tag{1.1}$$

where J(t) is a finite state Markov (jump) process. We shall consider the undiscounted, infinite horizon case of finding, within a given class U of stationary control functions, a control u which maximises the long-term expected average return per unit time

$$G = \lim_{T\to\infty} E\{\frac{1}{T}\int_0^T g\{x(t),u(x(t),J(t)),J(t)\}dt\}, \tag{1.2}$$

where g(x,u,j) is the return rate.

Problems of this type have been considered by Krassovskii and Lidskii [1]. A linear case has been considered by Wonham [2]. The undiscounted finite horizon case has been discussed in a general setting by Rishel [3] and Goor [4]. Related problems have been considered by Sworder [5]. The techniques used in the above references draw mainly upon the ideas of optimal control theory, with reliance placed much on maximum principles and the formulation of adjoint differential conditions. A criticism of this type of approach, from the practical point of view, is that it leads to conditions that are implicit in character, and which are difficult to implement into practical algorithms for actual numerical calculations. Difficulties of using this type of approach are mentioned by Cheng [6].

An alternative approach is to note that under suitable conditions, an optimal control problem of the above type can be represented approximately by a Markov decision process (MDP) It is then possible to obtain an approximation of the optimal control policy by solving the MDP problem. It turns out that the policy iteration method as described by Howard [7], is a particularly efficient method to use in this situation.

In order to motivate what follows, we give a rough outline of the procedure, before proceeding to details. First, define the conditional profit

$$\psi(x,i,t) = E\{\int_t^T g ds \mid x(t)=x, J(t)=i\}.$$

A differential-difference argument similar to that used by Wonham [2] relating the state of the system at t and t+dt then yields

$$\{\frac{\partial\psi(x,i,t)}{\partial t}\}dt + \{\frac{\partial\psi(x,i,t)}{\partial x}\}dx + \lambda_{ii}\psi(x,i,t)dt +$$

$$\sum_{j\neq i} \lambda_{ij}\psi(x,j,t)dt + g(x,i)dt = 0$$

where λ_{ij} is the transition intensity matrix defining the jump process. If we anticipate the result and assume that to first order in x and t, $\psi(x,i,t) = v(x,i) + (T-t)G$, (where

G is as defined in (1.2)) then $\frac{\partial \psi}{\partial t} = -G$, $\frac{\partial \psi}{\partial x} = \frac{\partial v}{\partial x}$, and the differential condition can be re-arranged in the form

$$Gdt + v(x,i) = (1 + \lambda_{ii}dt)v(x+fdt,i) + \sum_{j \neq i} (\lambda_{ij}dt)v(x,j) + g(x,i)dt \tag{1.3}$$

This equation is the continuous version of the well-known equation in MDP theory used in the value determination step. To see this, suppose that some appropriate discretization of the x-space can be made and that $\alpha = (x,i)$ denotes the discretized overall state of the system. Moreover suppose the system is observed only at points in time spaced dt apart. The (1.3) is of the form

$$G + v_{\alpha} = g_{\alpha} + \sum_{\beta} p_{\alpha\beta}v_{\beta}, \text{ all } \alpha, \tag{1.4}$$

with the interpretation, in terms of MDP terminology, that if we are in state $\alpha = (x,i)$ at time t, then at the next step we either (i) change to jump state j when we move to state $\beta = (x,j)$, this occurring with probability $p_{\alpha\beta} = \lambda_{ij}dt$, or else (ii) remain in jump state i, when we move to state $\beta = (x+fdt,i)$, this occurring with probability $p_{\alpha\beta} = 1 + \lambda_{ii}dt$.

The validity of the above interpretation is dependent on a discretization being possible which yields (1.4) as being an appropriate approximation of (1.3). A considerable amount of care is required if the discretization is to be truly valid. In section 2 conditions will be given under which such a discretization can be made, and some consequences of these conditions are given in section 3. In section 4 we show that the behaviour of the MDP resulting from such a discretization closely approximates that of the continuous control process. Our main result is Theorem 4.4 which gives a sufficient condition under which a sequence of MDP's will converge to yield the solution of the continuous control problem. Section 5 concludes with some brief comments on the computational advantages of the approach.

2. ASSUMPTIONS

2.1 Continuous Control Process

The assumptions are similar to those given in Rishel [3] in that the behaviour of $x(t)$ is assumed to be piecewise smooth. However as we consider the infinite horizon case we require certain positive recurrence conditions to ensure that a particular discretization remains valid in the long term. Also, the conditions have to hold in some uniform way to ensure that the discretization is valid over the class of stationary controls to be considered. The first set of assumptions, A, gives the general setting under which the process is to be observed.

A(i) The overall state of the system at t is described by $(x(t),J(t))$, where the point position $x(t)\in\Omega$, a compact region of some finite dimensional Euclidean space. The jump process $J(t)$, a finite state Markov process, takes values in $I = \{0,1,\ldots,\ell\}$ and is characterised by the transition matrix (λ_{ij}), $i,j \in I$, where

$$pr(J(t+h)=j|J(t)=i) = \lambda_{ij}h + o(h),\ pr(J(t+h)=i|J(t)=i) = 1 - \lambda_i h + o(h),$$

with $\lambda_i = -\lambda_{ii} = \sum_{j\neq i} \lambda_{ij}$. The process is assumed to be irreducible with all jump states positive recurrent.

A(ii) For each jump state i, $\Omega = \bigcup_n C_{n,i}$, a finite union of disjoint sets, to be called cells. Each cell $C_{n,i}$ is a relatively open subset of $\{x \mid \theta_{n,i}(x)=0\}$, where $\theta_{n,i}:\Omega \to E_{n-d_{n,i}}$ is a continuously differentiable mapping. Call $d_{n,i}$ the dimension of the cell.

A(iii) We make no specific assumption concerning the form of f or the form of restrictions on u in (1.1). As we shall only be considering stationary control policies, we use an alternative formulation that is more compact for our purposes. Let $\frac{dx}{dt} = u(x,i)$, $x \in \Omega$, $i \in I$, $u \in U$, where U is a given subset of $F(B_0,L_0)$ the set of all functions which are bounded by B_0

and are Lipschitz continuous on each $C_{n,i}$, with Lipschitz constant L_O. i.e. $|f(x,i)| < B_O \ \forall\ x,i$, and $|f(x,i) - f(y,i)| < L_O|x - y| \ \forall\ x,y \in C_{n,i}$, each i.

A(iv) The return rate is $g(x,u(x,i),i)$, when $u \in U$. We assume that $g \in F(B_1,L_1)$ whenever $u \in U$.

The next set of conditions places restrictions on the individual trajectories and incorporates the condition of positive recurrence. For given $i \in I$ and $u \in U$ the position vector x traces out a trajectory as time passes. The position of x at time t+h given $x(h) = x_O$ will be denoted by $y_t(x_O)$; the notation suppresses reference to the particular u giving rise to the trajectory as this should be clear from the context. The assumptions are

B(i) For given $i \in I$, each $u \in U$ induces a partial ordering of the cells $C_{n,i}$ so that each trajectory on leaving a cell must enter one of higher order; moreover the dimension of cells is non-increasing as the order increases.

B(ii) For given $i \in I$, all trajectories terminate in a cell of dimension zero, called a rest point, after a finite time T_O, which is independent of $u \in U$ and $i \in I$.

The final assumption concerns the stability of trajectories. Its effect is similar to the "positive angle" condition (8) of Rishel [3]. It ensures that trajectories which start out close together, remain so, by requiring that two trajectories close together but in different cells reach the same cell within a short time.

C There is a constant K_O such that for any $\varepsilon > 0$, $u \in U$, if the jump state remains at i, and if $x_1 \in C_{m_1,i}$, $x_2 \in C_{m_2,i}$ and $|x_1 - x_2| < \varepsilon$, then a cell $C_{m,i}$ can be found with $y_{t_1}(x_1) \in C_{m,i}$, $y_{t_2}(x_2) \in C_{m,i}$ for some $t_1, t_2 < K_O\varepsilon$; where $d_m \leqslant \min(d_{m_1}, d_{m_2})$.

This completes the specification of our assumptions of how the process behaves. The assumptions might appear at first sight to be somewhat involved and artificial. It may be of help therefore, before proceeding further, to give an example to show how they actually arise quite naturally in certain applications at least. The example concerns the operation of a certain industrial chemical plant. We do not go into details here of the genesis of the model as this is described in detail in [6]. We simply give the final formulation to illustrate the applicability of the assumptions.

In the example, the state variable $x = (x_1, x_2)$ is two-dimensional and represents the stock levels of two chemicals, whilst the jump process J takes one of two values 0 or 1 according to whether a certain processor is broken down or working. Capacity constraints restrict x to lie in the rectangle $[0,K_1] \times [0,K_2]$. This is depicted as OABC in Fig. 1.

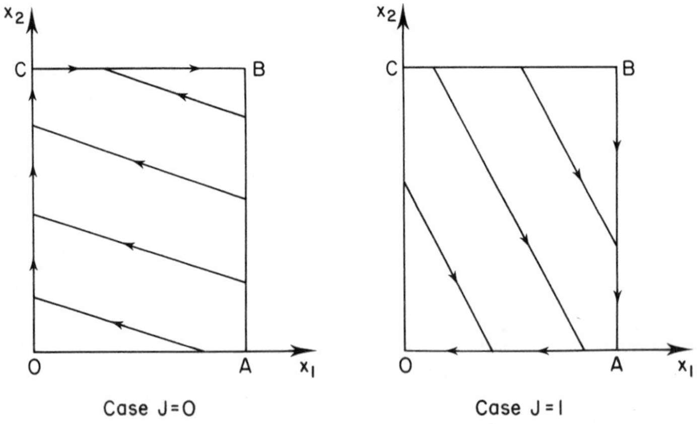

Fig. 1 Typical Trajectory Paths in the Chemical Plant Example

The system equation (1.1) turns out to be

$$\begin{aligned} f(x,u,0) &= (0, \tfrac{1}{2}a_1)' && \text{on OC,} \\ &= (u(x,0),\ 0)' && \text{on BC,} \\ &= (0,\ 0)' && \text{at B,} \\ &= (u(x,0)-a_3,\ \tfrac{1}{2}a_3)' && \text{elsewhere.} \\ f(x,u,1) &= (0,\ 0)' && \text{at O,} \\ &= (u(x,1)-\tfrac{1}{2}a_3,\ 0)' && \text{on OA,} \\ &= (0,\ -\tfrac{1}{2}a_3)' && \text{on AB,} \\ &= (u(x,1)+a_2-a_3,\ \tfrac{1}{2}a_3-a_2)' && \text{elsewhere,} \end{aligned}$$

where a_1, a_2, a_3 are constants satisfying $a_1 < a_3 < a_2$, $2a_1 < a_3$, and u satisfies $a_0 \leqslant u \leqslant a_1$. These restrictions ensure that the direction of movement is as shown in the Figure. The objective is to maximise the output rate of final product which turns out to be

$$
\begin{aligned}
g(x,u,0) &= a_1 \text{ on } OC, & g(x,u,1) &= 2a_1 \text{ at } O,\\
&= 0 \text{ on } BC, & &= a_3 \text{ elsewhere.}\\
&= a_3 \text{ elsewhere,} & &
\end{aligned}
$$

It will be seen from the Figure that if J = 0, and x is initially in the interior of OABC, x will run into either OC or BC and will finally reach B. This is in accordance with the assumptions if we think of the regions OABC, OC, BC and B as the cells corresponding to J = 0, with B as a rest point. Similarly when J = 1 we can take the cells as being OABC, AB, OA and A, with A as a rest point.

2.2 MDP Approximation of the Continuous Process

We turn now to the discretization procedure which yields the MDP approximation of the continuous process. The steps are as follows:

(i) For each $C_{n,i}$ we define a Δ_x-dissection as consisting of a set $\{X_a\}$ of disjoint regions such that $C_{n,i} = \cup X_a$, where $x_1, x_2 \in X_a$ implies that $|x_1 - x_2| < \Delta_x$. For each X_a we identify a point $x_a \in X_a$ to represent it in the discrete process. We let $C^{\Delta}_{n,i} = \{x_a | X_a \in C_{n,i}\}$. For simplicity we shall usually write a for x_a.

(ii) For the jump process we still use I to denote the set of jump states. However changes in jump state only occur at discrete time points spaced Δ_t apart. Let J^{Δ}_m denote the state of the discrete jump process at time $m\Delta_t$. The discrete transition probabilities are $pr(J^{\Delta}_{m+1}=j|J^{\Delta}_m=i) = \lambda_{ij}\Delta_t$, and $pr(J^{\Delta}_{m+1}=i|J^{\Delta}_m=i) = 1 - \lambda_i\Delta_t$, where λ_{ij}, λ_i are as defined in A(i).

(iii) Combining the notation of (i) and (ii) we shall write (a,i) for a typical overall state of the discrete process. To define transitions between overall states we first find, for each $i \in I$, all regions X_b that can be reached from x_a in time Δ_t for some $u \in U$:

$$A(a,i) = \{b \mid y_{\Delta_t}(a) \in X_b \text{ some } u \in U, \text{ with jump state fixed at } i\}. \tag{2.1}$$

A(a,i) represents the set of all actions possible at (a,i). If, at (a,i), decision $b \in A(a,i)$ is taken, then at the next step the overall state transition probabilities are defined to be:

$$\begin{aligned} (a,i) &\to (b,i) \text{ with probability } 1 - \lambda_i \Delta_t \\ &\to (a,j) \text{ with probability } \lambda_{ij} \Delta_t,\ j \neq i. \end{aligned} \tag{2.2}$$

(iv) The discrete process function g^Δ giving the one-step returns is defined as

$$\begin{aligned} g^\Delta(a,b,i) &= g(x_a,u_{ab},i)\Delta_t \quad \text{if } b \in A(a,i) \\ &= 0 \qquad\qquad\qquad \text{otherwise,} \end{aligned} \tag{2.3}$$

where u_{ab} is the (fixed) value of any $u \in U$ for which $y_{\Delta_t}(a) \in X_b$, with jump state fixed at i, evaluated at (a,i).

This completes the discretization process, with the resulting MDP defined by (2.1), (2.2) and (2.3). A stationary control for the MDP is identified if a unique action b(a,i) is assigned to each (a,i). For each $u \in U$ we can define a corresponding discrete control, denoted by u^Δ, if for any (a,i) we take b(a,i) as that b which satisfies

$$y_{\Delta_t}(a) \in X_b, \text{ with jump state fixed at } i. \tag{2.4}$$

In analogy to continuous trajectories we let $y_m^\Delta(a)$ denote the position of the discrete process at time $(m+n)\Delta_t$ given that it is at a at time $n\Delta_t$. A feature of assumptions B and C is that they induce similar properties in the trajectories of u^Δ. These will be discussed in detail in section 4.

3. PROPERTIES OF THE CONTINUOUS PROCESS

For the continuous control process an important implication of assumptions A, B and C is that trajectories depend continuously on initial position.

<u>Lemma 3.1</u> Let $u \in U$ and consider the trajectories $y_t(x_1)$, $y_t(x_2)$ for $t \in [0,T]$. If the jump state remains fixed at i in $[0,T]$ then $\exists K$ such that $|x_1 - x_2| < \varepsilon$ implies $|y_t(x_1) - y_t(x_2)| < K\varepsilon$, where K depends on T but not on u or ε.

<u>Proof.</u> From assumption B(i), we can divide $[0,T)$ into successive intervals $[t_n, t_{n+1})$, in each of which $y_t(x_1)$ remains in a given cell $C_{n,i}$ say. Consider $y_t(x_2)$ and suppose that $|y_{t_n}(x_1) - y_{t_n}(x_2)| < k_n\varepsilon$, some k_n. If $t_{n+1} - t_n < K_0k_n\varepsilon$ then the boundedness of u implies

$$|y_t(x_1) - y_t(x_2)| < (1 + B_0K_0)k_n\varepsilon \text{ for } t \in (t_n, t_{n+1}).$$

If $t_{n+1} - t_n > K_0k_n\varepsilon$ then by assumption C $\exists\, t' < t_n + K_0k_n\varepsilon$ such that $y_{t'}(x_1)$, $y_{t'}(x_2)$ both belong to $C_{n,i}$. Let $t'' = \sup\{t|y_\tau(x_1), y_\tau(X_2) \in C_{n,i}\ \forall\, t' < \tau < t\}$. It follows from the Lipschitz condition on u that up to t"

$$|y_t(x_1) - y_t(x_2)| < (1 + B_0K_0)k_n\varepsilon(1 + e^{(t''-t')L_0}).$$

Consequently $t_{n+1} - t'' < K_0(1 + B_0K_0)k_n\varepsilon(1 + e^{(t''-t')L_0})$ as if not, by assumption C, $\exists\, t'''$ such that $t''<t'''<t_{n+1}$ when both $y_t(x_1)$, $y_t(x_2) \in C_{n,i}$ again, contradicting assumption B(i). Hence applying the boundedness of u again in the period $[t'', t_{n+1})$ we get

$$|y_{t_{n+1}}(x_1) - y_{t_{n+1}}(x_2)| < (1 + B_O K_O)^2 k_n (1 + e^{(t''-t')L_O})\varepsilon$$

$$< (1 + B_O K_O)^2 k_n (1 + e^{TL_O})\varepsilon = k_{n+1}\varepsilon \text{ say.}$$

The result now follows by induction on n.

Another consequence of assumptions B and C is:

Lemma 3.2 Let $u \in U$ and $i \in I$. Then there is a unique point, ξ_i, the rest point of i, such that all trajectories reach ξ_i within time T_O, irrespective of starting position, provided the jump process remains fixed at i.

Proof. By assumption B, all trajectories reach a rest point with time T_O. Suppose for $u \in U$, $i \in I$, ξ_i is a rest point. Ω is bounded, so $\exists\ \delta$ such that $|x - \xi_i| < \delta\ \forall\ x \in \Omega$. By assumption C, $y_t(x)$ must reach ξ_i within time $K_O\delta$. Hence the rest point is unique.

We now concentrate attention on one particular jump state: O say, with rest point ξ_O. Suppose, just prior to time t, the overall state of the process is (ξ_O, O) and that at t there is a jump state change. We call such a moment a departure time. Let $t_O < t_1 < t_2 < \ldots$ be a sequence of such departure times and suppose that we measure time from the first such moment (i.e. let $t_O = 0$). Let $w_i = t_i - t_{i-1}$ be the intervals between successive departure times. Let $r_i = \int_{t_{i-1}}^{t_i} g\,dt$ be the total return between successive departure times. We have:

Lemma 3.3 The w_i are independently and identically distributed, as are the r_i. Moreover there exist $0 < K_1, K_2, K_3 < \infty$, independent of u, such that $K_1 < E(w) < K_2$, $E(r) < K_3$.

Proof. That the w_i are independently and identically distributed is an immediate consequence of assumption A(i), that the jump process is Markov. Moreover let $\ell_1, \ell_2, \ldots$ be the lengths of time between successive jumps out of state O. As the jump process is an irreducible finite state Markov process we have $E(\ell) < \infty$. Let N be the first such interval when the stay in jump state O is greater than T_O. As by B(ii), return to ξ_O in such an interval is certain, it follows that $w < \ell_1+\ell_2+\ldots+\ell_N = W$ say. Now N is a stopping rule (see Ross [8]) and as clearly $E(N) < \infty$ it follows that $E(w) \leq E(W) = E(\ell)E(N) = K_2$. As ℓ and N are associated with the jump process only this bound on E(w) holds for all $u \in U$. A similar bound applies to r as by A(iv), $r < B_1 w$, so that $E(r) < B_1 K_2 = K_3$. In a similar way $E(w) > K_1$.

The implication of the previous lemma is that the times $t_1, t_2, \ldots$ form a regenerative process. It follows that (Ross, Proposition 5.9).

Lemma 3.4

$$\lim_{t\to\infty} \frac{1}{t}\int_0^t g\,dt = G = \frac{E(r)}{E(w)} \text{ with probability 1.}$$

The continuous problem thus reduces to finding that $u \in U$ which maximises E(r)/(E(w).

In the next section we show that the above results extend to the MDP case, thus allowing a comparison to be made of the behaviour of the continuous process and the MDP approximation. Before we can do this we shall need to be able to evaluate E(w) and E(r) to any given accuracy. With this aim in mind we characterise those trajectories which contribute the most to E(w).

Consider the jump process in the interval $[0, \tilde{T}]$ where $\tilde{T}$ is assumed fixed, and suppose that at $t = 0$, a jump from state O to some other state has just occurred. A typical realisation of the ensuing jump process in $[0, \tilde{T}]$ will be written as

$$s = (t_0, t_1, \ldots t_{N-1}; j_1, j_2, \ldots, j_N)$$

where $t_0 = 0 < t_1 < \ldots < t_{N-1}$ are the moments when succeeding changes of jump state occur and $j_1, j_2, \ldots, j_N$ are the states jumped into. We can calculate $E(w)$ to any given accuracy by considering only those interdeparture times $w \leq T$, for T sufficiently large. In all that follows we assume $T \leq \tilde{T}$. Let

$$S_1(T) = \{s \mid w(s) \leq T\}$$

and let $dF(s)$ be the distribution of the jump realisations. Moreover let W be as defined in the proof of Lemma 3.3. As $E(W) < \infty$, we can find a T such that $\int_{W \geq T} W dH(W) \leq \varepsilon$ for any $\varepsilon > 0$ (where $dH(W)$ is the distribution of W). As, for all $u \in U$, $w \leq W$ it follows that

Lemma 3.5 Given $\varepsilon > 0$, $\exists\, T(\varepsilon)$ such that

$$E(w) - \int_{S_1(T)} w(s) dF(s) \leq \int_{W \geq T} W dH(W) \leq \varepsilon, \ \forall\, u \in U.$$

We can further restrict attention to the set of jump realisations defined by

$$S_2(\delta) = \{s \mid \text{In } s, \text{ all intervals between jump changes} > \delta\}.$$

In view of Assumption A(i) all intervals between jumps are strictly positive with probability 1. Consequently we have

Lemma 3.6 Given $\varepsilon > 0, \exists\, \delta(\varepsilon) > 0$ such that $\Pr\{s \in S_2(\delta)\} > 1 - \varepsilon$.

One further restriction can be made on the jump realisations that need be considered. Letting $B_\delta(x_0)$ denote $\{x \mid |x - x_0| < \delta\}$ we define

$S_3(\delta) = \{s \mid$ The trajectory, under s, reaches $(\xi_0, 0)$ again by re-entering $B_\delta(\xi_0)$ just once while in state 0, and on reaching $(\xi_0, 0)$ stays at least a further time $\delta\}$.

In view of assumption C, any trajectory, once it enters $B_\delta(\xi_0)$ whilst in state 0, will reach ξ_0 within a further time $K_0\delta$. Consequently trajectories not in $S_3(\delta)$ must change jump state within $(K_0 + 1)$ of entering $B_\delta(\xi_0)$ whilst in state O. In view of assumption A(i) that the jump process is Markov, the probability of this occurring can be made arbitrarily small by making δ small. We have therefore:

Lemma 3.7 Given $\varepsilon > 0$, $\exists\ \delta > 0$ such that $\Pr\{s \in S_3(\delta)\} > 1 - \varepsilon$, $\forall\ u \in U$.

The above results combine to give:

Lemma 3.8 Give $\varepsilon > 0$, $\exists\ T, \delta_1, \delta_2 > 0$ such that

$$E(w) - \int_S w(s)dF(s) < \varepsilon, \ \forall\ u \in U,$$

where $S = S_1(T) \cap S_2(\delta_1) \cap S_3(\delta_2)$.

Proof From Lemma 3.5 we can select T so that $E(w) - \int_{S_1(T)} wdF < \varepsilon/3$ $\forall\ u \in U$. But $\int_S wdF > \int_{S_1} wdF - T\left[\Pr\{s \notin S_2(\delta_1)\} + \Pr\{s \notin S_3(\delta_2)\}\right]$.
By Lemmas 3.6 and 3.7 we can make both probabilities less than $\varepsilon/(3T)$. Combining the two inequalities then gives the result.

An identical result applies to the calculation of E(r). This completes our investigation of the continuous control process, and we are now in a position to compare it with a MDP approximation.

4. COMPARISON OF THE CONTINUOUS PROCESS WITH A MDP APPROXIMATION

We show that the MDP approximation is a close representation of the continuous process in terms of the jump state behaviour, of the point trajectories and of the long term average return. We do this by deriving a set of results for the MDP that are analogous to Lemmas (3.3) - (3.8). As most of the arguments follow the same lines as those of the previous section, we only give an outline of most of the proofs.

Throughout the section we assume that the continuous process and its MDP approximation have both just previously been in overall state $(\xi_0, 0)$ and that at time $t = 0$, a jump change takes place.

Take first the jump process and consider the discretization where jump changes occur only at $m\Delta_t$, $m = 0,1,2,\ldots,M$, (with $M\Delta_t = \tilde{T}$). A typical realisation of this process will be written, in direct analogy to the definition of the jump realisation s in the continuous case, as

$$s^{\Delta} = (m_0, m_1, \ldots, m_{N-1}; j_1, j_2, \ldots, j_N)$$

where the m_n denote times when jump changes occur.

We need only consider those continuous jump realisations s for which there is at most one jump change within each time period Δ_t; these are grouped into subsets as follows:

$$\sigma(s^{\Delta}) = \{s \mid (m_n - 1)\Delta_t \leq t_n < m_n\Delta_t, \text{ for } n = 1,2,\ldots,N-1, \text{ and}$$
$$j_n\text{th jump state of } s = j_n\text{th jump state of } s^{\Delta}, \text{ for } n = 1,2,\ldots,N\}.$$

This grouping associates an s^{Δ} with each s; we denote this s^{Δ} by $s^{\Delta}(s)$. With these definitions we can make some slight modifications to some of the results in section 3 which will enable the comparison of the continuous process and a MDP approximation to be made much simpler. We define sets $S_1^{\Delta}(T)$, $S_2^{\Delta}(\delta)$ and $S_3^{\Delta}(\delta)$ corresponding to the sets $S_1(T)$, $S_2(\delta)$ and $S_3(\delta)$ defined in section 3. Let

$$S_i^{\Delta} = \{s^{\Delta}(s) \mid s \in S_i\}, \; i = 1, 2, 3.$$

For comparison purposes it proves simpler not to associate the S_i^{Δ} with S_i but instead to replace S_i by

$$R_i = \{s \mid s \in \sigma(s^{\Delta}), \text{ some } s^{\Delta} \in S_i^{\Delta}\}.$$

As from definition $R_i \supseteq S_i$ it follows immediately that Lemmas (3.5) - (3.8) hold a fortiori with S_i replaced by R_i. Moreover the definition of R_i implies that $S_i^\Delta = \{s^\Delta(s) \mid s \in R_i\}$ allowing a more direct comparison to be made of the s and s^Δ realisations.

We show next that the probability of a given s^Δ occurring in the MDP is close in value to the probability of $\sigma(s^\Delta)$ occurring in the continuous process. We have from the definitions in 2.2(ii) and assumption A(i):

$$P(s^\Delta) = \Pr\{s^\Delta \text{ occurring}\} = (\lambda_{Oj_1}/\lambda_O)(1-\lambda_{j_1}\Delta_t)^{m_1-1}(\lambda_{j_1j_2}\Delta_t)(1-\lambda_{j_2}\Delta_t)^{m_2-m_1-1}\cdots$$

$$\ldots(1-\lambda_{j_N}\Delta_t)^{M-m_{N-1}}$$

and

$$p(\sigma(s^\Delta)) = \Pr\{\sigma(s^\Delta) \text{ occurring}\} = (\lambda_{Oj_1}/\lambda_O)e^{-\lambda_{j_1}(m_1-1)\Delta_t}(\lambda_{j_1j_2} \times$$

$$\{e^{-\lambda_{j_1}\Delta_t} - e^{-\lambda_{j_2}\Delta_t}\}/\{\lambda_{j_2}-\lambda_{j_1}\})e^{-\lambda_{j_2}(m_2-m_1-1)\Delta_t}\ldots e^{-\lambda_{j_N}(M-m_{N-1})\Delta_t}.$$

We can establish a bound on $p(\sigma(s^\Delta))$ in terms of $p(s^\Delta)$ by repeatedly using the inequality $1 - x < e^{-x} < 1 - x + x^2(1 + x)^{-1}$ which holds for $x > 0$ but sufficiently small. Straightforward calculation then yields the following

<u>Lemma 4.1</u> Given $\varepsilon, \delta > 0 \ \exists \ \Delta^0 > 0$ such that for $s^\Delta \in S_1^\Delta(\delta)$

$$|p(s^\Delta) - p(\sigma(s^\Delta))| < \varepsilon p(s^\Delta) \ \forall \ 0 < \Delta_t < \Delta^0$$

Next we consider trajectories under the MDP. We use a slightly different notation from that used in section 3. For the continuous process let $y_m(0;s)$ denote the position of x at time $m\Delta_t$, given that it is initially at ξ_0 and that the jump realisation is s; let $y_m^{\Delta}(O;s^{\Delta})$ be similarly defined for the MDP. If it is clear from the context, we write these as y_m and y_m^{Δ} for short. The key relation between y_m^{Δ} and y_m is contained in the next Lemma which shows that not only does y^{Δ} remain close to y, but the cells in which each trajectory lies also match for most time points.

Lemma 4.2 Give T, $\delta > 0$, $\exists$ K, k, Δ^{O} such that if $s \in R_2(\delta)$, $0 < \Delta_t < \Delta^{O}$ and $\Delta_x = \rho\Delta_t^2$, with ρ constant; then with M defined by $M\Delta_t \leqslant T < (M+1)\Delta_t$:

(a) $|y_m(O;s) - y_m^{\Delta}(O;s^{\Delta}(s))| < K\Delta_t \quad \forall\ m \leqslant M,\ u \in U,$

(b) y_m, y_m both belong to the same cell (the cell may be different for different m) for at least $(1 - k\Delta_t)M$ of the m's, $\forall$ u $\in$ U.

Here K, k depend on T, δ but are independent of u, s and Δ_t.

Proof. Details of the proof are omitted as it follows fairly directly from the definitions of y_m and y_m^{Δ}, using an argument similar to the proof of Lemma 3.1. We consider a period (t_n, t_{n+1}) containing the time points $m\Delta_t$, $m = m_n, m_n+1,\ldots,m_{n+1}$ (with m_n, m_{n+1} dependent on Δ_t) where y_m remains in some given cell $C_{n,i}$. Similar considerations to those used in the proof of Lemma 3.1 show that if $\Delta_x = \rho\Delta_t^2$ and $|y_{m_n} - y_{m_n}^{\Delta}| < k_n\Delta_t$, some k_n independent of u and Δ_t, then there exists h_n, ℓ_n independent of u and Δ_t such that y_m, $y_m^{\Delta} \in C_{n,i}$ for $m = m_n+h_n$, $m_n+h_n+1,\ldots, m_{n+1}-\ell_n$. As there is a bound on the number of cells passed through in time T when $s \in R_2(\delta)$ both results (a) and (b) then follow by induction as in the proof of Lemma 3.1.

Now let w^{Δ} be the interval between successive departure times from $(\xi_O,0)$ for the MDP and r^{Δ} the total return in w^{Δ}.

Though the previous lemma is restricted to finite time intervals, combined with assumption B it is nevertheless sufficient to ensure the positive recurrence of departure times.

Lemma 4.3 $\exists\, \Delta^o$, K_4 such that, with $\Delta_x = \rho\Delta_t^2$:

$$E(w^\Delta),\ E(r^\Delta) \leqslant K_4, \quad \forall\, u \in U,\ \Delta_t \leqslant \Delta^o$$

Proof. Only an outline is given. For the continuous process, if the jump state remains fixed at O, by assumption B all trajectories will reach ξ_o within T_o irrespective of u. This result together with Lemma 4.2(b) implies that MDP trajectories have the same property, provided Δ_t is sufficiently small. An argument similar to that used in Lemma 3.3 for the continuous process then gives the result.

This shows that we can calculate the long term average return for the MDP in the same way as we calculated G. We have under the conditions of Lemma 4.3:

Lemma 4.4

$$\lim_{M\to\infty}(M\Delta_t)^{-1}\sum_{i=1}^{M} g^\Delta = G^\Delta = \frac{E(r^\Delta)}{E(w^\Delta)}, \text{ with probability 1.}$$

We now compare E(w) and $E(w^\Delta)$. The inter-departure time corresponding to jump realisation s^Δ will be written as $w^\Delta(s^\Delta)$ (provided $w^\Delta(s^\Delta) \leqslant \tilde{T}$, the time interval used to define s^Δ).

Lemma 4.5 Given ε, T, δ_1, $\delta_2 > 0$, $\exists\, \Delta^o$ such that if $s \in R = R_1(T)\cap R_2(\delta_1)\cap R_3(\delta_2)$ and $\Delta_x = \rho\Delta_t^2$, then $|w(s) - w^\Delta(s^\Delta(s))| < \varepsilon \quad \forall\, \Delta_t \leqslant \Delta^o$.

Proof. The lemma is a consequence of the definitions of $R_3(\delta_2)$ and Lemma 4.2. We give an outline only. Suppose t = O is a departure time from (ξ_o, O) and let t' be the time when (ξ_o,O) is again reached. From the definition of $R_3(\delta)$ and assumption C either $|y_m(O;s) - \xi_o| > \delta_2$ or $J(m\Delta_t) \neq O$ $\forall$ m such that $m\Delta_t < t' - K_o\delta_2$. Using Lemma 4.2(a) we can ensure that y^Δ

satisfies a similar property, i.e. that $|y_m^\Delta(0;s) - \xi_0| > \delta_2/2$ or $J_m^\Delta \neq 0$ $\forall$ m such that $m\Delta_t < t' - K_0\delta_2$. Once y_m satisfies $|y_m(0;s) - \xi_0| \leq \delta_2$ and $J(m\Delta_t) = 0$ we know from the definition of $R_3(\delta)$ that $y_m = \xi_0$ and $J(m\Delta_t) = 0$ within a further time $K_0\delta_2$. By Lemma 4.2(b) y^Δ has a similar property provided Δ_t is sufficiently small, and the result follows.

The above lemmas combine to give the following result.

Theorem 4.1 Given $\varepsilon > 0$, $\exists\, \Delta^0$ such that, with $\Delta_x = \rho\Delta_t^2$:

$$|E(w) - E(w^\Delta)| < \varepsilon, \quad \forall\, u \in U, \ \Delta_t \leq \Delta^0.$$

Proof. Given $\varepsilon_1 > 0$ we can choose by Lemma 3.8 (with R replacing S) T, δ_1, δ_2 so that $E(w) - \int_R w(s)dF(s) \leq \varepsilon_1$, $\forall\, u \in U$.

Now we use Lemmas 4.1, 4.3 and 4.5. Letting $R^\Delta = \{s^\Delta | s^\Delta = s^\Delta(s)$, some $s \in R\}$ $(= S_1^\Delta \cap S_2^\Delta \cap S_3^\Delta)$, these give for suitably chosen Δ^0:

$$\int_R w(s)dF(s) \leq \sum_{s^\Delta \in R^\Delta} (w^\Delta(s^\Delta) + \varepsilon_1)p(\sigma(s^\Delta)), \text{ by Lemma 4.5}$$

$$\leq \sum_{s^\Delta \in R^\Delta} w^\Delta(s^\Delta)(1 + \varepsilon_1)p(s^\Delta) + \varepsilon_1, \text{ by Lemma 4.1}$$

$$\leq \sum_{s^\Delta \in R^\Delta} w^\Delta(s^\Delta)p(s^\Delta) + (1 + K_4)\varepsilon_1, \text{ by Lemma 4.3}$$

$$\leq E(w^\Delta) + (1 + K_4)\varepsilon_1,$$

this inequality holding for $\Delta_t \leq \Delta^0$. Combining the two inequalities and setting $\varepsilon_1 = \varepsilon/(2 + K_4)$ shows that $E(w) \leq E(w^\Delta) + \varepsilon$.

We now follow an argument similar to that used in the proof of Lemma 3.8 to show that T, δ_1, δ_2 can be chosen to make

$$\sum_{s^\Delta \in R^\Delta} w^\Delta(s^\Delta)p(s^\Delta) \geqslant E(w^\Delta) - \varepsilon_1, \quad \forall\, u \in U.$$

A similar argument to that above, making use of this result, shows that we also have $E(w) \geqslant E(w^\Delta)-\varepsilon$, and the theorem follows.

A similar result can be obtained relating E(r) and $E(r^\Delta)$. The proof is along completely analogous lines.

<u>Theorem 4.2</u> Given $\varepsilon > 0$, $\exists\, \Delta^o$ such that, with $\Delta_x = \rho\Delta_t^2$,

$$|E(r) - E(r^\Delta)| < \varepsilon, \quad \forall\, u \in U,\ \Delta_t < \Delta^o.$$

The two theorems together with Lemma 3.3 combine to give

<u>Theorem 4.3</u> Given $\varepsilon > 0$, $\exists\, \Delta^o$ such that, with $\Delta_x = \rho\Delta_t^2$:

$$|G - G^\Delta| < \varepsilon, \quad \forall\, u \in U,\ \Delta_t < \Delta^o.$$

The theorem shows that G^Δ converges to G uniformly in u. Our final result is a straightforward consequence of this.

<u>Theorem 4.4</u> Let $\Delta^n > 0$, $n = 1, 2, \ldots$ be a sequence of numbers decreasing to zero as $n \to \infty$. For each Δ^n suppose there corresponds a MDP approximation with $\Delta_x = \rho\Delta_t^2$, $\Delta_t = \Delta^n$. Let v_n^Δ be the maximal solution of the MDP problem corresponding to Δ^n. Suppose that $\exists\, v_n \in U$ such that v_n^Δ corresponds to v_n in the sense of (2.4) and that v_n converges to $v^* \in U$. Then v^* solves the continuous optimal control problem.

The theorem is not of course an existence result. What it does provide is a sufficiency condition that is useful for practical purposes. Roughly speaking it guarantees that if the technique, as summarised in the theorem, can be made to work, then it gives the correct solution.

5. CONCLUSIONS

We end with some remarks on the practical aspects of solving the approximating MDP's. It is hoped to discuss this more fully elsewhere, but it is worth observing here that the policy iteration method of Howard [7] seems to be a particularly effective solution method. The main reason is that the one-step probability transition matrix as defined in (2.2) is sparse with all the non-zero entries limited to a small number of fixed positions, irrespective of the action policy taken. This enables efficient conjugate gradient methods to be used at the value determination step. To date problems involving two dimensional x with up to 1,000 overall states have been easily handled using this technique (see Cheng [6]) without recourse to special computer storage facilities for the matrix arrays.

6. REFERENCES

1. Krassovskii, N. N. and Lidskii, E. A. "Analytical Design of Controllers in Stochastic Systems with Velocity Limited Controlling Action", *Appl. Math. Mech.*, **25**, 627-643, (1961).

2. Wonham, W. M. "Random Differential Equations in Control Theory", in Probabilistic Methods in Applied Mathematics, Vol. 2, A.T. Bharucha-Raid, ed., Academic Press, New York, 191-199, (1970).

3. Rishel, R. "Dynamic Programming and Minimum Principles for Systems with Jump Markov Disturbances", *SIAM J. Control and Optn.*, **13**, 338-371, (1975).

4. Goor, R. M. "Existence of an Optimal Control for Systems with Jump Markov Disturbances", *SIAM J. Control and Optn.*, **14**, 899-918, (1976).

5. Sworder, D. D. "Feedback Control of a Class of Linear Systems with Jump Parameters", *IEEE Trans. Aut. Control*, **AC–14**, 9-14, (1969).

6. Cheng, R. C. H. "Optimal Control of Chemical Plants subject to Random Breakdowns", in Control of Distributed Parameters Systems: Proceedings of the 2nd IFAC Symposium, S. P. Banks and A. J. Pritchard, eds., Pergammon, Oxford, 413-422, (1977).

7. Howard, R. A. "Dynamic Probabilistic Systems, Vol II: Semi-Markov and Decision Processes", Wiley, New York, (1971).

8. Ross, S. M. "Applied Probability Models with Optimisation Applications", Holden-Day, San Francisco, (1970).

WEAK CONVERGENCE OF DECISION PROCESSES

A. Hordijk

(University of Leiden)

and

F. A. Van der Duyn Schouten

(Free University of Amsterdam)

1. INTRODUCTION

In this paper decision processes with countable state space and continuous time parameter are considered. The standard way in discrete-time dynamic programming to prove the existence of an optimal policy with a special structure (for instance monotone or (s,S)-type policies) is by induction on the length of the time horizon and passing to limits for the infinite horizon case. In continuous-time decision processes, however, the action may be changed on a non-countable set of time epochs, which makes an inductive approach inappropriate. In this paper we present a method to analyse continuous-time processes by approximation with discrete-time processes. For the approximating discrete-time processes we can apply the inductive method in order to prove structural results with respect to the optimal policy. The corresponding result for the continuous-time processes can then be obtained by arguments of weak convergence of the underlying stochastic processes.

In section 2 we will give definitions of countable-state decision processes with continuous as well as discrete time parameter (notation: CT - resp. DT-process). Moreover for every CT- and DT-process we define the class of permitted policies and for every policy a probability measure on the set of possible sample paths is defined.

In section 3 we give sufficient conditions for the weak convergence of a sequence of DT-processes controlled by fixed policies to a CT-process under a fixed policy.

We conclude with an application of this convergence result in section 4.

In this paper the proofs of the theorems are omitted. For details the reader is referred to Hordijk and Van der Duyn Schouten [2] and other forthcoming publications. For a more extensive analysis of the set of possible sample paths, introduced in section 2, we refer to Billingsley [1], Lindvall [3] and Whitt [5].

Finally we mention papers by Lippman [4] and Whitt [6] in which continuous-time decision processes are treated in a way similar to ours.

2. DEFINITIONS

In this section definitions are given of countable state decision processes with continuous (CT) as well as discrete (DT) time parameter. Next we describe the set of possible paths of these processes and we define the class of permitted policies. We conclude this section with the definition of a probability measure on the set of possible sample paths for any CT- and DT-process and any policy.

2.1 Definition

A countable-state continuous-time decision process (CT-process) is a five-tuple (S,A,H,q,c) where

(i) S is a countable set with discrete metric σ (the *state space*).

(ii) A is a complete metric space with Borel-σ-field $\mathcal{A}$ (the *action space*).

(iii) H is a measurable subset of $S \times A$ (the *constraint set*) i.e. $A(i) := \{a \in A: (i,a) \in H\}$ is the set of possible actions in state i.

(iv) q is a real-valued measurable function on $H \times S$, with $q(i,a,j) \geq 0$ for $j \neq i$, $(i,a) \in H$ and $\sum_{j \in S} q(i,a,j) = 0$ for $(i,a) \in H$. ($q(i,a,j)$ denotes for $j \neq i$ the *transition rate* from state i into state j when action $a \in A(i)$ is chosen in state i).

(v) c is a real-valued measurable function on H. ($c(i,a)$ denotes the *cost rate* when the system is in state i and action $a \in A(i)$ is chosen).

<u>Assumption</u> (i) $q(i,a,i) \geq -b$ for some $b>0$ and all $(i,a) \in H$.
(ii) $q(i,.,j)$ is continuous on $A(i)$ for all $i,j \in S$.

As a set of possible sample paths for CT-processes we introduce $D := \{x: [0,\infty) \to S$, right continuous and with left hand limits for every $s \in (0,\infty)\}$

and for $t>0$,

$D[0,t] := \{x: [0,t] \to S$, right continuous and with left hand limits in every $s \in (0,t]\}$.

It is well-known that every element in D and $D[0,t]$ has a finite number of discontinuities on every finite interval. We will make use of the following functions:

for $t>0$:

$$e_t : D[0,t] \to D, \text{ defined by } (e_t x)(s) = \begin{cases} x(s), & s<t \\ x(t), & s \geq t \end{cases}$$

$r_t : D \to D[0,t]$, defined by $(r_t x)(s) = x(s)$, $0 \leq s \leq t$.

$\pi_t : D \to S$, defined by $\pi_t x = x(t)$.

and

$\pi : D \times [0,\infty) \to S$, defined by $\pi(x,t) = x(t)$.

The space D is endowed with the Skorohod-metric ρ defined by

$$\rho(x,y) := \inf_{\lambda \in \Lambda} \int_0^\infty e^{-u}\left[\sup_{0 \leq t \leq u} \{|\lambda(t)-t| \vee \sigma(x(\lambda(t)),y(t))\} \wedge 1\right] du$$

where

$$\Lambda := \{\lambda : [0,\infty) \to [0,\infty),\ \lambda \text{ continuous, strictly incr.;}\ \lambda(0) = 0,\ \lambda(\infty) = \infty\}$$

The space $D[0,t]$ is endowed with the metric ρ_t defined by

$$\rho_t(x,y) = \rho(e_t x, e_t y)$$

The Borel-σ-fields of D and D$[0,t]$ are denoted by $\mathcal{D}$ resp. $\mathcal{D}[0,t]$. Note that the definition of ρ_t makes e_t continuous, while $r_t(x)$ is continuous at x if x is continuous at t.

For a more extensive study of the spaces D and D$[0,t]$ we refer to Billingsley [1], Lindvall [3] and Whitt [5].

2.2 Definition

Let (S,A,H,q,c) be a CT. A *policy* for (S,A,H,q,c) is a function R: D × $[0,\infty) \to$ A such that

(i) $R(x,t) \in A(x(t))$.

(ii) R(.,t) is r_t-measurable i.e. for all $B \in \mathcal{A}$ and all $t>0$ there exists a $C \in \mathcal{D}[0,t]$ such that $\{x \in D: R(x,t) \in B\} = r_t^{-1}(C)$.

(iii) R is $\mathcal{D} \times \mathcal{B}$ measurable, where $\mathcal{B}$ denotes the Borel-σ-field of $[0,\infty)$.

(iv) $\lambda(B(x)) = 0$, where $B(x) := \{t: R(.,.)$ is discontinuous at $(x,t)\}$ for all $x \in D$.

2.3 Definition

(i) A policy R is *memoryless* if R(.,t) is π_t-measurable, i.e. for all $B \in \mathcal{A}$ and all $t>0$ there exists a $C \subset S$ such that $\{x \in D: R(x,t) \in B\} = \pi_t^{-1}(C)$.

(ii) A policy R is *stationary* if R(.,.) is π-measurable, i.e. for all $B \in \mathcal{A}$ there exists a set $C \subset S$ such that $\{(x,t) \in D \times [0,\infty): R(x,t) \in B\} = \pi^{-1}(C)$.

In order to define for a given CT-process and a given policy R a probability measure on D we introduce a sequence of functions on D as follows. Let ξ be a fictitious state not belonging to S. For n = 1,2,... we define

$$T_n(x) := \begin{cases} \text{time epoch of the } n^{th} \text{ discontinuity of } x, \text{ if this exists} \\ \infty, \text{ else.} \end{cases}$$

$$J_n(x) := \begin{cases} \text{state occupied by } x \text{ immediately after its } n^{th} \text{ disc.} \\ \text{if this exists} \\ \xi, \text{ else.} \end{cases}$$

$J_0(x) = x(0)$ and $T_0(x) \equiv 0$.

Remark. Note that T_n is a function from D into $[0,\infty]$ and J_n from D into $S \cup \{\xi\}$. If $[0,\infty]$ is endowed with metric d defined by $d(s,t) = |e^{-s}-e^{-t}|$ and $S \cup \{\xi\}$ with the discrete metric, then T_n is continuous but J_n not.

For instance put

$$x_n(t) = \begin{cases} s_1, & 0 \leq t < n \\ s_2, & t \geq n \end{cases}, \text{ with } s_1 \neq s_2$$

and

$$x(t) = s_1, \ t \geq 0.$$

Then $\rho(x_n,x) = e^{-n}$, $J_1(x_n) = s_2$, $T_1(x_n) = n$, $T_1(x) = \infty$ but $J_1(x) = \xi$.

To define a probability measure on D in a convenient way we put

$$F: = \{ \bigcap_{n=0}^{m} J_n^{-1}(A_n) \cap T_n^{-1}(B_n) : m=0,1,..; A_n \subset S, B_n \text{ finite interval,}$$

$n=1,...,m-1$; $A_m \subset S$ and B_m finite interval ór B_m infinite interval including ∞ and $A_m = S \cup \{\xi\}\}$.

Theorem. F is a convergence determining class in D.

Corollary. F is a measure determining class in D.

2.4 Definition

Let (S,A,H,q,c) be a CT and R a policy. For any initial

distribution π_O on S we define a probability measure P_R on $\mathcal{F}$ (and hence on $\mathcal{D}$) by

(i) $P_R\{J_O \in A; T_O \in B\} = i_B(O) \cdot \pi_O(A)$, where $i_B(.)$ denotes the indicator function of the set B.
By induction on n we define

(ii) $P_R\{J_n \in A,\ T_n \leq t \mid J_O = j_O, T_O = O, \ldots, J_{n-1} = j_{n-1},\ T_{n-1} = t_{n-1}\} =$

$$= \int_{s=t_{n-1}}^{t} \exp\left(\int_{u=t_{n-1}}^{s} -q(j_{n-1}, R(x,u), j_{n-1})\,du\right) . q(j_{n-1}, R(x,s), A)\,ds$$

and

(iii) $P_R\{T_n \geq t \mid J_O = j_O,\ T_O = O, \ldots, J_{n-1} = j_{n-1}\ T_{n-1} = t_{n-1}\} =$

$$= \exp\left(\int_{u=t_{n-1}}^{t} -q(j_{n-1}, R(x,u), j_{n-1})\,du\right)$$

where x is defined by

$$x(t) = \begin{cases} j_i, & t_i \leq t < t_{i+1},\ i=O,\ldots,n-2\ (t_O = O) \\ j_{n-1}, & t \geq t_{n-1} \end{cases}$$

<u>Theorem</u>. P_R is a well defined probability measure on $\mathcal{D}$.

2.5 Definition

A countable-state discrete-time decision process (DT-process) is a six-tuple (S,A,H,k,p,c), where S, A and H have the same meaning as in definition 2.1. (i), (ii) resp. (iii) and

(iv) k is a natural number (k indicates that the *set of decision epochs* is $T_k := \{nk^{-1},\ n=O,1,2,\ldots\}$).

(v) p is a measurable *transition probability function* from H to S.
(p(i,a,j) denotes the probability that the state at the next decision epoch is j when action $a \in A(i)$ is chosen in state i).

(vi) c is a real-valued measurable function on H. (c(i,a) denotes the *direct costs* when action $a \in A(i)$ is chosen in state i).

As the set of possible sample paths for a DT-process (S,A,H,k,p,c) we introduce:

$$D_k := \{x \in D: x \text{ constant on } [nk^{-1}, (n+1)k^{-1}) \text{ for } n = 0,1,2,\ldots\}.$$

2.6 Definition

A <u>policy</u> for a DT-process (S,A,H,k,p,c) is a function R on $D_k \times T_k$ into A such that

(i) $R(x,t) \in A(x(t))$.

(ii) $R(.,t)$ is r_t-measurable.

(iii) R is $\mathcal{D} \times \mathcal{B}$-measurable.

2.7 Definition

Let (S,A,H,k,p,c) be a DT-process and R a policy. For any initial probability distribution π_0 on S we define a probability measure P_R on $\mathcal{D}$, concentrated on D_k, as follows:

(i) $P_R\{J_0 \in A;\ T_0 \in B\} = i_B(0).\pi_0(A)$ for $B \in \mathcal{B}$ and $A \subset S$.

(ii) $P_R\{J_n \in A,\ T_n = \ell k^{-1} | J_0 = j_0,\ T_0 = 0, \ldots J_{n-1} = j_{n-1},\ T_{n-1} = \ell_{n-1}k^{-1}\} =$

$$= \prod_{m=\ell_{n-1}}^{\ell-2} p(j_{n-1}, R(x,mk^{-1}), j_{n-1}).p(j_{n-1}, R(x,(\ell-1)k^{-1}), A).$$

(iii) $P_R\{T_n \geq \ell k^{-1} | J_0 = j_0;\ T_0 = 0, \ldots,\ J_{n-1} = j_{n-1},\ T_{n-1} = \ell_{n-1}k^{-1}\} =$

$$= \prod_{m=\ell_{n-1}}^{\ell-2} p(j_{n-1}, R(x,mk^{-1}), j_{n-1}),$$

where x is defined by

$$x(t) = \begin{cases} j_i, & \ell_i k^{-1} \leq t < \ell_{i+1} k^{-1}, \quad i=0,\ldots,n-2 \ (\ell_0=0) \\ j_{n-1}, & t \geq \ell_{n-1} k^{-1}. \end{cases}$$

3. WEAK CONVERGENCE

In this section we give sufficient conditions for the weak convergence of a sequence of probability measures on D induced by a sequence of DT-processes controlled by fixed policies to the probability measure induced by a CT-process controlled by a fixed policy.

Theorem. Let for k=1,2,.. $DT(k) := (S,A,H,k,p_k,c_k)$ be a DT-process and CT: = (S,A,H,q,c) a CT-process such that:

(i) $\lim_{k\to\infty} k^{-1} p_k(i,a,j) = q(i,a,j)$, uniformly in a, for all i, $j \in S$, with $i \neq j$.

(ii) $\lim_{k\to\infty} k^{-1}(p_k(i,a,i)-1) = q(i,a,i)$, uniformly in a, for all $i \in S$.

If R_k is a policy for DT(k), k=1,2,... and R for CT such that

(iii) for all $x \in D$ there exists a set $B(x) \subset [0,\infty)$ with $\lambda(B(x))=0$ and for all $t \notin B(x)$ we have $R_k(x_k,t_k) \to R(x,t)$ if

$\rho(x_k,x) \to 0$ and $t_k \to t$,

then $P_{R_k} \xrightarrow{W} P_R$ on D ($\xrightarrow{W}$ denotes weak convergence of probability measures).

3.1 Corollary

Let CT: = (S,A,H,q,c) be a CT-process and R a policy for CT. Define for k=1,2,.. DT-processes $DT(k) := (S,A,H,k,p_k,c_k)$ by

$$p_k(i,a,j) := k^{-1} q(i,a,j,), \quad j \neq i$$

$$p_k(i,a,i) := k^{-1} q(i,a,i) + 1$$

$$c_k(i,a) := k^{-1} c(i,a)$$

and policies R_k for DT(k) by

$$R_k(x,t) := R(x,t) \text{ for all } x \in D_k \text{ and } t \in T_k.$$

Then $P_{R_k} \xrightarrow{W} P_R$ on D.

Definition

(i) Let DT: = (S,A,H,k,p,c) be a DT-process and R a policy for DT. The *β-discount costfunctional for policy R* is a real-valued function f_R on D_k, defined by

$$f_R(x) = \sum_{\ell=0}^{\infty} \beta^{\ell} c(x(\ell k^{-1}), R(x, \ell k^{-1})).$$

(ii) Let CT: = (S,A,H,q,c) be a CT-process and R a policy for CT. The *α-discount costfunctional for policy R* is a real-valued function f_R on D defined by

$$f_R(x) = \int_0^{\infty} e^{-\alpha t} c(x(t), R(x,t))dt.$$

Theorem. Let CT: = (S,A,H,q,c) be a CT-process with policy R. Define DT-processes DT(k) and policies R_k as in corollary 3.1. Let f be the α-discount costfunctional for R and f_k the β_k-discount costfunctional for R_k, with $\beta_k = 1-\alpha k^{-1}$. If c(.,.) is bounded and continuous on H, then

(i) $f_k(x_k) \to f(x)$ if $\rho(x_k,x) \to 0$.

(ii) $P_{R_k} f_k^{-1} \xrightarrow{W} P_R f^{-1}$ on R.

(iii) $\int_D f_k(x)\, dP_{R_k}(x) \to \int_D f(x) dP_R(x)$.

Theorem. Let CT: = (S,A,H,q,c) be a CT-process and DT(k) be DT-processes defined as in corollary 3.1. Let R_k^* be an optimal *stationary* policy for DT(k) with respect to the β_k-discount costfunctional, where $\beta_k = 1-\alpha k^{-1}$ for some $\alpha>0$, k=1,2,... .

(i) There exist functions $r_k^*: S \to A$ such that

$R_k^*(x,t) = r_k^*(x(t))$, $k=1,2,\ldots$

(ii) If $r_k^*(i) \to r(i)$ for all $i \in S$ then $R(x,t) := r(x(t))$ is an optimal stationary policy for CT with respect to the α-discount costfunctional.

4. APPLICATION

We conclude this paper with an example which illustrates how the obtained results can be used. Consider an inventory-production system where a certain commodity is produced and sold. Customers arrive according to a Poisson input stream with parameter $\lambda \in [0,\lambda^*]$. Each customer asks for one unit of the commodity. The production time of one unit is negative exponential distributed with parameter $\mu \in [0,\mu^*]$. Both parameters λ and μ can be controlled. No backlog is allowed. The following costs are incurred.

$c_1(\lambda) :=$ *return-rate* when arrival-rate λ is used.

$c_2(\mu) :=$ *cost-rate* when production-rate μ is used.

$h(i) :=$ *cost-rate* when i items are at stock, $i=0,1,2,\ldots$.

<u>Assumption.</u>

(i) c_1 and c_2 are continuous.

(ii) h is convex.

Define CT: = (S,A,H,q,c) by

(i) $S := \{0,1,2,\ldots\}$ ($i \geq 0$ denotes that i items are at stock).

(ii) $A = [0,\lambda^*] \times [0,\mu^*]$.

(iii) $A(i) = A$ for $i \in \{1,\ldots,U\}$. (U represents the maximum capacity of the storage-facility)

$= \{0\} \times [0,\mu^*]$ for $i=0$

$= [0,\lambda^*] \times \{0\}$ for $i>U$.

(iv) $q(i,(\lambda,\mu),j) = \mu$ for $j=i+1$

$= -(\lambda+\mu)$ for $j=i$

$= \lambda$ for $j=i-1$.

(v) $c(i,(\lambda,\mu)) = h(i) - c_1(\lambda) + c_2(\mu)$.

Define DT-processes $DT(k) := (S,A,H,k,p_k,c_k)$ by

$$p_k(i, (\lambda,\mu),j) = k^{-1}\mu \quad \text{for } j=i+1$$
$$= 1-k^{-1}(\lambda+\mu) \quad \text{for } j=i$$
$$= k^{-1}\lambda \quad \text{for } j=i-1.$$

$$c_k(i,(\lambda,\mu)) = k^{-1}(h(i)-c_1(\lambda)+c_2(\mu)).$$

Note that $p_k(.,.,.)$ is a transition probability if $k \geq \lambda^* + \mu^*$.

Theorem.

(i) There exists an optimal stationary policy R_k for $DT(k)$ with respect to β_k-discount costfunctional ($\beta_k = 1-\alpha k^{-1}$).

(ii) There exists a function r_k: $S \to A$ such that $R_k(x,t) = r_k(x(t))$.

(iii) The first component of $r_k(i)$ is non-decreasing in i.

(iv) The second component of $r_k(i)$ is non-increasing in i.

Theorem.

(i) CT has an optimal stationary policy R with respect to the α-discount cost functional.

(ii) There exists a function r: $S \to A$ such that $R(x,t)=r(x(t))$.

(iii) The first component of $r(i)$ is non-decreasing in i.

(iv) The second component of $r(i)$ is non-increasing in i.

5. REFERENCES

1. Billingsley, P. "Convergence of probability measures", John Wiley, New York, (1968).

2. Hordijk, A. and Van der Duyn Schouten, F. A. "Countable state continuous time decision processes", Research report 12, Free University, Amsterdam, (1977).

3. Lindvall, T. "Weak convergence of probability measures and random functions in the function space $D[0,\infty)$", *J. Appl.*

Prob., **10**, 109-121, (1973).

4. Lippman, S. A. "Countable-state, continuous-time dynamic programming with structure", *Oper. Res.*, **24**, 3, 477-490, (1976).

5. Whitt, W. "Continuity of several functions on the function space D", Technical Report, Yale University, (1974).

6. Whitt, W. "Continuity of Markov processes and dynamic programs", Technical Report, Yale University, (1975).